T0400037

Handbook of Sustainable Development through Green Engineering and Technology

Green engineering involves the designing, innovation, and commercialization of products and processes that promote sustainability without eliminating both efficiency and economic viability. This handbook focuses on sustainable development through green engineering and technology. It is intended to address the applications and issues involved in their practical implementation.

A new range of renewable-energy technologies, modified to provide green engineering, will be described in this handbook. It will explore all green technologies required to provide green engineering for the future. These include, but are not limited to, green smart buildings, fuel-efficient transportation, paperless offices, and many more energy-efficient measures.

Handbook of Sustainable Development through Green Engineering and Technology acts as a comprehensive reference book to use when identifying development for programs and sustainable initiatives within the current legislative framework. It aims to be of great interest to researchers, faculty members, and students across the globe.

Green Engineering and Technology: Concepts and Applications

Series Editors: Brojo Kishore Mishra, *GIET University, India* and Raghvendra Kumar, *LNCT College, India*

Environment is an important issue these days for the whole world. Different strategies and technologies are used to save the environment. Technology is the application of knowledge to practical requirements. Green technologies encompass various aspects of technology which help us reduce the human impact on the environment and creates ways of sustainable development. Social equability, this book series will enlighten the green technology in different ways, aspects, and methods. This technology helps people to understand the use of different resources to fulfill needs and demands. Some points will be discussed as the combination of involuntary approaches, government incentives, and a comprehensive regulatory framework will encourage the diffusion of green technology, least developed countries and developing states of small island requires unique support and measure to promote the green technologies.

Convergence of Blockchain Technology and E-Business
Concepts, Applications, and Case Studies
Edited by D. Sumathi, T. Poongodi, Bansal Himani, Balamurugan Balusamy, and Firoz Khan K P

Big Data Analysis for Green Computing
Concepts and Applications
Edited by Rohit Sharma, Dilip Kumar Sharma, Dhowmya Bhatt, and Binh Thai Pham

Handbook of Sustainable Development Through Green Engineering and Technology
Edited by Vikram Bali, Rajni Mohana, Ahmed Elngar, Sunil Kumar Chawla, and Gurpreet Singh

Integrating Deep Learning Algorithms to Overcome Challenges in Big Data Analytics
Edited by R. Sujatha, S. L. Aarthy, and R. Vettri Selvan

For more information about this series, please visit: https://www.routledge.com/Green-Engineering-and-Technology-Concepts-and-Applications/book-series/CRCGETCA

Handbook of Sustainable Development through Green Engineering and Technology

Edited by
Vikram Bali, Rajni Mohana, Ahmed A. Elngar,
Sunil Kumar Chawla, and Gurpreet Singh

CRC Press
Taylor & Francis Group
Boca Raton London New York

CRC Press is an imprint of the
Taylor & Francis Group, an **informa** business

First edition published 2023
by CRC Press
6000 Broken Sound Parkway NW, Suite 300, Boca Raton, FL 33487-2742

and by CRC Press
4 Park Square, Milton Park, Abingdon, Oxon, OX14 4RN

CRC Press is an imprint of Taylor & Francis Group, LLC

© 2023 Taylor & Francis Group, LLC

Library of Congress Cataloguing-in-Publication Data
Names: Bali, Vikram, editor.
Title: Handbook of sustainable development through green engineering and technology / edited by Vikram Bali, Rajni Mohana, Ahmed A. Elngar, Sunil Kumar Chawla, and Gurpreet Singh.
Description: First edition. | Boca Raton : CRC Press, 2022. | Series: Green engineering and technology : concepts and applications | Includes bibliographical references and index.
Identifiers: LCCN 2021014483 (print) | LCCN 2021014484 (ebook) | ISBN 9780367650926 (hbk) | ISBN 9780367650957 (pbk) | ISBN 9781003127819 (ebk)
Subjects: LCSH: Sustainable engineering.
Classification: LCC TA170 .H364 2022 (print) | LCC TA170 (ebook) | DDC 628--dc23
LC record available at https://lccn.loc.gov/2021014483
LC ebook record available at https://lccn.loc.gov/2021014484

ISBN: 978-0-367-65092-6 (hbk)
ISBN: 978-0-367-65095-7 (pbk)
ISBN: 978-1-003-12781-9 (ebk)

DOI: 10.1201/9781003127819

Typeset in Times
by MPS Limited, Dehradun

Contents

Preface

Greetings!

Due to huge and rapid-pace socio-environmental happenings all around the world, several metropolitan cities have directed their development efforts toward smart policies for sustainable mobility, energy upgrading of the building stock, increased energy production from renewable sources, improved waste management, inclusion of solar energy systems, and energy-efficient implementations of ICT infrastructures. With an increased urbanization and energy consumption, one of the main responsibilities of a smart city is to focus on low energy use, renewable energy, and small carbon footprints. Green engineering is a conscious effort to minimize the risk of advancement projects to human health, the environment, and nature.

Green engineering encompasses many aspects in real life. Green engineering comprises design, innovation, and commercialization of products and processes, which promotes sustainability without eliminating efficiency and economic viability of the processes involved.

This book is a collective effort toward providing a range of green engineering perspectives for sustainable development of society. This book acts as a link between innovations and strategies researchers have proposed for society's overall welfare in context to the applications of green engineering and technology. This ranges from Sustainability Entrepreneurship to Environmental Performance Index Score, Smart Grid to Smart Farming, Smart Factories to Electrical Vehicles, Energy Efficient Buildings to Renewable Energy, Sustainable and Green Materials to Smart Construction Materials, and Environmental Protection Applications to Solar Thermal Power Generation.

In Chapter 1, "Achieving Sustainable Development through Sustainable Entrepreneurship and Green Engineering," the concept of sustainable entrepreneurship is defined, discussed, and elaborated. This chapter reviews green engineering principles for key decision making from the sustainable entrepreneurship perspective; it encourages the use of green engineering by entrepreneurs and managers and the application of green entrepreneurship for sustainable development.

In Chapter 2, "Environmental Performance Index Score: A Driving Force toward Green Business Model and Process Innovations," the authors confine the scope of this EPI index study for 2020 against that of last the ten years' longitudinal data series driven baseline index. The findings of the study reveal innovative green business models and processes as best practices to improve environmental performance.

In Chapter 3, "Precursors and Impediments of Green Consumer Behaviour: An Overview," the authors provide an insight to manufacturers and marketers regarding the precursors and impediments of green consumer behaviour. It will facilitate them in designing their future production and marketing strategies accordingly.

In Chapter 4, "Green Smart Farming Techniques and Sustainable Agriculture: Research Roadmap toward Organic Farming for Imperishable Agricultural Products," highlights the requirement of a suitable Farm Management System to look after the important parameters like productivity, profitability, ecosystem, and human health for practicing and managing green smart sustainable agriculture.

In Chapter 5, "Toward Circular Product Lifecycle Management through Technologies 4.0," authors present the potential of technologies 4.0 for smooth transition from linear to circular business models in Industry 4.0. The perspective shows the importance of Technologies 4.0 toward Circular Product Lifecycle Management through a set of multi-sector case studies.

In Chapter 6, "A Taxonomy on Smart Grid Technology," authors provide a comprehensive review of terminology related to smart grid technology in context of a smart and robust power system with significance of using natural resources for power generation. Emphasis is on the use of smart meters, smart card systems, and other smart solutions through the application of advanced technologies like cloud computing, Internet of Things, and machine learning.

In Chapter 7, "Toward Urban Sustainability: Impact of Blue and Green Infrastructure on Building Smart, Climate Resilient, and Livable Cities," the author identifies a set of recommendations that can help stakeholders adapt Blue and Green Infrastructure practices and their integration into the urban-planning policy and development practices for urban sustainability as the desired outcome.

In Chapter 8, "Smart Factories: A Green Engineering Perspective," the author covers the perspective of smart factories by using various technologies that empower the concept, and thereby the utilization of resources, energy, and space required, thus enhancing productivity, reducing wastage of materials, and employing better sustainable and cleaner methods in the view of Industry 4.0.

In Chapter 9, "Electric Vehicle Research: Need, Opportunities and Challenges," the authors make a sincere effort to cover the fundamental aspects in the field of Electric Vehicle Research. They begin with an overview of electric vehicles, with their need and relevance, and highlight the advantages of electric vehicles as a primary transport mode. The authors have also discussed embedded system modelling of electric vehicles, with special emphasis on the sustainability and safety aspects of electric vehicles.

In Chapter 10, "Sustainable Developments through Energy-Efficient Buildings in Smart Cities: A Biomimicry Approach," focuses on understanding and application of biomimicry principles to make a building self-aerated, self-ventilated, and energy-efficient. A case study of Oxalis Oregana Leaf is presented to understand the property and working mechanism of the leaf to achieve solar capturing and controlling. Bio-inspired strategies were followed to retrofit a prototype, and several simulations were performed to examine the results and analyze the amount of energy saved and reduction in carbon emission.

In Chapter 11, "Biomass-derived Activated Biochar for Wastewater Treatment," the authors discuss the role of the activated biochar produced as a bio-waste in wastewater treatment.

In Chapter 12, "A Study on the Cause and Effects of Paddy Straw Burning by Farmers in Fields and Proposal of using Rice Husk as a Novel Ingredient in Pottery Industry," the authors discuss the various causes and effects of farmers burning paddy straw, as well as the alternate options of using paddy straw or rice husks in making useful products or using it as an alternate fuel source.

In Chapter 13, "Clay and Ceramics as Sustainable and Green Materials to Remove Methylene Blue from Water: A Critical Analysis," the focus is on a big challenge of developing a novel and improved method for removing a toxic and harmful dye known as methylene blue. The authors use clay and ceramic materials for methylene blue removal from aqueous systems. An in-depth analysis shows the significance of different reaction parameters like contact time, percentage removal, adsorbent dosage, size of adsorbent, pH co-ions, temperature and concentration of methylene, and providing reaction kinetics and mechanism involved for methylene blue adsorption by clays and ceramic material simultaneously.

In Chapter 14, "Metal Chalcogenides-based Nanocomposites for Sustainable Development with Environmental Protection Applications," author presents a successful attempt for the efficient preparation of MoS2/Graphene nanocomposites through a simple hydrothermal process to study the photo-catalytic performance of methylene blue degradation.

In Chapter 15, "Renewable Energy in Smart Grid: Futuristic Power System," the authors explore different renewable energy resources, smart grid systems, and challenges associated with the integration of a smart grid with renewable energy. The findings of this study are fruitful for new-generation researchers working in this area.

In Chapter 16, "Experimental Investigation on Different Supplementary Cementitious Materials as Smart Construction Materials to Produce Concrete: Towards Sustainable Development," authors prepare a design mix to prepare higher-grade M50 to M90 by using Supplementary Cementing Materials like Fly Ash and Micro Silica with different proportions to replace cement to save natural resources. Different tests are carried out by authors to study their resistance against acid, chloride, and sulphate attacks to understand durability & sustainability concepts in depth. The outcomes of these test results are given with their effects on High Performance Concrete mixes and examined to give sustainable solutions by using smart construction materials.

In Chapter 17, "Solar Thermal Power Generation: Application of Internet of Things for Effective Control & Management," the author explores the application of Internet of Things for Effective Control & Management in the context of solar thermal-power generation providing multiple benefits of the proposed scheme to help enhancing the performance and reliability of solar thermal-power plants. This study can prove a revolutionary step in this direction.

We wish all our readers and their family members good health and prosperity.

**Vikram Bali, Rajni Mohana, Ahmed A. Elngar,
Sunil Kumar Chawla, and Gurpreet Singh**

Editors

Dr. Vikram Bali is director at IMS Engineering College, Ghaziabad, Uttar Pradesh, India. He graduated from REC, Kurukshetra – BTech (CSE), Post-Graduation from NITTTR, Chandigarh – ME (CSE) and Doctorate (PhD) from Banasthali Vidyapith, Rajasthan. He has more than 20 years of rich academic and administrative experience. He has published more than 50 research papers in International Journals/ Conferences and edited books. He has authored five textbooks. He has published six patents. He is on the editorial board and on the review panel of many international journals. He is series editor for three book series of CRC Press/Taylor & Francis Group. He is a lifetime member of IEEE, ISTE, CSI and IE. He was Awarded Green Thinker Z-Distinguished Educator Award 2018 for remarkable contributions in the field of Computer Science and Engineering duringthird International Convention on Interdisciplinary Research for Sustainable Development (IRSD) at Confederation of Indian Industry (CII), Chandigarh. He has attended faculty enablement programs organized by Infosys and NASSCOM. He is a member of boards of studies for different Indian universities and a member of organizing committee for various national and international seminars/conferences. He is working on four sponsored research projects funded by TEQIP-3 and Unnat Bharat Abhiyaan. His research interests include Software Engineering, Cyber Security, Automata Theory, CBSS and ERP.

Dr. Rajni Mohana is currently working as associate professor in the Department of Computer Science and Engineering, at Jaypee University of Information Technology (JUIT), Waknaghat, Solan, India. She has over ten years of experience in academia. She has various national and international publications to her credit. She is also a reviewer for various renowned international journals and international conferences. She has completed her BTech (CSE) form MITS, Orissa, MTech (CSE) from DAVIET, Jalandhar and PhD (CSE) from Jaypee University of Information Technology, Wakhnaghat, Solan. Her interest areas are software engineering, information systems, service-oriented architecture, aspect-oriented programming, NLP, sentiment analysis, text mining and opinion mining. She is currently pursuing research in Cloud Computing, NLP and Software Engineering. She has supervised various MTech Students and PhD Scholars.

Dr. Ahmed A. Elngar is assistant professor of Computer Science at the Faculty of Computers and Artificial Intelligence, Beni-Suef University, Egypt. Dr. Elngar is the founder and head of Scientific Innovation Research Group (SIRG). Dr. Elngar is a director of the Technological and Informatics Studies Center (TISC), Faculty of Computers and Artificial Intelligence, Beni-Suef University. Dr. Elngar has more than 55 scientific research papers published in prestigious international journals and over 25 books covering such diverse topics as data mining, intelligent systems, social networks, and smart environment. Dr. Elngar is a collaborative researcher. He is a member in Egyptian Mathematical Society (EMS) and International Rough Set Society (IRSS). His other research areas include Internet of Things (IoT), network security, intrusion detection, machine learning, data mining, artificial intelligence, big data, authentication, cryptology, healthcare systems, and automation systems. He is an editor and reviewer of many international journal around the world. Dr. Elngar won several awards including the Young Researcher in Computer Science Engineering, from Global Outreach Education Summit and Awards 2019, on 31 January 2019 in Delhi, India. Also, was awarded the Best Young Researcher Award (Male), Global Education and Corporate Leadership Awards (GECL-2018), Plot No-8, Shivaji Park, Alwar 301001, Rajasthan.

Dr. Sunil Kumar Chawla is working as an assistant professor in the Department of Computer Science and Engineering, Chandigarh University, Mohali, Punjab, India. He received his doctorate from IKG Punjab Technical University. He has taught for more than 17 years. His research interests lie in digital image processing, biometrics, image segmentation, and machine learning. He is a member of IEEE and other professional societies like ISTE, CSTA, IAENG, and IETE. He has more than 30 publications in International journals of repute indexed in Scopus, ESCI, Google Scholar, and other eminent databases. He has been associated with many reputed international conferences in the capacity of Technical Programme Committee member. He is serving as a reviewer of several reputed journals. He has published a book entitled *Segmentation and Normalization of Iris for Human Recognition* with Lambert Academic Publishing, Germany. He has filed two patents. He is the guest editor of two books published by CRC Press/Taylor and Francis Group and many special issues of reputed international journals.

 **Dr. Gurpreet Singh** completed his BTech from Punjabi University, Patiala, and MTech and PhD from IKG Punjab Technical University Kapurthala. His areas of interest are cloud computing, green computing, the Internet of Things, next generation networks, databases and data mining. He has written more than ten papers in various national and international referred journals. He has presented more than 15 papers in various national and international conferences. Dr. Singh has also filed a patent on electronic payment systems using Internet of Things to secure direct payments through wifi cards. He has written one book with International Publisher Lambert Publishers, Germany.

Contributors

Hadeel Fahad Alharbi
College of Computer Science and
   Engineering
University of Ha'il
Ha'il, Kingdom of Saudi Arabia

Damyanti G. Badagha
Civil Engineering Department
S. N. Patel Institute of Technology &
   Research Centre
Umrakh, Gujarat, India

Pankaj Bhambri
Department of Information Technology
Guru Nanak Dev Engineering College
Ludhiana, Punjab, India

Hima C. S.
School of Architecture
KLE Technological University
Hubli, India

Bhawna Chahar
Department of Business Administration
Manipal University
Jaipur, Rajasthan, India

Pradeep Singh Chahar
Department of Physical Education
Banaras Hindu University
Uttar Pradesh, India

Pratik G. Chauhan
Civil Engineering Department
Dr. S. & S. S. Ghandhy Government
   Engineering College
Surat, Gujarat, India

Sunil Kumar Chawla
Department of Computer Science &
   Engineering
Chandigarh University
Gharuan, Mohali, Punjab, India

Amit Chopra
Guru Nanak Dev University
Amritsar, Punjab, India

Subhra Das
Solar Engineering Department
Amity University Haryana
Gurugram, Haryana, India

Bansari N. Dave
Nascent Associates
Rajkot, Gujarat, India

Ahmed A. Elngar
Computers & Artificial Intelligence
Beni-Suef University
Banī Suwayf, Beni-Suef, Egypt

O. P. Gupta
Department of Electrical Engineering &
   Information Technology
College of Agriculture Engineering &
   Technology
Punjab Agricultural University
Ludhiana, Punjab, India

Anurag Jain
Virtualization Department
School of Computer Science
University of Petroleum and Energy Studies
Dehradun, Uttarakhand, India

Indu K.
Department of Electronics and
   Communication Engineering
CHRIST (Deemed to be University)
Bangalore, Karnataka, India

Vinod Karar
CSIR-Central Scientific Instruments
  and Organisation
Chandigarh, India
and
Academy of Scientific and Innovative
  Research (AcSIR)
Ghaziabad, India
and
Chandigarh University
Gharuan, Mohali, Punjab, India

Balween Kaur
Department of Commerce
DAV College for Women
Ferozepur, India

Kritika Khandelwal
Department of Business Administration
Manipal University
Jaipur, Rajasthan, India

Rajneesh Kumar
Department of Computer Science and
  Engineering
Maharishi Markandeshwar (Deemed to
  be University)
Mullana, Ambala, Haryana, India

Aswatha Kumar M.
Department of Electronics and
  Communication Engineering
CHRIST (Deemed to be University)
Bangalore, Karnataka, India

Rizwana M.
Department of Management Studies
Ramaiah Institute of Technology
Bangalore, India

Meenakshi Mittal
Department of Commerce
DAV College for Women
Ferozepur, India

Mohamed Jaffer Sadiq Mohamed
Department of Chemistry
School of Chemical Sciences and
  Technology
Yunnan University
Kunming, China
and
National Center for International Research
  on Photoelectric and Energy Materials
Yunnan Province Engineering Research
  Center of Photocatalytic Treatment of
  Industrial Wastewater
Yunnan Provincial Collaborative
  Innovation Center of Green Chemistry
  for Lignite Energy
Yunnan University
Kunming, China

Barbara Ocicka
Collegium of Business Administration
Department of Logistics
SGH Warsaw School of Economics
Warsaw, Poland

Ashish Pawar
Department of Renewable Energy
  Engineering
College of Technology and Engineering
Maharana Pratap University of
  Agriculture and Technology
Udaipur, Rajasthan, India

N. L. Panwar
Department of Renewable Energy
  Engineering
College of Technology and Engineering
Maharana Pratap University of
  Agriculture and Technology
Udaipur, Rajasthan, India

Kushal Qanungo
Division of Chemistry
University Institute of Science
Chandigarh University
Gharuan, Mohali, Punjab, India

Sita Rani
Department of Computer Science and
    Engineering, Gulzar
Group of Institutions
Khanna, Punjab, India

P. V. Raveendra
Department of Management Studies
Ramaiah Institute of Technology
    University
Bangalore, India

Prabha Roy
Ernest & Young, Global Delivery
    Services - IA
New Delhi, India

Divya Sharma
School of Architecture
KLE Technological University
Hubli, India

Priyanka Sharma
Division of Chemistry
University Institute of Science
Chandigarh University
Gharuan, Mohali, Punjab, India

Ravinder Tonk
Chandigarh University
Gharuan, Mohali, Punjab, India

Shikha Tuteja
CSIR-Central Scientific Instruments
    and Organisation
Chandigarh, India
and
Academy of Scientific and Innovative
    Research (AcSIR)
Ghaziabad, India
and
Chandigarh University
Gharuan, Mohali, Punjab, India

Grażyna Wieteska
Faculty of Management
Department of Logistics
University of Lodz
Lodz, Poland

Beata Wieteska-Rosiak
Faculty of Economics and Sociology
Department of Investment and Real
    Estate
University of Lodz
Lodz, Poland

Kusum Yadav
College of Computer Science &
    Engineering University of Ha'il
Ha'il, Kingdom of Saudi Arabia

1 Achieving Sustainable Development through Sustainable Entrepreneurship and Green Engineering

P. V. Raveendra and Rizwana M.
Department of Management Studies, Ramaiah Institute of
Technology, Bangalore, India

CONTENTS

1.1 INTRODUCTION

Since 1987, the term sustainable development has become a significant notion in the language of various policy makers like politicians, practitioners, and planners (Burton, 1987). One of the oldest definitions for the term sustainable development was published in 1987 in the Brundtland Commission report; its main aim was to help world nations toward sustainable development to encourage harmony between humankind and nature. According to the report, the term sustainable development

DOI: 10.1201/9781003127819-1

1

has been aimed at "satisfying the present need not be at the cost of future." In simple terms, development should be a continuous process of meeting the current needs and considering the requirements of future generations also (WCED, 1987). The term "needs" in the definition emphasizes the indispensable requirement of the poor. In spite of the existence of many definitions for the term sustainable development, the most repeatedly used definition is the one suggested by the Brundtland Commission (Dernbach, 1998). From its inception until now, the concept of sustainable development has made incessant progress as it has been augmented with new variables, coordinates, hypothetical, procedural, and practical valances (Kardos, 2012). "Conserving resources for future generations" is how the term sustainable-development policy stays distinct from conventional environmental policy. Further, the comprehensive objective of sustainable development (SD) emphasizes long-term survival of the economy by integrating and acknowledging the concerns of macro environmental factors throughout the decision-making process (Emas, 2015).

The role of developing nations is vital in maintaining sustainability as they have better experience in the practical application of responsible innovation to serve individuals and society. One of the critical components of sustainable development is designing scientific and technological solutions that are accustomed with the nature of energy and materials society uses (Mihelcic 2005). In the process of achieving sustainable development to maintain and enhance the quality of life without disturbing the planet, sustainability into products, procedures, processes, and systems must be accomplished in an accessible way. This concept is the main theme of green engineering (Anastas & Zimmerman, 2003). To make the resources available in the long run, sustainable development aims to use resources at an optimum level. The society, the economy, and the environment are the three essential pillars of sustainable development, and all the three are interrelated and interdependent; the actions of one pillar may impact the other directly or indirectly, either in the short run or long run (Hoyer & Naess, 2001).

1.2 ENTREPRENEURSHIP AND SUSTAINABLE DEVELOPMENT

The association between entrepreneurship and sustainable development has been communicated by multiple researchers from various streams of thought, theories, and literature. For instance, ecopreneurship focuses on providing solutions to environmental problems, social entrepreneurship focuses on providing solutions to social issues, sustainable entrepreneurship focuses on providing solutions to economic issues, and social and environmental concerns provide solutions through the understanding of successful business goals (Zahedi & Otterpohl, 2015). Entrepreneurship plays a crucial role in addressing unending socio-economic challenges and also contributes toward the attainment of sustainable development goals, especially related to food, security, hygiene and sanitation, and trustworthy sources of energy, which are the pressing priority of the developing and underdeveloped countries (UNCTAD, 2017). In addition to that, entrepreneurs have been strengthened and encouraged to execute ideas on green entrepreneurship that stimulates economic and environmental benefits (Hockerts & Wüstenhagen, 2010). Green entrepreneurship is a broad concept that connects environmental entrepreneurship and sustainable development, and obeys to the "triple bottom line" of the environment, society, and the economy (Ye et al., 2020). With numerous tycoons advocating entrepreneurship as a solution for many social-environmental problems, entrepreneurship is highly instrumental in bringing out positive transformation toward sustainable products and processes (Hall, Daneke, & Lenox, 2010). Developing entrepreneurship for sustainable development is a phenomenon that exhibits the linkages of three aspects, namely, environmental, social, and economic dimensions among entrepreneurial procedures, market revolution, and societal developments (Johnson & Schaltegger, 2019). The paradigm shift toward a sustainable business model, along with the implementation of environment-friendly business practices, can, possibly, provide a surplus range of opportunities for entrepreneurs (Jeevan & Priti, 2014). Many research papers that focus on social, environmental, and sustainable entrepreneurship emphasize that entrepreneurs can ascertain, generate, and capitalize on the opportunities for sustainable development by encouraging major reforms through social and technological innovations leading to market transformations (Cohen & Winn, 2007). Sustainable development is a multifaceted phenomenon that addresses five key attributes: financial capital, physical capital, social capital, human capital, and natural capital. A sustainable society is one that is long lasting, resilient, and intelligent enough for the sustenance of its physical or social systems of support (Meadows, Meadows, & Randers, 1992). Long-term productive capacity coupled with long-term wealth and comfort derived from substitute resources are the key elements of sustainable development (Ahmed & McQuaid, 2005). Thus, sustainability is a thought-provoking process for all society, and particularly for business. It should be a commitment and a fundamental driver of the corporate strategy for the majority of businesses (Jeevan & Priti, 2014). Sustainability requires creative thinking across the field of human endeavor, which is not limited to scientists and technologists (Ahmed & McQuaid, 2005). Undeniably, sustainable development has been viewed as one of the important tools to progress toward a more inclusive model by promoting an exemplary relationship among economic, social, and environmental systems for both existing and future generations (Cobbinah, Black, & Thwaites, 2011). The contribution of entrepreneurship and small business holds an important role in addressing the challenges involved in the implementation of sustainable development (Omri, 2018). Though researchers have agreed that entrepreneurship plays a key role in the economic development of the nation, however, the majority of the literature emphasizes that both sustainable development and entrepreneurship are considered as solutions to guarantee the overall development of the entire society (Hall, Daneke, & Lenox, 2010). In addition to that, the green engineering principles also align with sustainable development.

Considering the concepts of entrepreneurship, sustainability, and sustainable development, sustainable entrepreneurship can be stated as a fusion concept stemming from the concept of both business entrepreneurship and sustainable development (Katsikis & Kyrgidou, 2007).

1.3 SUSTAINABLE DEVELOPMENT AND SUSTAINABLE ENTREPRENEURSHIP

To justify the concepts, an intensive review of literature that emphasizes sustainable development, principles of sustainable development, sustainable entrepreneurship, and green engineering has been carried out and is presented below.

(Austin, Stevenson, & Wei-Skillern, 2016) stated that while establishing the venture, the founders of the company should remember sustainable business models. Accordingly, when sustainability is linked with entrepreneurship, "sustainable entrepreneurship" will fall in place and the same will become the driving force for the integration of a sustainable economic-environmental-social system. Thus, sustainability shifts the focus from the development of technology, innovations, and living conditions to ensure endurance with greater responsibility regarding social, environmental, and economic factors. Thus, sustainable entrepreneurship aims at a perfect blend of people, planet, and profit (Nejoua, Fateh, & Charaf, 2017). Identification of a sustainable innovation and its implementation, either through dynamic reorientation of existing companies or a start-up, achieves the desired ecological or social objectives (Hockerts, 2003). It is the way opportunities are identified, created, and exploited for productions of goods and services, for the present and future, by stakeholders knowing the environmental and ecological consequences (Cohen & Winn, 2007). The definition given by (Tilley & Young, 2009) focuses on ethical dimensions and future dimensions. It aims to improve the quality of life of all stakeholders locally and globally, for both the present and future generations, by doing business through ethical and economic development. However, (Brundtland, 1987) is the first to give a formal definition of sustainable development "which meet the needs of the present without affecting the future generations' ability to meet their own needs."

Various common elements among the many definitions include equity among public, public health, water management, education system for all, renewable energy management, chemical engineering, waste management, environmental protection, optimum utilization of resources, and protection of natural resources. Different disciplinarians stated their own interpretations of sustainable development and tried their best, from their own point of view, without missing the focus on the themes of sustainable development. Among all the disciplinarians, engineering discipline is the one that has applied this concept most effectively compared to any other disciplines. This might be because they are in mechanical (products and processes) or civil (use of natural resources) or chemical (environment protection), and interesting aspects all fall under the engineering or economics domain. This might be the input for the development of the concept of green engineering. Community development and environment should be given equal importance for sustainable development. Development of one without the other will not solve the purpose of sustainability (Khosla, 1987). Sustainable development should focus on ethical principles and future interest with scientific realities (Repetto, 1986). The present generation's development should not be at the cost or survival of future human generations. There must be equal consideration for present and future human health. Of course, this goal should be achieved with satisfaction of immediate and future subsistence needs with a low degree of risk (Norgaard, 1988). Sustainable entrepreneurship should aim for community development, without which the purpose of it will not be achieved (Patzelt & Shepherd, 2011). Sustainable entrepreneurship shall also concentrate on transforming sectors toward sustainability (Schaltegger & Wagner, 2011). Accordingly, the authors have made a comparison between sustainable development and sustainable entrepreneurship based on the available literature.

1.3.1 SIMILARITIES BETWEEN SUSTAINABLE DEVELOPMENT AND SUSTAINABLE ENTREPRENEURSHIP

Authors	Identified Areas in Sustainable Development	Authors	Identified Areas in Sustainable Entrepreneurship
(Norgaard, 1988)	Survival of future human generations and future human health are core concepts. In addition to that, the present and future needs are to be satisfied. All the above should be achieved with minimum risk.	(Nejoua, Fateh, & Charaf, 2017), (Schaltegger & Wagner, 2011)	Innovations, social and environmental dimensions, sustainable technological innovations, transforming sectors toward sustainability
(OECD, 2006)	Developing and sharing the economic benefits to all the stakeholders by converting brownfields into educational and housing projects. Development of innovative industrial processes with ecological balance and environmental protection.	(Hockerts, 2003)	Ecological and social objectives Sustainability of natural resources
(Anastas & Zimmerman, 2003)	Enhancement of quality of life without disturbing ecological balance by proper design of products, systems, processes for today and tomorrow.	(Cohen & Winn, 2007)	Products for present and future with environmental, social, economic consequences
(Repetto, 1986)	Focus on ethical principles and future interest with scientific realities.	(Tilley & Young, 2009)	Ethical dimensions and future dimension along with new opportunities
(Khosla, 1987)	Community development with due consideration of the environment.	(Patzelt & Shepherd, 2011)	Community development

1.4   BUILDING SUSTAINABLE ENTREPRENEURSHIP ECOSYSTEM

The term "entrepreneurship ecosystem" has gained its cynosure in the last decade. It was first defined by Cohen (2005) as "interrelated group of performers in a geographic community who are dedicated to seed sustainable development" by providing assistance and creating novel sustainable ventures. According to (Van De Ven, 1993), entrepreneurial ecosystems progress through an interaction of a myriad of mutually supporting components, which may create new ventures over time. (Figure 1.1).

The fundamental idea of an entrepreneurial ecosystem is to provide a favorable environment that supports innovation and aids in the formation of new ventures. Therefore, sustainable economic growth can bring it within a specific geographic boundary (Garud, Kumaraswamy, & Karnøe, 2010). The application of multiple components of entrepreneurial ecosystem was first examined by (Neck, Meyer, Cohen, & Corbett, 2004). According to (Neck et al., 2004), entrepreneurial ecosystem components, namely informal networks like entrepreneur family members and friends, formal networks like government, professional services, universities, support services, capital services, and access to the talent pool are the predominant variables in fostering a sustainable entrepreneurial ecosystem.

The main objective of sustainable development ventures in the purview of the sustainable entrepreneurial ecosystem arises to create economic, social, and

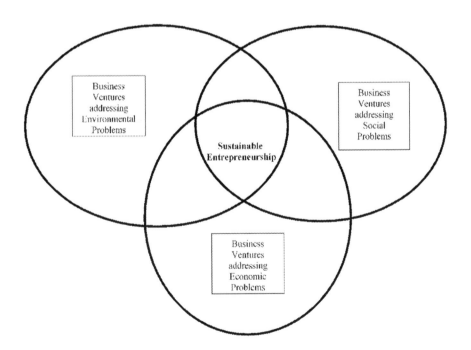

FIGURE 1.1   Conceptual framework of sustainable entrepreneurship and green engineering.

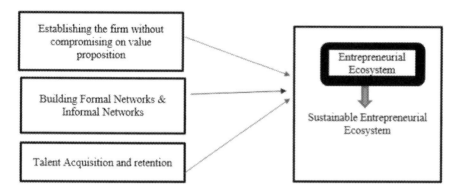

FIGURE 1.2 Components of building sustainable entrepreneurial ecosystem.

environmental value (Pankov, Velamuri, & Schneckenberg, 2019). While establishing the business venture only, it is imperative for the entrepreneurs to have a sustainable business idea, and the actions of sustainable business ventures must be grounded by considering the concerns that are particular to economy, environment, and society.

In addition to that, to build a sustainable ecosystem, an entrepreneur should ensure supportive talents and not compromise on the value of the long-term vision, culture, employees, collaborations, and product planning (Forbes Business Council, 2020). Similarly, entrepreneurs must focus on developing a sustainable business model that stays afloat in the long run by considering environmental impacts. To execute a sustainable business plan, entrepreneurs must understand the key elements of a sustainable business model as a value proposition, long-term vision, and usage of sustainable resources (Hendricks, 2018).

Well-developed entrepreneurship ecosystems are highly instrumental in building a sustainable economy. They look at the micro-level overview. For corporate entrepreneurs, it is very important to focus on the key actions like drafting to support guidelines for an underprivileged segment of people, establishing the ventures that support the local community, and collaborating with other private partners (Endeavor Insight, 2015). For example, the UAE is an oil-driven economy, and to materialize the knowledge driven and entrepreneurial economic model from the oil-driven economy, they have considered three core components, like "getting corporates involved to use the competencies, investors' participation, and community education" (Jagannathan, 2017) (Figure 1.2).

1.5 SUSTAINABLE DEVELOPMENT AND GREEN ENGINEERING

Innovations play an important role in designing the products, processes, and systems with sustainability as a core theme. It is the right time to think about social innovations that will aim to protect the environment, value generation to all the stakeholders, recycle waste, and use renewable-energy resources in the process. Many companies are focusing on sustainability missions, considering the above factors, with optimum utilization of resources for today and tomorrow and benefit to all stakeholders in the society. One should not forget that, ultimately, the companies have to compete in a dynamic market. There are many other challenges for the companies to play safe in the competitive world (Jenck, Agterberg, & Droesher, 2004).

Different researchers had given different principles of sustainable development. Of course, all the principles have commonalities. However, Veselaj (2019) has identified the following ten principles to achieve sustainable development.

a. Optimum utilization of resources: Resources are not abundant but scarce. In such a situation, the resource utilization should be careful and productive in such a way that it will lead to the development of the present and future. It includes protection of natural resources. Some European countries have adopted these principles as an integrated part of their laws so that all will follow automatically.

b. "Prevention is better than the cure" is a well-known saying. It holds well in environment degradation, too. Optimum utilization of resources should be attained without affecting the pollution level and environmental degradation.

c. Social, economic, and environmental development is incomplete without the public's participation. Awareness of sustainable development among the public is not sufficient; involvement of the public in decision making is important.

d. Manufacturers have to pay real costs for the activities involved in the consumption of natural resources, as well as the activities engaged that will harm the environment. Now many governments have National Pollution Boards in place to address this need. Integration of natural resources of the nations and globe is essential to achieve sustainable development. In simple terms, national water policy, forest policy, energy policy, agricultural policy, industrial policy, trade policy, and other related policies should have a common vision.

e. Equity among the generations is the core principle.

f. A systems approach is also known as a holistic approach. A system is a group of interrelated and interdependent elements that are working toward a common goal. Similarly, social, ecological, and economic concepts are interrelated and interdependent and working toward the common goal of sustainable development.

g. Social reasonability of the producers of the goods and services is always an important element of sustainable development. Use of limited resources should be compensated in some way. This principle also highlights the need for waste management and reforestation.

h. Conservation and sustainable use of nature and local diversity needs to be maintained.

i. A healthy environment should be made a fundamental right in society.
j. The factors to be sustained for sustainable development are i) Nature ii) Sources of life support and iii) Communities.

Current engineering evolution is due to increased concerns regarding the degradation of the environment, global warming, and the decrease in natural resources and increase in population growth. This all leads to new design development, with sustainability as a performance factor. The goal of wealth creation for society is concerned about the current generation as well as the future generation. Optimum utilization of resources is needed for society both today and tomorrow. Utilization of resources largely depends on the design, development, and commercialization of industrial processes that are not only sustainable but also economically feasible with protecting the environment. Through green engineering, the above will be met with little risk to human health and environment. The right decision-making process at the design and development phase of a product or process will impact the cost effectiveness with proper protection of environment and human health (Patel, Kellici, & Saha, 2013).

The 12 green engineering principles (Anastas & Zimmerman, 2003) are collectively guiding principles for engineers. However, one should not forget that engineers are more involved in successful implementation rather than making policy decisions. Entrepreneurs play a vital role in decision making. Both engineers and entrepreneurs should be aware of these principles. These principles are focused much beyond quality and safety, and they should be considered in designing processes, products, and systems. These principles of green engineering are discussed in detail.

a. Focusing on nonhazardous production in society by proper utilization of entire materials and energy inputs.
b. Focusing on prevention of waste rather than recycling of waste.
c. Designing framework with purification and separation operations.
d. Increase production with systems components, including maximizing mass and energy.
e. Pulling output rather than pushing inputs through system components.
f. Planning of the recycling process and reuse of resources should be considered from an investment point of view that will lead to sustainability.
g. Long-lasting results should be the goal for design.
h. Using "one size fits all" solutions rather than building unnecessary capacity.
i. Minimizing material diversity and value retention for multi-component product development.
j. Integration and interconnectivity of design of process and system with available energy and material flows.
k. Performance will not be output unless it is designed. The design should also aim for the performance after the commercial life.
l. Usage of reusable and largely available inputs throughout the life cycle.

1.5.1   SIMILARITIES BETWEEN SUSTAINABLE DEVELOPMENT AND GREEN ENGINEERING

Authors	Identified Areas in Sustainable Development	Authors	Identified Areas in Green Engineering
(Norgaard, 1988)	Survival of future human generations, future human health, satisfaction of immediate and future subsistence needs with low degree of risk	(Patel, Kellici, & Saha, 2013).	Factors in consideration with optimum utilization of resources for today and tomorrow and aim to benefit all the stakeholders in the society.
(OECD, 2006)	Developing and sharing the economic benefits to all stakeholders by converting brownfields into educational and housing projects. Development of innovative industrial process with ecological balance and environmental protection.	(Veselaj, 2019)	Performance will not be output unless it is designed. The design should also aim for the performance after the commercial life. There should be clear cut plants for reuse of the resources throughout its life cycle.
(Anastas & Zimmerman, 2003)	Enhancement of quality of life without disturbing ecological balance by proper design of products, systems, processes for today and tomorrow.	(Anastas & Zimmerman, 2003)	Principles are focus beyond quality and safety, and should be considered in designing process, products, and systems.
(Repetto, 1986)	Focus on ethical principles and future interest with scientific realities.	(Veselaj, 2019)	Planning of recycling process and reuse of resources should be considered from investment point of view that will lead to sustainability
(Khosla, 1987)	Community development with due consideration of the environment.	(Ye et al., 2020).	Environment, society, and economy.

The following conceptual framework has been developed. (Figure 1.3).

From the above picture, it is clear that sustainable development is the common area among social entrepreneurship and green engineering.

After studying the 12 principles of green engineering and ten principles of sustainable development, one can understand that there exist some commonalities among these principles.

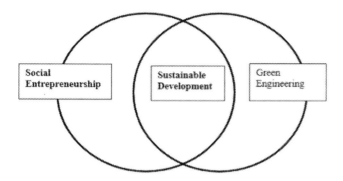

FIGURE 1.3 Inter relationships between social entrepreneurship, sustainable development, and green engineering.

Mere knowledge of these principles is not sufficient. One has to plan for implementation of these principles. After gaining awareness of these principles, the biggest challenge lies with the implementation of these principles. The following table gives commonalities in the implementation of sustainable development and green engineering principles.

1.6 COMMONALITIES IN THE IMPLEMENTATION OF SUSTAINABLE DEVELOPMENT AND GREEN ENGINEERING PRINCIPLES

Sustainable Development Principles	Green Engineering Principles	How to Implement
Optimum utilization of resources: Resources are not abundant but scarce. In such a situation, the resource utilization should be careful and productive, leading to the development of the present and future. It includes protection of natural resources. Some European countries have made these principles as part of their laws so that all will follow automatically.	Focusing on non-hazardous production in society by proper utilization of entire material and energy inputs.	Both the principles are aiming for the same but with different approaches. Optimum utilization of all the resources is the main theme of sustainable development and not to produce non-hazardous production in the society is added input from the green engineering. The scope of engineering is wide and different disciplines of engineering started practicing it. Green construction is one among them. All the countries, irrespective of developing or under-developing, who are planning to construct houses for the needy should plan only for green houses. Similarly, all the major project

(Continued)

Sustainable Development Principles	Green Engineering Principles	How to Implement
		approvals should be based on green engineering principles. When the governments take the initiative of implementation of these principles, private organizations also will come forward to practice the same. Of course, the financial resources might be a consideration in the short run, but in the long run, it is definitely economically viable and beneficial to the society. Similarly, green chemistry is one of the emerging areas, if the government's plan to give licenses to those industrial houses which plan for green production. Its implementation becomes easy. Governments should plan to give financial incentives for those projects, industries, startups which are going to plan these principles.
Prevention is better than cure is a well-known proverb. It holds well in environment degradation also. Optimum utilization of resources should be attained without affecting the pollution level and environmental degradation.	Focusing on prevention of waste rather than recycling of waste.	Use of renewable energy resources will avoid production of waste itself. Depending upon the country's geographical conditions it has to plan whether to use wind energy, solar energy, hydro power generation. The electric power lost in transportation is around 20% in many countries. If they reduce these distribution losses, it will be great savings for them. If sufficient numbers of dams are built, water scarcity can be avoided in many countries. By building proper storage facilities across the nation, the wastage of grains and other items can be minimized. The main principle is prevention of waste is a suitable strategy rather than recycling the waste after its production.
Manufacturers have to pay real costs for the activities involved in consumption of natural resources as well as	Increased production with systems components, including maximizing mass and energy.	The competition in the markets forced them to change from mass production to mass customization. The price of the product is no more

Sustainable Development Principles	Green Engineering Principles	How to Implement
the activities engaged that will harm the environment.	Minimizing material diversity and value retention for multi-component product development.	cost plus product, but it is the value for money or the price affordable by the customer. Both the concepts focus on the value for money. The resources utilized directly and indirectly to be considered. At the same time, the functions offered by the product directly and indirectly to be considered. The cost of inputs includes usage of all environmental resources to be borne by the producers. To do this, many governments had national boards to take care of this. Integration of natural resources of the nations and globe is essential to achieve sustainable development. In simple terms, national water policy, national forest policy, national energy policy, national agricultural policy, national industrial policy, national trade policy, and other related policies should have a common vision of sustainability
Systems approach is also known as holistic approach to be practiced. Similarly, social, ecological, and economic concepts are interrelated and interdependent, working toward the common goal of sustainable development.	Integration and interconnectivity of design of process and system with available energy and material flows. Pulling output rather than pushing inputs through system components.	A system is a group of elements that are interrelated and interdependent and working toward a common goal. The use of natural resources productively leads to economic development and if it. The integrated approach toward the use of resources shall be considered in production systems.
Conservation and sustainable use of nature and local diversity to be maintained.	Long lasting should be the goal for design.	We produce what we design. Sustainability should be considered while designing the products and use of resources. Government should encourage those companies that are going to produce long-lasting products and the companies that plan to manufacture with sustainable goals.
Social reasonability of the producers of the goods and services is an important	Usage of reusable and largely available inputs throughout the life cycle.	One-time use of resources should be discouraged. Instead manufacturers should be encouraged to use

(Continued)

Sustainable Development Principles	Green Engineering Principles	How to Implement
element of sustainable development. Use of limited resources should be compensated in some way. This principle also highlights need for waste management and reforestation.	Performance will not be output unless it is designed. The design should also aim for the performance after the commercial life.	renewable resources throughout the life cycle of the products.

1.7   CONCLUSION

From the above reviews, it can be inferred that the following are the key dimensions of sustainable entrepreneurship, sustainable development, and green engineering:

- Preservation of natural resources
- Development of goods and services for present and future
- Economic gains along with social gains
- Balancing both environmental and social concerns
- Long term integrated approach for people, planet and profit
- Transforming sectors towards sustainability
- Sustainable venture performance and cultural consideration

Apart from exhibiting similarities between sustainable development and sustainable entrepreneurship, the summary clearly provides the simple difference between conventional and sustainable entrepreneurs where the former one focuses on creating customer value for profit maximization (Ludeke-Freund, 2019) and the latter one focuses on solving environmental and social problems through their business. The term "sustainability" could be viewed expansively as three goals of economic development, ecological balance, and societal equity. In addition to that, the development of sustainable entrepreneurship in the world depends on the action of passionate people who have that internal drive to create a business model that has environmental, social, and economic impact. As this concept is in the development stage, all the stakeholders of the world should be involved in the development of sustainable entrepreneurship.

1.8   LIMITATIONS OF THE STUDY

Sustainable development requires huge amounts of investment by governments and enterprises. Whether they are ready for it is a big challenge. Some rich countries can afford this type of investment. However, the underdeveloped nations and developing countries will have constraints. Underdeveloped nations are more interested in the present community development rather than focusing on both present and future. At the same time, it is a big challenge for all the companies in both developed nations and developing nations to pay the price for the natural resources they have used. If they start paying the price for all the resources they have used, it will be tough for them to compete in the market. In the completion, "price of the product or service" plays an important role. Already, there are huge disparities among the developed nations and underdeveloped nations; forcing any compulsions may further increase the disparities among them. For example, the true spirit of the World Trade Organization was difficult to implement successfully due to various conflicts among the member countries. There are so many benefits to society through green engineering, sustainable development, and sustainable entrepreneurship. Making all stakeholders aware of benefits and involving them all in the decision making will minimize some of the limitations. The discussions on concepts and principles of sustainable development, sustainable entrepreneurship, and green engineering are not new. They have occurred since the 1980s onward; however, there is not much progress in the implementation of these principles. There are some exceptions to this idea. For example, Bhutan had some of the principles of these concepts in its constitution itself. The Oman government had made sustainability in its vision document with strict compliance. The push should come from big countries rather than smaller ones. Some universities are offering degrees and certificate courses on green engineering and sustainable development. Universities offering entrepreneurship courses should have made green engineering, sustainable development, and sustainable entrepreneurship as part of the course. There are journals in the name of green engineering and sustainable development. However, this study is restricted to the discussion of the conceptual framework and relationship among them.

1.9   FUTURE SCOPE OF THE STUDY

There are 12 principles for green engineering and ten principles of sustainable development, which were discussed in the above paragraphs. There is scope of further research in each of these principles in detail. The following list of the principles is rewritten from the point of view of identifying further research.

    a. Focusing on nonhazardous production in society by proper utilization of entire material and energy inputs.

    b. Focusing on prevention of waste rather than recycling of waste.

    c. Designing a framework with purification and separation operations.

    d. Increasing production with systems components, including maximizing mass and energy.

    e. Pulling output rather than pushing inputs through system components.

    f. Planning of recycling process and reuse of resources should be considered from an investment point of view that will lead to sustainability.

    g. Long lasting should be the goal for design.

    h. Using "one size fits all" solutions rather than building unnecessary capacity.

    i. Minimizing material diversity and value retention for multi-component product development.

    j. Integration and interconnectivity of design of process and system with available energy and material flows.

    k. Performance will not be output unless it is designed. The design should also aim for the performance after the commercial life.

    l. Usage of reusable and largely available inputs throughout the life cycle.

For example, renewable energy itself is a wide area. Waste management and recycling, green chemistry, and green construction are some of the emerging areas that have green engineering and sustainability in their roots. Involvement of community in sustainable development is vital. Without the support of all stakeholders, no program can be successfully implemented. Government should plan for bringing awareness about the use of natural resources for sustainability. For example, integration of former lands with optimum utilization of water resources will be beneficial. Researchers can focus on their local areas' natural resources and plan how to involve local stakeholders for the benefit of all. Researchers can also study how to make sustainability a social movement rather than any other government schemes. Some countries make sustainability principles as part of their constitutions so the programs will be implemented irrespective of the parties who run the government. Researchers can focus more on developing nations and underdeveloped nations since these countries are going to make huge investments in infrastructure in the near future. With additional investments, these countries can plan to implement sustainable principles easily. Researchers can also plan how sustainable development can be made part of the syllabus in the university curriculum. The authors are suggesting that these sustainability principles can be part of primary and secondary education. Researchers can focus on how to utilize the water, soil, and labour optimally. They can also focus on how artificial intelligence, block-chain technology, design thinking, and other latest developments will be integrated with the sustainability principles. Usage of latest technologies will be helpful, particularly for those countries with huge natural resources but below the poverty line. Researchers can focus on the outcomes of corporate social responsibility activities and can suggest sustainability can be made part of CSR. Some countries are not aware of how much natural resources are wasted. In developing nations, electricity transmission losses would be around 30% to 40%. Wastage of food grains due to lack of storage facilities would be around 20% to 30%. Soil erosion is the problem in some of these countries. Use of unnecessary fertilizers has badly impacted the fertility of the soil throughout the world. Again, the traditional farming is back within the name organic farming. However, this concept is not prevalent in many countries, but proper implementation of organic farming will lead to some of the components of sustainability. Research discusses the reasons behind this and how to overcome the problems. Many studies restrict their studies in finding the reasons for the problems rather than suggesting strategies to overcome the problems. Involvement of the researchers in finding the solutions will be great support to the society in suggesting suitable strategies. Engineering should encourage and design only sustainable products for the future. They have to plan for sustainable processes and products with limited resources. There is huge scope for the researchers also to study in detail each principle in theory as well as its implementation in various domains.

REFERENCES

Ahmed, & McQuaid, R (2005). Entrepreneurship, management, and sustainable development. *World Review of Entrepreneurship, Management and Sustainable Development*, 1, 6–13. doi:10.1504/WREMSD.2005.007750

Anastas, P, & Zimmerman, J (2003). Design through the twelve principles of green engineering. *Environmental Science and Technology*, 37, 94A–101A.

Austin, J, Stevenson, H, & Wei-Skillern, J (2016). *Social and Commercial Entrepreneurship: Same, Different, or Both?* Baylor University, pp. 1042–1258.

Brundtland, GH (1987). World commission on environment and development. *Our com-mon future*. Oxford, New York: Oxford University Press.

Burton, I (1987). Our common future: The world commission on environment and development. *Environment*, 29(5), 25–29.

Cobbinah, P, Black, R, & Thwaites, R (2011). Reflections on six decades of the concept of development: Evaluation and future research. *Journal of Sustainable Development in Africa*, 13(7), 143–158.

Cohen, B (2005). Sustainable valley entrepreneurial ecosystems. *Business Strategy and the Environment*, 15(1), 1–14. doi:10.1016/j.jbusvent.2004.12.001

Cohen, B, & Winn, MI (2007). Market imperfections, opportunity and sustainable entrepreneurship. *Journal of Business Venturing*, 22(1), 29–49.

Dernbach, JC (1998). Sustainable development as a framework for national governance. *Case Western Reserve Law Review*, 49(1), 1–103.

Emas, R (2015). *The Concept of Sustainable Development: Definition and Defining Principles*. Florida International University, pp. 134–350.

Endeavor Insight (2015, August 25). How to create a sustainable business ecosystem 101? *Entrepreneurship Ecosystem Insight*. Retrieved December 17, 2020, from Entrepreneurship Ecosystem Insights website: http://www.ecosysteminsights.org/how-to-create-a-sustainable-business-ecosystem-101/

Forbes Business Council (2020, February 24). Council post: 12 ways entrepreneurs can build a sustainable business. Retrieved December 16, 2020, from Forbes website: https://www.forbes.com/sites/forbesbusinesscouncil/2020/02/24/12-ways-entrepreneurs-can-build-a-sustainable-business/?sh=33fe9a6d3392

Garud, R, Kumaraswamy, A, & Karnøe, P (2010). Path dependence or path creation? *Journal of Management Studies*, 47(4), 760–774. doi:10.1111/j.1467-6486.2009.00914

Hall, J, Daneke, G, & Lenox, M (2010). Sustainable development and entrepreneurship: Past contributions and future directions. *Journal of Business Venturing*, 25(5), 439–448. doi:10.1016/j.jbusvent.2010.01.002

Hendricks, D (2018). What is a sustainable business model? – business.com. Retrieved December 16, 2020, from business.com website: https://www.business.com/articles/how-to-create-a-sustainable-business-model/

Hockerts, K (2003). *Sustainability Innovations, Ecological and Social Entrepreneurship and the Management of Antagonistic Assets*. Bamberg: Difo-Druck. http://books.google.dz/books?

Hockerts, K, & Wüstenhagen, R (2010). Greening Goliaths versus emerging Davids—Theorizing about the role of incumbents and new entrants in sustainable entrepreneurship. *Journal of Business Venturing*, 25(5), 481–492. doi:10.1016/j.jbusvent.2009.07.005

Hoyer, KG, & Naess, P (2001). The ecological traces of growth: Economic growth, liberalization, increases consumption – and sustainable urban development. *Journal of Environmental Policy and Planning*, 3(3), 177–192.

Jagannathan, R (2017, February 7). Three steps to building a sustainable entrepreneurial economy in the UAE. Retrieved December 16, 2020, from Entrepreneur website: https://www.entrepreneur.com/article/288800

Jeevan, & Priti (2014). Green entrepreneurship – A conceptual framework (April 28, 2017). National Conference on Change and Its Contemporary Social Relevance – Department of Social Work, SIMS, 27th September 2014, ISBN No. 978-81-929306-1-9, Available at SSRN: https://ssrn.com/abstract=2960064

Jenck, JF, Agterberg, F, & Droesher, MJ (2004). Products and processes for a sustainable chemical industry: A review of achievements and prospects. *Green Chem, 6*, 544–556.

Johnson, MP, & Schaltegger, S (2019). Entrepreneurship for sustainable development: A review and multilevel causal mechanism framework. *Entrepreneurship Theory and Practice, 44*(6), 1141–1173. doi:10.1177/1042258719885368

Kardos, M (2012). The relationship between entrepreneurship, innovation and sustainable development. Research on European union countries. *Procedia Economics and Finance, 3*, 1030–1035. doi:10.1016/s2212-5671(12)00269-9

Katsikis, IN, & Kyrgidou, LP (2007). The concept of sustainable entrepreneurship: A conceptual framework and empirical analysis. *Academy of Management Proceedings, 2007*(1), 1–6. doi:10.5465/ambpp.2007.26530537

Khosla, A (1987). Alternative strategies in achieving sustainable development, pollution prevention as corporate entrepreneurship. *Journal of Organizational Change Management, 11*(1), 86–89. doi:10.1108/09534819810369554

Ludeke-Freund, F (2019). Sustainable entrepreneurship, innovation, and business models: Integrative framework and propositions for future research. *Business Strategy and The Environment, 29*(2), 665–681. doi:10.1002/bse.2396

Meadows, D, Meadows, DL, & Randers, J (1992) *Beyond the Limits.* Chelsea Green Pub. Co.

Mihelcic, J, Ramaswami, A, & Zimmerman, J (2005). Integrating developed and developing world knowledge into global discussions and strategies for sustainability. 1. Science and technology. *Submitted to Environmental Science and Technology, 41*(10), 3415–3421. https://doi.org/10.1021/es060303e

Neck, HM, Meyer, GD, Cohen, B, & Corbett, AC (2004). An entrepreneurial system view of new venture creation. *Journal of Small Business Management, 42*(2), 190–208. doi:1 0.1111/j.1540-627x.2004.00105.x

Nejoua, DS, Fateh, DM, & Charaf, DB (2017). A conceptual overview of sustainable entrepreneurship. *JFBE*, II(I), 370–394.

Norgaard, RB (1988). Sustainable development: An evolutionary view. *Futures, 20*(6), 606–620.

OECD (2006). *Advancing Sustainable Development.* Paris: OECD Policy Briefs OECD Publishing.

Omri, A (2018). Entrepreneurship, sectoral outputs and environmental improvement: International evidence. *Technological Forecasting and Social Change, 128*, 46–55.

Pankov, S, Velamuri, VK, & Schneckenberg, D (2019). Towards sustainable entrepreneurial ecosystems: Examining the effect of contextual factors on sustainable entrepreneurial activities in the sharing economy. *Small Business Economics, 56*(2021), 1073–1095. https://doi.org/10.1007/s11187-019-00255-5

Patel, D, Kellici, S, & Saha, B (2013). Some novel aspects of green process engineering. *Chim. Oggi*, 31, 57–61.

Patzelt, H, & Shepherd, D (2011). Recognizing opportunities for sustainable development. *Entrepreneurship Theory and Practice, 35*(4), 54–56.

Repetto, R (1986). *Economic Policy Reforms for Natural Resource Conservation.* Washington, DC: World Resources Institute, pp. 16–17.

Schaltegger, S, & Wagner, M (2011). Sustainable entrepreneurship and sustainable innovation: Categories and interactions. *Business Strategy and The Environment, 20*(4), 227–237.

Tilley, F, & Young, W (2009). Sustainability entrepreneurs – Could they be the true wealth generators of the future? *Greener Management International*, 55, 79–92.

UNCTAD (2017). Promoting entrepreneurship for sustainable development: A selection of business cases from the empretec network. Retrieved December 1, 2020, from: https://unctad.org/system/files/official-document/diaeed2017d6_en.pdf

Urban, B, & Kujinga, L (2017). The institutional environment and social entrepreneurship intentions. *International Journal of Entrepreneurial Behavior & Research*, 23(4). doi:10.1108/IJEBR-07-2016-0218

Van De Ven, H (1993). The development of an infrastructure for entrepreneurship. *Journal of Business Venturing*, 8(3), 211–230. doi:10.1016/0883-9026(93)90028-4

Veselaj, Z (2019). Principles of sustainable development as norms of the current legislative framework in Kosovo. *European Journal of Sustainable Development Research*, 3(4), em0099. doi:10.29333/ejosdr/5878Research

WCED (UN World Commission Environment and Development) (1987). Our common future: Report of the World Commission Environment and Development, p. 43.

Ye, Q, Zhou, R, Anwar, MA, Siddiquei, AN, & Asmi, F (2020). Entrepreneurs and environmental sustainability in the digital era: Regional and institutional perspectives. *International Journal of Environmental Research and Public Health*, 17(4), 1355. MDPI AG. Retrieved from: 10.3390/ijerph17041355

Zahedi, A, & Otterpohl, R (2015). Towards sustainable development by creation of green social entrepreneur's communities. *Procedia CIRP*, 26, 196–201. doi:10.1016/j.procir.2014.07.037

2 Environmental Performance Index Score: A Driving Force towards Green Business Model and Process Innovations

Kritika Khandelwal and Bhawna Chahar
Department of Business Administration, Manipal University, Jaipur, Rajasthan, India

Pradeep Singh Chahar
Department of Physical Education, Banaras Hindu University, Uttar Pradesh, India

CONTENTS

DOI: 10.1201/9781003127819-2

2.1   INTRODUCTION

Sustainability is a broad concept meaning that people cannot ignore the environmental impact and business impact, both positive and negative senses (Couckuyt & Van Looy, 2019). Currently, numerous business models have eco-efficiency as their objective. Eco-efficiency is defined as the production of products and services at competitive prices that meet human requirements and provide the worth of life. However, the environmental consequences and the use of numerous resources during the life cycle are progressively reduced-level equivalent, at least, to the estimated capacity of the planet.

The issue of environmental pressure on firms is increasing day by day so that this part of the domain must have recent strategic planning (Javed, Yasir, & Majid, 2018). The fast pace of industrialization toward achieving economic development has had a lasting and detrimental impact on the environment. Environmental damage has created a danger to environmental health and caused an ecological imbalance. Moreover, a large part of the pollution is caused by substantial wasteful utilization of resources (Chen, 2008). Therefore, damage to the environment is irrefutable.

In the above Figure 2.1, WMO has highlighted accelerating adverse effects of environmental change on public wellbeing due to environmental health deterioration and impact on nature due to ecosystem volatility. The significant impact of environmental deterioration is rising temperature. Global climate study of the year 2020 provides authoritative evidence of an incremental increase in global temperature (Figure 2.2).

From 1880–2020, global temperatures in March have shown an incremental pace and a significant increase in resulting impacts, for example, accelerating sea-level rise, contracting ocean ice, glacier retreat, and heat waves. As a result, the global average surface temperature in 2020 has been around 1.16°C over the pre-industrial baseline.

2.2   SIGNIFICANCE OF THE STUDY

According to the Climate Action Summit (2019), "There is no longer time for delay." The effects of environmental change must be emphasized, and there is an urgent need to accomplish the sustainable development goals for environmental safety. The crusade for safeguarding the environment with the self-doctrine of sustainable development has become one of the biggest challenges and most important targets of the present time. Environmental performance as a sustainable move to control the growth rate of environmental hazards has become an urgent need. Global green transition can only be a savior in this alarming situation. Greening of value-chain processes may help improve resource productivity (Bisgaard et al., 1995). Green policies supporting eco-innovation and green business model development may positively impact sustainable

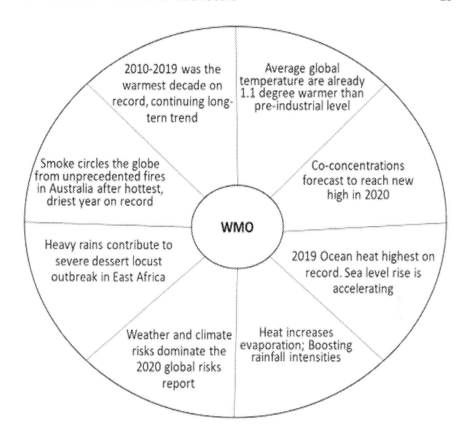

FIGURE 2.1 Impacts of environmental health deterioration and ecosystem volatility.

Source: World Meteorological Organization (2020) Climate action is a priority and a driver of world affairs: UN chief.

development (Henriksen et al., 2012). A stepping point toward environmentally sustainable practices calls for an industrial paradigm shift from conventional business models and processes to green business models and processes. To reinforce the green dimension, many nations have imposed a penalty mechanism through a green policy framework. The companies are measured on the environmental indexes and are enforced to improve the index for their long-term sustainability. Therefore, we must identify green practices and their relative contribution to avoid environmental damage. The growing importance of saving the environment by regulation gave impetus to the study to measure and identify the best green practices for environmental performance.

2.3 LITERATURE REVIEW

2.3.1 GREEN BUSINESS MODEL AND PROCESS INNOVATION

Green innovations include developing an entirely new process or modification in the existing processes, products, and overall operating systems that are environmentally

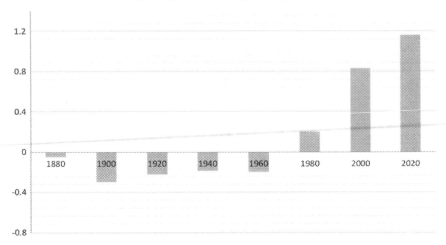

FIGURE 2.2   Global March temperature from the Year 1880 to 2020.

Source: Global Climate Report – NOAA (March 2020).

free and help drive environmentally sustainable operation (Oltra & Saint Jean, 2009). More business models that are innovated or re-engineered with a subsequent green effect have reflected a more significant green change. The greener the business model innovation, the greater the likelihood of forming essential green innovation (Bisgaard et al., 1995).

2.3.2 ENVIRONMENTAL PERFORMANCE INDEX

Green transformation is desirable to preserve the enduring development of the economy. Green business models focusing on sustainability in environmental and resource matters are broad contributors to environmental protection (Ernst & Young, 2008). The 2020 environmental performance index (EPI) ranks 180 countries on 32 performance indicators across 11 issue categories covering environmental health and ecosystem vitality. This metric measures how closed countries are to well-known environmental policy goals at a national scale. The EPI bids a scorecard that highpoints leaders and laggards in environmental performance (Wendling et al., 2020). Some countries have designed very encouraging green policies and robust penalty mechanisms for non-green companies (Seychelles National Climate Change Committee, 2009). For example, Libya aims to meet 10% of energy requirements from renewable energy by 2030 (Nachmany et al., 2017). The countries that are excelling in EPI have revolutionized business practices by innovative green models and processes. The ever-increasing environmental threat has lately but awakened the laggards' countries to go green.

2.3.3   Lead Generation

EPI provides complete insights on most acceptable practices and provides direction for nations seeking to be privileged in sustainability. EPI uses 32 indicators of environmental performance, which guide other nations to improve their overall environmental performance. However, which indicator can be improved if the innovative green process model is not strongly analysed? This was identified as a gap, and research is directed toward analysing green models and process innovations about the specific indicator on EPI.

2.4   RESEARCH METHODOLOGY

The research aims to explore innovative green practices and identify the impact of green practices on environmental performance specific to each indicator. First, the exploratory research design is used to get insights into green practices and technological advancements. After that, the EPI improvement score is identified with a descriptive design. The scope of the research has been confined to measuring the 2020 EPI index v/s last ten years longitudinal data series drove baseline index. The top ten indicators of EPI contributing to environmental health and ecosystem volatility are taken for impact analysis of green practices. Green practices are identified from the top five ranked countries on EPI as Denmark, Luxembourg, Switzerland, United Kingdom, and France.

The hypothesis of the study has been defined as **H0:** green business model and process innovations do not have a significant impact on indicators of EPI.

2.4.1   Environmental Performance Index Measurement Methodology

The measurement of EPI has a hierarchical approach. The below table describes the methodology used to construct the 2020 environmental performance index. The measurement is based upon two objectives as environmental health and ecosystem vitality. It includes 32 total variables grouped into 11 major indicators. Four indicators explain environmental health and seven ecosystem vitality (Table 2.1).

2.5   ANALYSIS

The following Table 2.2 identifies the top five countries aggressive in environmental performance with the highest score attainment

The green business innovation model and business-process innovation analysis are done concerning the top five EPI-listed countries set as benchmarking green practices.

2.5.1   Circular Economy

The circular economy is a closed cycle with a 3R approach – reduce, reuse, and recycle. The circular economy is analysed as an alternative to the traditional linear economy as a more resilient and future-fit strategy to manage waste. Denmark is the

TABLE 2.1

Environment Performance Indicator Metrics

Environment Performance Index (EPI) Indicators	Environmental Health (40%) (It measures threat to human health)	Air quality (20%)	PM$_{25}$ Exposure (11%)
			Household solid fuels (8%)
			Ozone (1%)
		Sanitation & Drinking Water (16%)	Drinking water (9.6%)
			Sanitation (6.4%)
		Heavy metals (2%)	Lead (2%)
		Waste management (2%)	Solid waste (2%)
	Ecosystem vitality (60%) (It measured threat to natural resources and eco system services.)	Climate change (24%)	CO$_2$ (11.6%)
			CH$_4$ (3.15%)
			SNMI (3%)
			F-Gas (2.4%)
			So$_2$ (1.5%)
			NO$_x$ (1.5%)
			Black C (1.2%)
			GHG Int. (1.2%)
			N$_2$O (1.05%)
			GHG pop (0.6%)
			Land cover (0.6%)
		Biodiversity & Habitat (15%)	Marine protect (3%)
			Biome protect [Nat'l] (3%)
			Biome protect [Global] (3%)
			SPI (1.5%)
			PARI (1.5%)
			SHI (1.5%)
			BHI (1.5%)
		Fisheries (6%)	Stock status (2.1%)
			MTI (2.1%)
			Trawling (1.8%)
		Ecosystem services (6%)	Tree cover (5.4%)
			Wetlands (0.3%)
			Grasslands (0.3%)
		Water resources (3%)	Waste water (3%)
		Pollution emission (3%)	SO$_2$ (1.5%)
			NO$_x$ (1.5%)
		Agriculture (3%)	SNMI (3%)

Source: Compiled by the researcher from the 2020 ranking methodology derived from Nardo et al. (2009).

world's largest recycler with a maximum number of start-ups models on waste-to-energy mechanisms. E.g. CopenHill is the cleanest waste-to-energy plant with annual processing of 440,000 tons of combustible waste. Moreover, it has inbuilt the idea of the world's first steam-ring generator and crowdfunding through Kickstarter.

TABLE 2.2

Top Five Countries on EPI

Rank	Country	EPI Score
1	Denmark	82.5
2	Luxembourg	82.3
3	Switzerland	81.5
4	United Kingdom	81.3
5	France	80.0

Source: The 2020 EPI rankings.

2.5.2 CARBON CAPTURE TECHNOLOGY

Step-change Company was established with a green business model to purify the air. Step-change is the most significant industrial-scale carbon capture plant. The plant removes carbon dioxide from the air directly and keeps the air clean (The Guardian, 2020).

2.5.3 GREEN DIGITAL FINANCE

Switzerland is soundly recognized as a worldwide centre for green digital finance. It is accredited to the following fintech business models that have speedily scaled up and heavily contributed toward environmental performance (Table 2.3).

2.5.4 ECO-REFINING SYSTEM TECHNOLOGY

France has introduced a new industrial dimension of refinement. OLVEA group has come up with the business model of eco-refining technology. They have set up a plant having two refining lines with 100 tonnes of capacity daily. The technology uses equipment and processes that purify the oils and leave no possibility of cross-contamination. The plant runs on utterly renewable energy and refines other fuels.

2.5.5 BIO-MASS ENERGY TECHNOLOGY

Denmark had provided 70% of renewable energy from biomass. It targets to grow bio-mass technology to become a significant source of bringing heat and power to residential and commercial buildings.

2.5.6 REGASIFICATION LNG TO POWER

In 2015, France swapped fuel production from oil to natural gas and made all infrastructure facilities available for the gasification process. As a result, oil consumption cut down by 35% in 1973 and 2015, although natural gas usage augmented.

TABLE 2.3

Fintech Business Model in Switzerland to Assist Green Digital Finance

Fintech Company and Its Business Model	Business Process

Fintech companies:
  Carbon Delta AG
  Net guardians
	It collects publicly available and proprietary data and uses computer modelling to correlate its value and climate change risks and opportunities.

Business model: Data power MLAI to inform greener	**Result:** **Switzerland enjoys leadership in green investment screening through MLAI**
Fintech company: WePower	It mediates wind or solar energy projects to increase capital by selling energy tokens that signify energy they are obligated to deliver through Blockchain technology.
Business model: Blockchain for Social Impact Coalition (BSIC)	**Result:** **Switzerland ranked in the top three countries for blockchain technology, with many fintech startups**
Fintech company: Green match	It provides tools to value and evaluate renewable energy project proposals and compute situations on a software-as-a-service (SaaS) basis
Business model: Matchmaking funding platform	**Result:** **Matchmaking Platforms in Switzerland have become a common way of getting new sources of finance, especially for renewable energy**
Fintech company: Sela-labs	It eliminates funding barriers to sustainable projects by creating a marketplace for them, by connecting them using DLT to unlock funds.
Business model: Distributed Ledger Technology as cryptocurrency	**Result:** **Switzerland is recognized as a green Crypto-financial center and aiming to become Crypto valley**

Source: Developed by the researcher from published data (Ries, 2011).

2.5.7 HYDROGEN FUEL CELL TECHNOLOGY-ELECTRO MOBILITY

To remove the storage barrier of renewable energy sources, solar or wind, hydrogen fuel cell technology experiments at Luxembourg. Hydrogen is an abundant source in the universe. For the further development of this technology, billions of Euros are currently being invested. Hydrogen is not the energy source like wind energy, or solar energy splits water molecules like hydrogen and oxygen and then uses hydrogen as a transport fuel industrial fuel. It becomes a storable energy source.

2.5.8 CONSTRUCTION REENGINEERING: PASSIVE HOUSE SYSTEM

Luxembourg is attributing its second rank in EPI majorly to introducing the "Passive house" construction system. It is a construction standard that is energy resourceful and reasonable. It captures available energy sources inside the building structure, e.g. the body heat from the residents in the house, from employees in the corporate/factories. It also absorbs solar heat toward the inside of the building. Suitable windows with ventilation and building shell with decently protected outside walls, roof, and floor slab keep the space warm throughout wintertime and allow heat to go out during summer. Construction21 is designed on the Passive House System. Passive houses restrict the use of traditional energy sources and control emissions into the air. Though Luxembourg is very high in energy consumption, it regulates greenhouse gas emissions through innovative renewable energy techniques. Over the years, Luxembourg increased its renewable energy (Thomas & Piron luxembourg. "Certifications", n.d).

2.6   FINDING AND DISCUSSION

It has been found that innovative green business models and green processes have remarkably helped control the specific emission and have become significant contributors to improving the country's EPI score and thereby ranking well. Green innovative models and processes include green digital finance initiatives with blockchain technology, distributed ledger technology, cryptocurrency, data power MLAI, and other green initiatives like passive house system, hydrogen fuel cell technology, regasification LNG to power, bio-mass energy, eco-refining system have yielded good improvement in EPI, which is presented in the below Tables 2.4 and 2.5.

In the indicator environmental health, it can be seen that Switzerland, Luxembourg, and France are excelling in air quality as per current status but observing the growth rate of EPI, France environmental health indicator is increased by 7.39%, which is quite a significant improvement. In environmental health, Switzerland has topped in air quality due to its major investments in operating through green digital finance. However, Luxembourg has shown an 11.50% highest improvement score in air quality due to hydrogen-fuel technology and passive house system initiatives. All five countries have reached a significant high in sanitation and drinking water, in which Switzerland and the United Kingdom have the top rank with 100% attainment. Denmark scores highest in metal exposure due to complete regulation on GHG emission through circular economy and compulsion on renewable sources. Luxembourg, United Kingdom, followed by Switzerland, have shown significant improvement scores due to emphasizing green industrial projects. Ecosystem Vitality, France, and the United Kingdom have earned top scores in the indicator biodiversity & habitat, primarily due to the eco-refining system of France and carbon-captured technology initiatives. The United Kingdom has joined them in first place in the protection of terrestrial biomass. Denmark excels in climate change and in fisheries due to the circular economy. This assistance has helped move toward green processes. In water resources and agriculture, steady control is found with no fluctuation or incremental growth.

2.7   CONCLUSION AND IMPLICATIONS

In general, high scorers depict a committed approach toward defending public well-being, conserving natural resources, and controlling and dissociating GHG emissions

TABLE 2.4
2020 Current Score v/s Baseline Score of All Indicators on EPI

Top Five Countries on EPI and Their Indicators Score (%)

EPI Indicators	Comparison	Denmark	Luxembourg	Switzerland	UK	France
EPI	Current	82.5	82.3	81.5	81.3	80
	Baseline	75.2	70.7	72.9	72.3	74.2
Environmental Health	Current	91.7	92.6	95	91.7	91.5
	Baseline	86	86.8	90.7	88.3	85.2
Air quality	Current	85.5	87.2	90.6	84.7	88.1
	Baseline	76.7	78.2	82.7	79.2	79.8
Sanitation & Drinking Water	Current	97.4	98.6	100	100	96.2
	Baseline	94.9	96.6	–	99.2	92
Heavy metals (Lead)	Current	100	96.1	95	94.6	84
	Baseline	93.6	85.3	88.4	87	75.5
Waste management (Solid waste)	Current	99.8	96.2	99	92.9	94.8
	Baseline	–	–	–	–	–
Ecosystem vitality	Current	76.4	75.4	72.5	74.3	72.3
	Baseline	68	59.9	61	61.6	66.9
Climate change	Current	95	77.5	81.6	90	81.9
	Baseline	75.4	48.2	58.1	70.1	69.7
Biodiversity & Habitat	Current	81.7	85.5	63	88	88.3
	Baseline	81.1	83.4	59.5	68.7	87.7
Fisheries	Current	13.2	–	–	8.8	12.1
	Baseline	12.6	–	–	5.3	8.5

Ecosystem services	Current	30.2	34.3	46.4	28.3	36.1
	Baseline	25.8	31.3	43.7	28.9	33.9
Water resources (Wastewater treatment)	Current	100	98.5	96.7	98.5	88
	Baseline	–	–	–	–	–
Pollution emission	Current	100	100	100	100	100
	Baseline	–	68.8	–	–	–
Agriculture Sustainable N Mgmt Index	Current	73	42.2	47.6	54.3	65.2
	Baseline	73.6	46.2	51.7	61.9	69.1

Sources:
Current Score: EPI 2020 score of indicators.
Baseline score: Calculated from last ten years longitudinal data series.
Extracted and compiled by the researcher from secondary sources of data.

TABLE 2.5

Improvement Rate Score of EPI Indicators

EPI Indicators	Improvement Rate Score of Indicators on EPI (%)				
	Denmark	Luxembourg	Switzerland	United Kingdom	France
EPI	9.70744681	16.407355	11.7969822	12.4481328	7.81671159
Environmental Health	6.62780698	6.68202765	4.74090408	3.8505 0963	7.3943662
Air quality	11.4732725	11.5089514	9.55259976	6.94444444	10.4010025
Sanitation & Drinking Water	2.63435195	2.07039337	–	0.80645161	4.56521739
Heavy metals (Lead)	6.83760684	12.6611958	7.46606335	8.73563218	11.2582781
Waste management (Solid waste)	–	–	–	–	–
Ecosystem vitality	12.3529412	25.8764608	18.852459	20.6168831	8.07174888
Climate change	25.994695	60.7883817	40.4475043	28.380171	17.5035868
Biodiversity & Habitat	0.739827374	2.51798561	5.88235294	28.0931587	0.684150513
Fisheries	4.76190476	–	–	66.0377358	42.3529412
Ecosystem services	17.0542636	9.58466454	6.1784897	–2.07612457	6.48967552
Water resources (Wastewater treatment)	–	–	–	–	–
Pollution emission	–	45.3488372	–	–	–
Agriculture Sustainable N Mgmt Index	–0.815217391	–8.65800866	–7.9303675	–12.2778675	–5.64399421

Source: Developed by researcher.

from economic activity. Almost in all indicators, Denmark has excelled with the highest improvement rate, followed by Luxembourg. It can be said that green digital finance assistance initiatives have encouraged green models and processes to expand. For example, in Luxembourg, hydrogen-fuel cell technology-based models have significantly controlled harmful emissions.

In the EPI, Cote d'Ivoire and Sierra Leone and Afghanistan, Myanmar, and Liberia are close to the lowest positions. Shallow scores on the EPI reveal the urgency of sustainable efforts by the respective nation. Green models and processes like eco-refining system, renewable energy, bio-mass energy, circular economy, and re-gasification of LNG to power can be adopted immediately to initiate the move toward green and to protect the environment. As a future scope, hydrogen-fuel cell technology, passive house systems, and green digital finance can be targeted to excel in environmental performance. These techniques also hold true for other nations looking to improve the score to outshine the environmental performance index.

All five countries have implemented a feed-in tariff scheme to promote renewable energy sources. Feed-in tariffs are a continuing agreement and pricing attached to costs of production for renewable energy producers. Thus, these tariffs subsidize the cost of generating renewable energy, and producers are protected from nearly all of the intrinsic perils in renewable energy production. Moreover, as the feed-in tariff scheme has prominently encouraged green move in their respective countries, the other countries can incorporate it in their regulatory framework toward green.

REFERENCES

Bisgaard, T, Henriksen, K, & Bjerre, M   (1995). Green business model innovation: Definition, next practice, and nordic policy implications. *Sustainable Innovation 2012 Resource Efficiency, Innovation and Lifestyles*, 30.

Chen, YS (2008). The driver of green innovation and green image–green core competence. *Journal of Business Ethics*, 81(3), 531–543.

Climate Action Summit (2019) Press release, Retrieved from https://public.wmo.int/en/media/press-release/state-of-climate-2018-shows-accelerating-climatechangeimpacts#:~:text=%E2%80%9CThere%20is%20no%20longer%20any,WMO's%20contributions%20to%20the%20Summit

Couckuyt, D, & Van Looy, A (2019). A systematic review of green business process management. *Business Process Management Journal*. Environment Performance Index-Yale University. (2020). Retrieved from https://epi.yale.edu/

Di Nardo, F., Saulle, R., & La Torre, G. (2010). Green areas and health outcomes: a systematic review of the scientific literature. *Italian Journal of Public Health*, 7(4), 402–413.

Ernst, & Young, BERR (2008). "Comparative advantage and green business"; report for the department of business, enterprise, and regulatory reform.

Henriksen, K, Bjerre, M, Øster, J, & Bisgaard, T (2012). *Green Business Model Innovation-Policy Report*. Nordic Council of Ministers.

Javed, A, Yasir, M, & Majid, A (2018). Psychological factors and entrepreneurial orientation: Could education and supportive environment moderate this relationship? *Pakistan Journal of Commerce and Social Sciences (PJCSS)*, 12(2), 571–597.

Nachmany, M, Fankhauser, S, Setzer, J, & Averchenkova, A (2017). Global trends in climate change legislation and litigation: 2017 update.

NOAA National Centers for Environmental Information (2020). State of the Climate: Global Climate Report for March 2020, published online April 2020, retrieved on July 30, 2020 from https://www.ncdc.noaa.gov/sotc/global/202003

Oltra, V, & Saint Jean, M (2009). Sectoral systems of environmental innovation: An application to the French automotive industry. *Technological Forecasting and Social Change*, 76(4), 567–583.

Ries, E. (2011). The lean startup: How today's entrepreneurs use continuous innovation to create radically successful businesses: Currency.

Seychelles National Climate Change Committee. (2009). Seychelles national climate change strategy. Republic of Seychelles.

The Guardian (27 June, 2020). UK biggest carbon project is step change on emission. Retrieved from https://www.theguardian.com/environment/2019/jun/27/uks-biggest-carbon-capture-project-is-step-change-on-emissions

Thomas, & Piron Luxembourg: "Certifications". (n.d.). Retrieved from https://www.thomas-piron.lu/en/certifications-2/

Wendling, ZA, Emerson, JW, de Sherbinin, A, Etsy, DC, et al. (2020). Environmental Performance Index 2020. Yale Center for Environmental Law & Policy: 4 June 2020.

World Meteorological Organization (4 February, 2020). Climate action is a priority and a driver of world affairs: UN chief. Retrieved from https://public.wmo.int/en/media/news/climate-action-priority-and-driver-of-world-affairs-un-chief

3 Precursors and Impediments of Green Consumer Behaviour: An Overview

Balween Kaur and Meenakshi Mittal
Department of Commerce, DAV College for Women, Ferozepur, India

CONTENTS

DOI: 10.1201/9781003127819-3

3.1 INTRODUCTION

The world has realized that "going green" is the only option left for mankind to save planet Earth. The revolution in industries and technology and the communication sector has left no stone unturned to exaggerate the troubling issues of the environment. Climatic conditions have undergone drastic changes, and natural disasters are increasing rapidly. Rapid deforestation, global warming, greenhouse impact, contamination of air and water with harmful gases, acid rain, erosion, degradation of soil, and the depletion of ozone layer are the factors highlighting that environmental problems are the biggest and most detrimental problems mankind faces today. To save Mother Earth from this present turmoil, it becomes the duty of every human being to contribute towards healing the planet. These severe environmental threats have forced manufacturers and marketers to produce and market such products and services that do not add to environmental problems, such as air and water pollution and further destruction of natural ecosystems. Therefore, the sellers are shifting from selling conventional non-sustainable products to green products that do not harm the environment through the manufacturing process, consumption, and disposal. It is urgent to make consumers aware of such environmental threats and stimulate them to modify their buying habits and use only sustainable products that do not harm the environment in any way.

The green products are eco-friendly since they are manufactured in an environmentally conscious manner, cause minimal hazardous effects on nature, do not release chemical gases and solid wastes into the environment, use recycled sourced materials, do not use plastics and other nonbiodegradable materials, do not endanger the health of any living being, do not promote unnecessary waste, and do not lead to cruelty to animals. Green products are recyclable, reusable, biodegradable, refillable, carbon free, pro-environmental, ecologically friendly, and sustainable. The consumers who are aware of the environmental threats have started accepting green products to satisfy their needs because they desire to contribute to the healing process of the environment. A green consumer is an eco-friendly person, deeply concerned about contributing to the healing process of the environment; this consumer buys only green products made of chemical-free ingredients and intends to minimize the adverse effects on the environment by reducing the use of toxic and hazardous substances (Paul and Rana, 2012). The positive behaviour of consumers toward adopting green products for their routine needs is called green consumer behaviour. Green consumer behaviour is the outcome of positive intentions of consumers toward purchasing only green products and availing only green services (Sadiq et al., 2020). (Figure 3.1).

3.2   PRECURSORS OF GREEN CONSUMER BEHAVIOUR

3.2.1   PSYCHOGRAPHIC FACTORS

3.2.1.1   Environmental Knowledge

The pool of knowledge consumers possess regarding environment-related issues is called environmental knowledge. Consumers who have proper information and

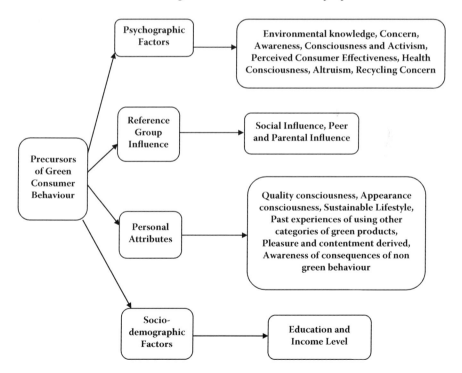

FIGURE 3.1   Precursors of green consumer behaviour.

scientific knowledge relating to the deteriorating effects of manufacturing processes of chemically composed products and harmful effects of the non-green products on air quality and overall environment are better equipped to adopt new behaviours. They are more inclined to behave in a pro-environmental manner and are motivated to accept and use sustainable products and services only, knowing that it seems to be one of the best options to save environment from further degradation (Srivastava and Chawla, 2017; Doorn and Verhoef, 2015).

3.2.1.2   Environmental Concern

The excessive usage and destruction of natural resources is a serious threat people are facing all over the globe. The only saviour of this alarming condition and unending destruction is making human beings aware of environmental wellness so that they can contribute to the curative process of the environment by adopting ways and techniques suggested by government agencies, ecologists, environmentalists, socialists, and researchers. Numerous research studies conducted recently in various zones of the world have explored the association of environmental concern, consumers' demographic and psychographic characteristics, and other methods marketers have adopted to inculcate sustainable buying habits among the masses. Reports have elucidated that the environmental concern of consumers is positively and significantly related with green purchasing intentions and behaviour of global consumers (Cheung et al., 2019). Hence, the regulatory authorities need to devise stringent regulations stimulating the consumers to become concerned for the deteriorating environment and contribute to the healing process by inculcating sustainable buying habits (Delafrooz et al., 2014).

3.2.1.3   Environmental Awareness and Consciousness

Environmental issues like air pollution, depletion of the ozone layer, deforestation, depletion of natural resources, water pollution, soil erosion, and global warming require the formation of strict and mandatory regulations by environmental bodies at the international level, which should be compulsorily adopted by the residents of all nations. It is a much needed concern to popularize environmental issues among the masses and the general public. When the consumers are aware of the environmental issues and problems, they themselves get inspired to contribute positively towards the solutions for these problems by behaving in an eco-friendly manner (Dagher and Itani, 2012; Lin and Niu, 2018; Shin et al., 2019). The awareness of such issues among consumers definitely motivates them to accept and buy only green products made by adopting eco-friendly methods, causing minimum hazardous effects on nature (Shin et al., 2019; Xu et al., 2020; Yadav and Pathak, 2016).

3.2.1.4   Environmental Activism

Environmental activism has been found as one of the antecedents of green buying behaviour of consumers. When consumers actively participate by working to advocate and motivate others to behave in a pro-environmental manner, they are surely choosing green and eco-friendly goods and services for daily personal and family needs. Therefore, environmental activism significantly impacts the buying intentions of consumers and inspires them to behave in a pro-environmental manner while opting for products and services (Lee, 2010).

3.2.1.5   Health Consciousness

Consumers have become more health conscious after they have realised the ravages a deteriorated environment take on the health of living beings. The drastic changes in the quality of the environment have created various health hazards for humans in the form of diseases and other infections and allergies. They have also realized the role of green products in saving them from such health problems. The green products are made up of chemical-free raw materials involving minimum usage of plastics and other harmful ingredients. Therefore, the increasing awareness of consumers to use the products involving health benefits motivates them to purchase and consume eco-friendly products and services (Hoque et al., 2019; Kim and Seock, 2009; Rana and Paul, 2020).

3.2.1.6   Perceived Consumer Effectiveness

Perceived consumer effectiveness represents the confidence of consumers to bring about the results and outcomes that they desire to achieve and value personally. This novel tool is used to build the positive stance of consumers towards purchasing intention and behaviour relating to green products and services. The researches done in cross-cultural contexts elucidate that this confidence level of consumers towards sustainable products motivates them to change their conventional buying habits and switch over to green buying, which is the need of the hour to save the environment from further destruction. Thus, it has been evident from the studies that more consumers trust green products, their perceived effectiveness increases, and they start adopting only sustainable products and services for routine and special needs (Heo and Muralidharan, 2019). Higher levels of perceived consumer effectiveness have been found to make the consumer environmentally concerned and increase a sense of contribution to augment the healing of the environment by purchasing green products and exhibiting sustainable behaviour.

3.2.1.7   Altruism

Altruism refers to the selfless concern for the well-being of others. The altruistic values of consumers are the strongest predictors of their green buying behaviour. The concern to save the environment and Mother Earth from further destruction and to repair the damage done to the environment and the plane propels consumers to behave selflessly and modify their purchasing intentions; they shift towards buying pro-environment products only, which reduces hazardous effects on nature and the Earth. Altruistic values modify the attitude of consumers and make them behave in a pro-environmental manner.

3.2.1.8   Recycling Concern

In the past few years, researchers have conducted several empirical studies to decipher the impact of recycling concerns on pro-environmental behaviour of consumers. The results of the studies done in cross-cultural contexts describe that the consumers who have positive intentions for recycling and reusing materials and finished products definitely have concern for the deteriorating quality of the environment and are worried for its healing. Recycling concerns mould the attitude of consumers for the environment and make them ecologically conscious and green consumers. These consumers generally exhibit green purchasing behaviour while making their routine purchasing. They tend to buy only eco-friendly, herbal, natural, and organic products. Therefore, a significant connection exists between recycling concern and green behaviour of consumers (Kautish et al., 2019).

3.2.2 Reference Group Influence

3.2.2.1 Social Influence

Subjective norms refer to the belief that an important person or group of persons will approve and support a particular behaviour (Persaud and Schillo, 2017). These are determined by the perceived social pressure from others for an individual to behave in a certain manner (Johnstone and Hooper, 2016). The Theory of Reasoned Action suggests that consumers' attitude and subjective norms predict purchasing intentions, which shape consumers' behaviour. This theory has been used by researchers as a theoretical model to confirm the impact of social influence on green buying behaviour of consumers. It explains that when consumers are influenced by social norms, they tend to behave in an eco-friendly manner and accept only sustainable products (Clark et al., 2019). Hence, significant impact is exerted by social influence on the green buying intentions and behaviour of consumers toward green products (Khare, 2015; Khare et al., 2013).

3.2.2.2 Peer and Parental Influence

The peer group of a person consists of family, close friends, and colleagues. A consumer spends maximum time with peer groups and parents. The suggestions, advice, recommendations, and values of the peer group generally influence the tastes and preferences of a person. When the parents and peers of a consumer are concerned and conscious of environmental degradation, the consumer will surely be impacted and have similar connotations regarding the environment and benefits of using green and eco-friendly products (Suki and Suki, 2019). Thus, it can be stated that the positive intentions of a consumer's parents and peer groups for green products moulds behaviour and encourages the purchase and use of green products and green behaviour.

3.2.3 Personal Attributes

3.2.3.1 Quality Consciousness

The characteristic of quality consciousness among consumers makes them very cautious for the quality of products they use and consume. Such consumers never compromise on quality of products and services; rather, they are always willing to pay premium prices and travel long distances to buy good quality products they have previously used and trusted. Such consumers now have shifted their preferences from conventional products to green, herbal, and organic products that are composed of chemically free raw materials and do not endanger the life of any living beings. The more consumers are quality conscious, the more they are motivated to shift towards behaving in a pro-environmental manner and buying only green and sustainable products.

3.2.3.2   Appearance Consciousness

Modern consumers have become very conscious for their appearance and looks. They spend a good amount of money on personal care products such as cosmetics, hair care products, skin care products, and other beauty-enhancing and make-up products. They have realized that the conventional and non-herbal products use large amounts of chemicals and other hazardous raw materials that satisfy consumers instantly but cause harmful side effects for the users in the long run. Consumers wish to have a younger look as compared to their age, so they tend to consume herbal, organic, and sustainable green cosmetics and other personal care products because these products are manufactured using only herbal and natural ingredients, causing negligible harm to the users in the long run. Thus, the appearance consciousness of consumers, especially women, provokes them to discard chemically composed products and use only natural, herbal, organic, and green products, causing minimum damage to nature and environment.

3.2.3.3   Sustainable Lifestyle

When consumer adopt a lifestyle that considers the environment before making any minor or major decision, they have adopted a sustainable lifestyle. This sustainable lifestyle consumers adopt provoke him to accept and purchase only eco-friendly products. Such a consumer is very concerned for environmental threats and is fully devoted to healing the environment. There is a significant favourable impact of the consumer's sustainable lifestyle on purchasing intentions and behaviour towards sustainable products.

3.2.3.4   Past Experience of Using Green Products of Other Category

The consumer may be using one category of green products and wishes to buy the same to satisfy needs for other categories of products. Consumers who are satisfied after using initial green products will be inspired to buy green products only for other needs also. Once the sustainable products develop the consumer's trust in its green claims, quality, and price, the consumer's demand for other categories of green products also rises. Hence, the past experiences of using green products predict the consumers' intentions and behaviour for future demand for other kinds of eco-friendly products (Ghazali et al., 2017).

3.2.3.5   Pleasure and Contentment Derived

Consumers who get satisfaction after using a product will promote the product by narrating its uses and benefits among their peer group. The satisfaction, contentment, and pleasure derived from green products motivate the user to continue these purchases and also convince others to consume the same products. The past researchers have also found a prominent relationship between contentment derived and consumers' green buying intentions and behaviour. Therefore, post-purchase satisfaction of consumers is strongly impacting their purchase intentions for green products in near future.

3.2.3.6   Awareness of Consequences of Non-green Behaviour

When consumers are aware of the harmful effects of non-green conventional products on health, as well as on the environment, they will avoid the risk of buying and consuming non-green products. The non-green products may be harmful in different ways. Sometimes they are chemically composed and harmful to produce, use, and dispose of. They may contain plastics and other non-biodegradable materials, which are harmful to air, water, and soil resources. They may be cruel to animals or may be causing excessive waste of natural resources in the production process. When consumers are aware of the consequences of using non-green products on their own health and on the environment, they automatically get more inclined toward using only green products and services.

3.2.4 SOCIO-DEMOGRAPHIC FACTORS

3.2.4.1 Education Level

Consumers' green purchasing intentions and behaviours are significantly affected by their education level. Researchers have determined that the well-qualified consumer is more inclined towards purchasing green products and wishes to contribute to healing of the environment (Nath et al., 2013; Jain and Kaur, 2006). The academic qualifications of the consumers also makes them aware of the deteriorating nature of conventional and non-green products, which use plastics, cause air, soil, and water pollution, and harm the natural environment. Therefore, global researchers have confimed that consumers' educational qualities encourage green qualities, meaning they always exhibit green behaviour while making their purchases.

3.2.4.2 Income Level

Consumers' level of income is directly related to green buying intentions and behaviour. Consumers who fall in a high income bracket are more conscious and concerned for the quality of products they use (Ifegbesan and Rampedi, 2018; Jermsittiparsert et al., 2019). They prefer to choose good quality green products made up of natural ingredients and negligible amounts of chemicals. They are inclined toward consuming organic, herbal, and natural products only. But a consumer with low income is more price sensitive than environmentally conscious. These consumers make all their purchases within their limited income, not able to care about quality and environmental impacts. Therefore, there is a positive link between income and green buying intentions of consumers. (Figure 3.2).

3.3 IMPEDIMENTS OF GREEN CONSUMER BEHAVIOUR

3.3.1 PERSONAL FACTORS

3.3.1.1 Own Ignorance

The consumers who are ignorant about the deteriorating environment and increasing threats to the existence of mankind do not have positive behaviour for sustainable products. Such consumers do not have any knowledge about the environment and lack environmental attitude, awareness, concern, and consciousness. Their choice and preferences of products are not governed by eco-friendly behaviour. Hence, these

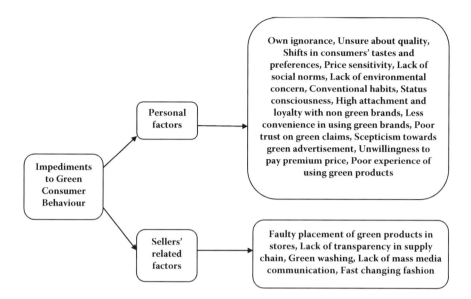

FIGURE 3.2   Impediments of green consumer behaviour.

consumers never buy green products and do not contribute to the healing of the environment by behaving in an ecologically conscious manner.

3.3.1.2   Unsure about Quality

When consumers are planning to buy green products for the first time, they hesitate to make quick decisions since they are unsure about product quality, having never used sustainable products. This factor creates suspicion in the minds of prospective buyers and hinders their decision-making process of buying green products to meet their specific needs. But where the seller gives guarantees and persuades the customer of the superior quality and environmental benefits of green products, the customers sometimes believe and make their initial purchases for sustainable products.

3.3.1.3   Shifts in Consumers' Tastes and Preferences

When there is a shift in tastes, preferences, and affordability of consumers, they tend to shift from green buying behaviour to conventional non-green buying behaviour. Therefore, modifications in needs, wants, desires, and paying capacity inhibit the further buying of green products.

3.3.1.4   Price Sensitivity

Consumers' price sensitivity affects the degree to which their buying behaviour is affected by price-level changes of the product. Moreover, consumers also relate the quality of products with their prices. It is generally believed that high-priced products are better in quality as compared to low-priced ones. Therefore, the consumers who are more sensitive to prices compare the prices and quality of green and non-green products (Bhutto et al., 2019). If the benefits derived are more than their sacrifice, then they prefer to pay more prices and behave in a pro-environmental manner. If the price gap is more and the benefits derived are less, then they opt to consume non-green conventional products. Therefore, the price-sensitive nature of consumers restricts them from behaving in an ecologically conscious manner and exhibiting green behaviour.

3.3.1.5   Lack of Social Norms

The informal understandings that govern the behaviour of consumers as members of society are social norms. These are the unwritten or unpublished rules that are considered acceptable in a group, and the individuals are expected to follow such norms of behaviour. The use of green products is one such social norm that consumers are expected to follow to save the environment from further degradation. But lack of social norms in consumers restricts them from behaving in a pro-environmental manner.

3.3.1.6   Lack of Environmental Concern

Lack of environmental concern, environmental awareness, environmental knowledge, and environmental consciousness impeded consumer buying intentions and behaviour for pro-environmental products. Such consumers are least concerned and bothered with environmental hazards and do not have any motivation to contribute to environmental healing.

3.3.1.7   Conventional Habits

Due to their nature, some consumers are very reluctant to change their previous habits and adopt new products. They are least concerned about the environment and are not willing to modify their conventional habits and adopt green products that are beneficial to the environment and the society at large.

3.3.1.8   Status Consciousness

When consumers associate their status with buying highly priced non-green products being sold by costly multinational brands, they are less interested in the concept of sustainability. Such status-conscious consumers use products of only costly brands, which prove their high status among their peer group. This characteristic of being status conscious restricts their ability to adopt pro-environmental habits and switch over to green products.

3.3.1.9   High Attachment and Loyalty with Non-green Brands

The consumers who are using conventional non-green products of certain categories and are very satisfied and content with those products, they have developed high attachment and loyalty with those brands (Mugge et al., 2010). Such consumers are least inspired to sacrifice their most trusted products and shift their preference for green products. The contentment derived from conventional non-green products prevents them from exhibiting green buying intentions and behaviour.

3.3.1.10   Less Convenience in Using Green Products

Consumers wish to remain in their comfort zone while making purchase decisions. When they have to travel a little longer to reach the green store or pay a premium price to buy green products, they shift their preference and buy low-priced non-green products available nearby. In a nutshell, they are not willing to sacrifice their conformity and convenience to buy sustainable products.

3.3.1.11   Poor Trust on Green Claims

The marketers sometimes exaggerate the green claims in advertisements pertaining to the green products to convince consumers to modify their preference and introduce green buying habits in their daily routine. But when these green claims turn out to be false or over exaggerated, the consumers lose their trust in marketers and shift back to conventional non-green buying intentions. The poor trust on green claims hinders the buying behaviour of consumers toward green products.

3.3.1.12   Scepticism towards Green Advertisement

Consumers, having doubt in their minds regarding manufacturers' claims about green products in advertisement, do not behave in a pro-environmental manner and continue their conventional buying habits of non-green products.

3.3.1.13   Unwillingness to Pay Premium Price

Green products are made up of natural ingredients causing minimum hazardous effects on nature and environment; therefore, the cost of producing such products is higher than conventional non-green products. That is why the selling price is more than the selling price of conventional non-green products. When the consumers are unwilling to pay the high price or premium price for buying green products, they develop unfavourable intentions for purchasing such green and eco-friendly products.

3.3.1.14   Poor Experience of Using Green Products

Consumers who have earlier used green products but did not attain any satisfaction and contentment from them lose their trust for such products. Such consumers do not try other categories of green products since the previous consumption did not provide any pleasure and satisfaction. Thus, the poor experience of using one category of green products plays the role of inhibitor and prohibits consumers from trusting the quality and green claims of marketers. It becomes very difficult to inculcate the green buying behaviour in such consumers in the future.

3.3.2   Sellers' Related Factors

3.3.2.1   Faulty Placement of Green Products in Stores

The green products should be placed in the stores to attract the attention of customers. The personal and environmental benefits of such products should also be highlighted by the seller to arouse the instinct of customers to shift their buying preference from non-green to green buying behaviour. But the faulty placement of sustainable products in stores inhibits customers from buying such sustainable products.

3.3.2.2   Lack of Transparency in Supply Chain

The transparent supply chain entails visibility and traceability not only on retail level but also at the supplier level, raw material level, and worker level. This transparency reduces reputational risk and enhances sellers' standing as a trustworthy enterprise.

Therefore, lack of transparency in the supply chain inhibits consumers from purchasing sustainable products and behaving in a pro-environmental manner.

3.3.2.3 Green Washing

The indulgence of marketers in green-washing activities inhibits the consumers from buying green and sustainable products. These activities convey a false impression and provide misleading information about the environmental soundness of sellers' so-called green products. Such environmental claims are unsubstantiated and deceive consumers into trusting that sellers' products are environmentally friendly. This is a significant impediment inhibiting consumers to behave in an ecologically conscious manner.

3.3.2.4 Lack of Mass Media Communication

The information regarding benefits of using sustainable products to the consumer, society, and environment have not been properly and effectively communicated by the media. The masses are still unaware of the hazardous consequences of using chemically composed products to their health and environment. Therefore, the lack of mass media communication restricts the consumers from behaving in an ecologically conscious manner and buying green products.

3.3.2.5 Fast Changing Fashion

Once consumers start using green products and are satisfied with their features, a new range of products and new varieties of non-green products are advertised and displayed by sellers of conventional products, impeding the consumers from repeating their purchase for green products (Joung, 2014).

3.4 CONCLUSION

The chapter is a reservoir of progenitors and repercussions of green buying behaviour of consumers. It provides a detailed account of diverse factors promoting and inspiring consumers to behave in an environmentally conscious manner and adopt green products and services to fulfil their routine needs. It provides an ocean of information to manufacturers and marketers regarding the antecedents and impediments of green consumer behaviour; it will facilitate them in designing their future production and marketing strategies accordingly (Hossain and Rahman, 2018). The authors have outlined diverse progenitors of green consumer behaviour, which motivate the consumer to behave in an eco-friendly manner and contribute to the healing of the planet by purchasing green products (Maichum et al., 2017; Maniatis, 2016). Consumers are affected by psychographic factors, socio-demographic factors, social and reference group influence, green marketing mix variables adopted by sellers and product attributes, and past experiences of using green products (Han et al., 2019). All these factors affect their decisions to purchase and repurchase eco-friendly products. As regards the barriers of green consumer behaviour, two categories have been outlined as impediments: personal factors and sellers' related factors. Both the categories of impediments play a stringent role of demotivating consumers to avoid buying green and eco-friendly products (Gleim et al., 2013).

REFERENCES

Bhutto, MY, Zeng, F, Soomro, YA, & Khan, MA (2019). Young Chinese consumer decision making in buying green products: An application of theory of planned behavior with gender and price transparency. *Pakistan Journal of Commerce and Social Sciences (PJCSS)*, *13*(3), 599–619.

Cheung, MF, & To, WM (2019). An extended model of value-attitude-behavior to explain Chinese consumers' green purchase behavior. *Journal of Retailing and Consumer Services*, *50*, 145–153.

Clark, RA, Haytko, DL, Hermans, CM, & Simmers, CS (2019). Social influence on green consumerism: Country and gender comparisons between China and the United States. *Journal of International Consumer Marketing*, *31*(3), 177–190.

Dagher, G, & Itani, O (2012). The influence of environmental attitude, environmental concern and social influence on green purchasing behavior. *Review of Business Research*, *12*(2), 104–111.

Delafrooz, N, Taleghani, M, & Nouri, B (2014). Effect of green marketing on consumer purchase behaviour. *QScience Connect*, *2014*(1), 5.

Ghazali, E, Soon, PC, Mutum, DS, & Nguyen, B (2017). Health and cosmetics: Investigating consumers' values for buying organic personal care products. *Journal of Retailing and Consumer Services*, *39*, 154–163.

Gleim, MR, Smith, JS, Andrews, D, & Cronin Jr, JJ (2013). Against the green: A multi-method examination of the barriers to green consumption. *Journal of Retailing*, *89*(1), 44–61.

Han, H, Eom, T, Chung, H, Lee, S, Ryu, HB, & Kim, W (2019). Passenger repurchase behaviours in the green cruise line context: Exploring the role of quality, image, and physical environment. *Sustainability*, *11*(7), 1985.

Heo, J, & Muralidharan, S (2019). What triggers young Millennials to purchase eco-friendly products?: The interrelationships among knowledge, perceived consumer effectiveness, and environmental concern. *Journal of Marketing Communications*, *25*(4), 421–437.

Hoque, MZ, Alam, M, & Nahid, KA (2018). Health consciousness and its effect on perceived knowledge, and belief in the purchase intent of liquid milk: Consumer insights from an emerging market. *Foods*, *7*(9), 150.

Hossain, MI, & Rahman, MS (2018). Measuring the impact of green marketing mix on green purchasing behaviour: A study on Bangladeshi consumers. *The Comilla University Journal of Business Studies*, *5*(1), 5–19.

Ifegbesan, AP, & Rampedi, IT (2018). Understanding the role of socio-demographic and geographical location on pro-environmental behaviour in Nigeria. *Applied Environmental Education & Communication*, *17*(4), 335–351.

Jain, SK, & Kaur, G (2006). Role of socio-demographics in segmenting and profiling green consumers: An exploratory study of consumers in India. *Journal of International Consumer Marketing*, *18*(3), 107–146.

Jermsittiparsert, K, Haseeb, M, & Dawabsheh, M (2019). Enhancing organic food identity through green marketing in Thailand: *Mediating* role of environmental, health, and social consciousness. *World Food Policy*, *5*(2), 74–91.

Johnstone, ML, & Hooper, S (2016). Social influence and green consumption behaviour: A need for greater government involvement. *Journal of Marketing Management*, *32*(9-10), 827–855.

Joung, HM (2014). Fast-fashion consumers' post-purchase behaviours. *International Journal of Retail & Distribution Management*, *42*(8), 688–697.

Kautish, P, Paul, J, & Sharma, R (2019). The moderating influence of environmental consciousness and recycling intentions on green purchase behavior. *Journal of Cleaner Production*, *228*, 1425–1436.

Kim, S, & Seock, YK (2009). Impacts of health and environmental consciousness on young female consumers' attitude towards and purchase of natural beauty products. *International Journal of Consumer Studies, 33*(6), 627–638.

Khare, A (2015). Antecedents to green buying behaviour: A study on consumers in an emerging economy. *Marketing Intelligence & Planning, 33*(3), 309–329.

Khare, A, Mukerjee, S, & Goyal, T (2013). Social influence and green marketing: An exploratory study on Indian consumers. *Journal of Customer Behaviour, 12*(4), 361–381.

Kim, HY, & Chung, JE (2011). Consumer purchase intention for organic personal care products. *Journal of Consumer Marketing, 28*(1), 40–47.

Lee, K (2010). The green purchase behaviour of Hong Kong young consumers: The role of peer influence, local environmental involvement, and concrete environmental knowledge. *Journal of International Consumer Marketing, 23*(1), 21–44.

Lin, ST, & Niu, HJ (2018). Green consumption: Environmental knowledge, environmental consciousness, social norms, and purchasing behavior. *Business Strategy and the Environment, 27*(8), 1679–1688.

Maichum, K, Parichatnon, S, & Peng, KC (2017). Factors affecting on purchase intention towards green products: A case study of young consumers in Thailand. *International Journal of Business Marketing and Management (IJBMM), 2*(3), 01–08.

Maniatis, P (2016). Investigating factors influencing consumer decision-making while choosing green products. *Journal of Cleaner Production, 132*, 215–228.

Mugge, R, Schifferstein, HN, & Schoormans, JP (2010). Product attachment and satisfaction: Understanding consumers' post-purchase behaviour. *Journal of Consumer Marketing, 27*(3), 271–282.

Nath, V, Kumar, R, Agrawal, R, Gautam, A, & Sharma, V (2013). Consumer adoption of green products: Modeling the enablers. *Global Business Review, 14*(3), 453–470.

Paul, J, & Rana, J (2012). Consumer behaviour and purchase intention for organic food. *Journal of Consumer Marketing, 26*(6), 412–422.

Persaud, A, & Schillo, SR (2017). Purchasing organic products: Role of social context and consumer innovativeness. *Marketing Intelligence & Planning, 35*(1), 130–146.

Rana, J, & Paul, J (2020). Health motive and the purchase of organic food: A meta-analytic review. *International Journal of Consumer Studies, 44*(2), 162–171.

Sadiq, M, Paul, J, & Bharti, K (2020). Dispositional traits and organic food consumption. *Journal of Cleaner Production*, 121961.

Shin, YH, Im, J, Jung, SE, & Severt, K (2019). Motivations behind consumers' organic menu choices: The role of environmental concern, social value, and health consciousness. *Journal of Quality Assurance in Hospitality & Tourism, 20*(1), 107–122.

Srivastava, K, & Chawla, D (2017). Demographic and psychographic antecedents of ecologically conscious consumer behaviour: An empirical investigation. *International Journal of Indian Culture and Business Management, 14*(4), 480–496.

Suki, NM, & Suki, NM (2019). Examination of peer influence as a moderator and predictor in explaining green purchase behaviour in a developing country. *Journal of Cleaner Production, 228*, 833–844.

Van Doorn, J, & Verhoef, PC (2015). Drivers of and barriers to organic purchase behaviour. *Journal of Retailing, 91*(3), 436–450.

Xu, X, Wang, S, & Yu, Y (2020). Consumer's intention to purchase green furniture: Do health consciousness and environmental awareness matter?. *Science of the Total Environment, 704*, 135275.

Yadav, R, & Pathak, GS (2016). Young consumers' intention towards buying green products in a developing nation: Extending the theory of planned behavior. *Journal of Cleaner Production, 135*, 732–739.

4 Green Smart Farming Techniques and Sustainable Agriculture: Research Roadmap towards Organic Farming for Imperishable Agricultural Products

Sita Rani
Department of Computer Science and Engineering, Gulzar Group of Institutions, Khanna, Punjab, India

Pankaj Bhambri
Department of Information Technology, Guru Nanak Dev Engineering College, Ludhiana, Punjab, India

O. P. Gupta
Department of Electrical Engineering & Information Technology, College of Agriculture Engineering & Technology Punjab Agricultural University Ludhiana, Punjab, India

CONTENTS

DOI: 10.1201/9781003127819-4

49

4.1   INTRODUCTION

The demand of agricultural products has drastically increased in the last few years, which puts agricultural sustainability at risk. There is a need for an adequate food system with a prime focus on sustainability (Yousuf, Titikshya, & Singh, 2018). Farmers all around the world are facing many issues. A few prominent ones are finding suitable solutions for efficient water utilization, minimizing soil erosion, and meeting the energy requirement at minimal cost. But, these issues can't be addressed successfully with traditional farming practices. With a continuous rise in the need for food products and sustainable agriculture, it becomes essential for farmers to adopt green smart-farming techniques (Roy, 2015). These advanced farming techniques are developing a transformation in agricultural practices and production processes (Wolfert et al., 2017). Most of the developed nations and developing countries have already initiated these practices, after understanding their usefulness. So far, the developed countries like Japan and China have already begun to use information & communication technology, along with the IoT, at rapid rate in developing precision agriculture methods (Walter et al., 2017). With an aim to augment the market, the governments of other countries are also recognizing the need for advanced agricultural practices to establish a culture of green smart agriculture (Glaroudis, Lossifieds, & Chatzimisios, 2020). Many centers are established to facilitate farmers with these state-of-art practices. These types of radical advancements in agricultural techniques are the seeds of opportunities, as well as challenges, in the agriculture sector. To overcome the various challenges in the developing countries, awareness about advanced agriculture techniques among the farmers is significant. So, the requirement of green smart agriculture in practice is highly needed.

Smart farming is a contemporary farming method that targets the adoption of technology to increase crop yield with significantly limited inputs (Walter et al., 2017). It is an information-driven approach. Usually, it makes use of the global positioning of data for the adoption of various measures, e.g., auto-piloted harvesters and tractors, drones, automated seeders, and many more (Charatsari & Lioutas, 2020). Various issues addressed by the smart-farming techniques include the requirement of water and fertilizers to the plants, time of application, land area where required, etc. Smart farming and sustainable agriculture are both data dependent (Meybeck et al., 2012). Smart farming contributes toward cost-effective sustainable agriculture using the amalgamation of observations collected from the Earth and navigation satellites. Data collected from these two sources help farmers to make various decisions and to obtain economic and ecological outputs. In short, smart farming is revolutionizing various agricultural practices. This movement reveals that the regular use of smart-farming practices will contribute to solving many food security issues worldwide.

In organic farming, crops are cultivated without using synthetic agrochemicals like fertilizers, pesticides, or genetically modified organisms (Yadav et al., 2013). These organisms can maintain the health of the soil, biodiversity, ecosystem, and people. It is an approach of farming focused on growing crops using organic wastes. Crops and animal wastes, along with other bio-fertilizers, provide the required nutrients to the crops for better production (Narayan, 2012). The NSSO published in the 66th round that a big transition is occurring in the consumption habits of people. Now, people have become health and diet conscious and are more interested in the consumption of organic products. Considering these facts, this work brings forth a thorough review that encompasses the technologies needed, along with gains of practicing green smart agriculture for organic farming.

4.2 SMART FARMING

Numerous modern techniques have been incorporated in the past to enhance the production of agricultural products. Still, there is a huge gap in need and the supply of agricultural products due to the high population. According to predictions, by the year 2050, the total world population will increase, tentatively by 25% (Ayaz et al., 2019). A major segment of population in all countries has drifted toward urban areas; so, a high rise in the urban population is also expected by 2050, too. It is also estimated that in the future, the income per family is going to increase, especially in advanced countries. All these factors will contribute to an increase in the demand for food.

People are giving more attention to diet and quality food. These factors will become the key drivers in the need for a higher yield, as shown in Figure 4.1. Along with food, other cash crops are also attracting equal attention because of their major role in the economy of a country (Tripathi et al., 2019). So, the rising need of agricultural products puts an additional burden on existing limited agricultural resources.

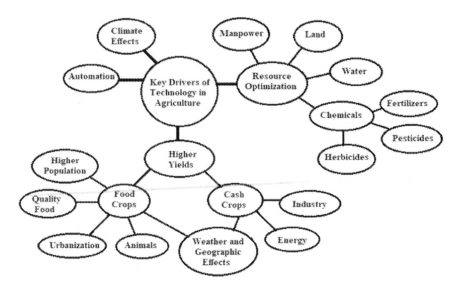

FIGURE 4.1   Main technology drivers in agriculture domain.

Regrettably, only a confined segment of the Earth's land is available for farming due to various environmental, physical, political, and economical factors. In the last decade, it has been analyzed that there is a downturn in the land used for the production of food items (Bruinsma, 2003). Over time, the difference between the requirement and the yield of the food items is evolving as a concern.

To counter the increasing demand and various challenges, farmers require the incorporation of advanced technologies in the agricultural processes to produce more yield from the limited land area and resources. Using the traditional methods of farming, most worker time is spent on observing and apprehending the crop conditions (Navulur & Prasad, 2017). Observing today's needs, advanced technology-based solutions are needed to handle the tasks like crop supervision and observation without being physically in the fields and with minimal effect on the environmental. Consequently, the concept of smart farming emerges, as shown in Figure 4.2. Along with advanced technologies, smart farming needs the support of communication networks (Mahbub, 2020), various government and research institutions, and a diversity of mechanisms.

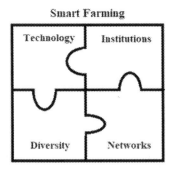

FIGURE 4.2   Conflux of smart farming.

The latest sensor-based technologies and communication networks have assisted farmers by enabling their virtual presence in the fields. Wireless sensors are also playing a vital role in crop observation and monitoring. These sensors have the competence to identify the undesirable occurrences of the crops accurately and early. Because of these advantages, technology-equipped instruments are used in almost all the phases of the crop life cycle. A variety of sensors are used in the various agricultural tools, which make the processes fast and economical. Different smart equipment, like automated tractors, spray machines, weeders, and harvesters, facilitate the various agriculture processes. With the usage of sensors, the process of data collection, communication, processing, and dissemination has become very fast. The involvement of the sensor technology has taken agriculture in a new dimension by supporting crop and location precise agriculture.

The Internet of Things (IoT) is benefitting a broad range of sectors like healthcare, manufacturing, communication, agriculture, and many more. The basic idea behind the integration of IoT is to improve the efficiencies and productivity of these sectors (Lin et al., 2017; Elijah et al., 2018; Sisinni et al., 2018). If we analyze the previous role of IoT in different domains, it can be easily predicted that IoT will also play a vital role in the different agriculture operations. Due to the various features supported by IoT, like automated remote data collection with the help of sensors, communication, cloud-based data analysis, and mechanization of different agriculture operations, these predictions are easily realizable and a lot of advancement is expected in the agriculture domain. To epitomize the above discussion, the various technology drivers are shown in Figure 4.1, whereas the various challenges faced for the incorporation of the technology are shown in Figure 4.3.

Numerous articles describe the contribution of IoT in the domain of agriculture; however, the major published content only discusses the applications (Zhang, Dabipi, & Brown, 2018; Khanna & Kaur, 2019). According to the content published

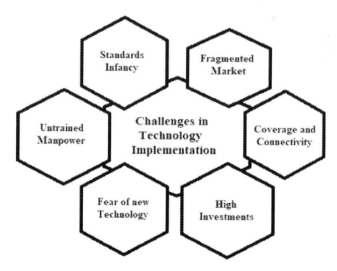

FIGURE 4.3   Major challenges in exertion of smart farming.

in these papers, IoT eases the process of data collection and management, helps to improve crop yield, reduces the labor cost and production cost considerably, and increases the profits. But, in the majority of the published papers, other important issues like IoT-based agricultural frameworks, the role of IoT for food quality check, prototypes, etc., are either overlooked or discussed with limited emphasis.

4.3 APPLICATIONS OF SMART FARMING

Traditional farming can be changed to the contemporary style by using advanced agricultural techniques. Smooth assimilation of IoT in agriculture can advance the domain to new heights, something not imagined in the past. Areas of applications, along with the services provided and different types of sensors used in smart agriculture, are shown in Figure 4.4. Different application areas of smart agriculture are discussed below.

4.3.1 Soil Analysis

The fundamental aim of the process of soil analysis is to identify the nutrient level in the soil. Experts in the domain recommend conducting, at least once a year, a soil

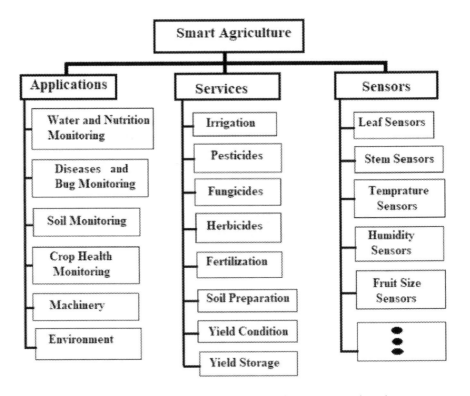

FIGURE 4.4 Smart agriculture: Main applications, various sensors and services.

analysis test for a piece of land. Either fall or winter is the assumed suitable time to conduct the soil analysis test (Dinkins & Jones, 2103). The various important aspects considered during soil analysis are the usage of fertilizers, water level, type of the soil, and the crops grown in the past year.

At present, various corporate houses are manufacturing a broad range of sensor-based equipment and toolkits to determine the quality of soil. This equipment is also helpful for suggesting the solutions to abstain soil quality deterioration. Labin-a-Box is an example of such a smart-testing tool. It is manufactured by AgroCares. With this toolkit, up to 100 samples of soil can be processed for quality analysis.

4.3.2 IRRIGATION

It is important point to stress that the agriculture sector uses, on its own, tentatively 70%–75% of the available fresh water (Berger, Pfister, & Motoshita, 2016). In some countries, this percentage even increases to 80% (Oliveira et al., 2017). For the sake of stressing the problem of the water crisis around the world, water usage must be reduced to only the needed level in the agriculture sector, too. Although several restrained irrigation techniques already are being practiced to save water, a high need remains for advanced irrigation techniques with the use of IoT technologies. One such IoT-based irrigation solution is VRI (variable rate irrigation) (LaRue & Fredrick, 2012). It advantageously supports the efficient usage of water.

4.3.3 DISEASE AND PESTS ANALYSIS

According to data provided by the Food and Agriculture Organization (FAO), approximately 20%–40% of the agricultural produce around the world is ruined because of pests and diseases. To regulate this damage, pesticides are developed as an integral component of the agriculture domain, to the contrary effect, which is very dangerous for human and animal lives (Greene, 2013). To counter this problem, IoT-based smart solutions provide guidance for the location of specific use of pesticides and insecticides. These intelligent devices also aid real-time observations of the crops and disease forecasting (Kim, Lee, & Shin, 2018) (Venkatesan, Kathrine, & Ramalakshmi, 2018). Three level disease and pest control mechanisms are adopted by IoT-based devices, i.e. analysis, evaluation, and treatment.

4.3.4 FERTILIZER

Fertilizers are used to enhance the level of nutrients in the soil. But exaggerated use of fertilizers not only deteriorates soil quality but also pollutes the ground water, resulting in harm to human life and the environment (D. Mcguire, 2014). The smart devices in the IoT framework help approximate the needed quantity of the nutrients accurately. They also aid to provide site-specific requirements more precisely. Using IoT-based equipment, the required quantity of fertilizers can be spread over the fields with minimum labor (Lavanya, Rani, & Ganesh Kumar 2019). The Normalized Difference Vegetation Index (NDVI) is an important IoT-based solution for the analysis of the crop nutrients (Benincasa et al., 2018) and (Liu, Wang, & Bing-kun,

2018). Many advanced technologies, like geo mapping (Suradhaniwar et al., 2018), GPS accuracy, etc., are playing a key role in IoT-configured fertilization.

4.3.5 YIELD ANALYSIS AND HARVESTING

Yield analysis is one of the most important agricultural processes. In this process, the various activities performed include moisture content analysis and quantity of the grain. Even the analysis of yield quality is addressed as one of the important tasks in the mechanism of yield analysis (Chung et al., 2016; Wietzke et al., 2018). It plays a vital role in estimating the right time for crop harvesting. Consequently, the best quality yield can be obtained. To reap the true benefits, it is very important to harvest the crops at the right time. A yield monitor was proposed, which can be integrated with the harvester. It is also equipped with a mobile app, FarmRTX. Through this app, the data related to the harvesting process automatically becomes live on a web-based platform. Manfrini et al. (2015) presented an IoT-based solution to estimate the growth of the fruits.

4.4 ADVANCED IOT-BASED AGRICULTURAL METHODS

New techniques to improve the quantity and quality of yield are always practiced in the agriculture industry. In the beginning, crop yield was improved by concentrating on higher quality seeds and fertilizers. Afterward, analysis showed that the adoption of technology would further improve the agricultural processes; hence, the same was incorporated to achieve the targets (Zhang, Wohlhueter, & Zhang, 2016). In this direction, research was carried out for a long time to analyze the advantages of IoT platforms in agricultural processes. The incorporation of IoT- and sensor-based technologies has made advanced agricultural methods feasible.

4.4.1 VERTICAL FARMING

While the requirement of food is increasing day by day, the auricular land is decreasing because of urbanization and soil erosion (Pimentel & Burgess, 2013). Another important issue is the usage of almost 70% of water in the agriculture sector. One possible solution to overcome these issues is vertical farming (VF). VF is practiced generally in urban areas with an aim to grow plants in a restrained environment and with limited resources. It helps to take more yield from a limited ground area. IoT systems play a key role in the VF technique to grow plants. Human intervention is not required in IoT-connected vertical farms at any stage of the agriculture process. A number of sensor-based solutions for waste collection, irrigation, and other processes are used.

4.4.2 GREENHOUSE FARMING

Greenhouse farming is the most traditional among smart-farming techniques. This agricultural technique became popular in the 19th century. Initially, it was most practiced in Italy and France. Later, in the 20th century, it was adopted in countries with severe climate conditions (Hind, 1988). With greenhouse farming, crops that could be grown only in certain regions or under certain climate conditions are grown anywhere. It was the beginning of the integration of sensor technology in agricultural processes. In a controlled environment, the farming of crops depends upon many factors, like structure and shape of the greenhouse shed, traits under observation, ventilation system of the area, etc. Shamshiri et al. (2018) presented an in-depth analysis of all these factors. The usage of the sensor technology in the monitoring of all these factors is also discussed. Akkaş and Sokullu (2017) proposed an IoT-based model to observe the various parameters like temperature, humidity, and other parameters in the greenhouses.

4.4.3 HYDROPONICS

Hydroponics is an extension of greenhouse farming in which crops are grown without soil using only water. In this type of farming, the vital nutrients the plants need are dissolved in the water and crops are grown into it. In some cases, where required, the roots of the plants are also made to stand in gravel dipped in water. With hydroponics, higher yields from the crops are obtained with fewer resources and pesticides. At present, a number of IoT-based solutions are available to measure the nutrient level in water for hydroponics. Theopoulos et al. (2018) discussed various IoT-based solutions for tomato hydroponics, soil-less growing, and water nutrient-level identifiers.

4.4.4 PHENOTYPING

Phenotyping is one of the most prominent among advanced agricultural techniques under experimentation. With this technique, quantitative analysis related to the crop parameters can be done (Tripodi et al., 2018). Various parameters that can be quantified are related to plant growth, quality, and quantity of yield, etc. An IoT-based platform was presented by Zhou et al. (2017) for crop breeding.

4.5 TOOLS AND TECHNOLOGIES FOR SMART AGRICULTURE

In the last few decades, the agricultural domain has revolutionized, like any other industry. This transformation expects the agricultural domain to be administered like any other industry. So, every trait of the domain that can be automated can reap the true benefit of the IoT platforms.

In the last few years, researchers and engineers have made many efforts to propose automated solutions for various agriculture tasks like planting, fertilizing, weeding, watering, spraying, transportation, etc. Various manufacturers are converting these solutions to the products to aid various applications; some are depicted in Figure 4.5. The various tools and technologies used in smart farming are discussed below.

4.5.1 WIRELESS SENSORS

Among the various technologies used in smart agriculture, wireless sensors are the most prominent ones. Wireless sensors play a very important role in collecting the

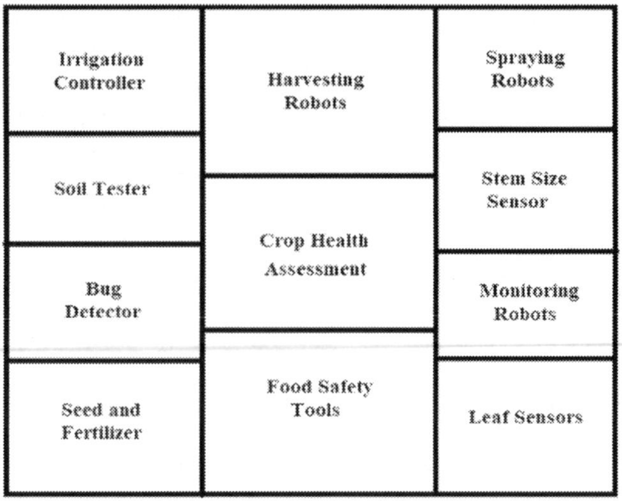

Irrigation Controller	Harvesting Robots	Spraying Robots
Soil Tester	Crop Health Assessment	Stem Size Sensor
Bug Detector		Monitoring Robots
Seed and Fertilizer	Food Safety Tools	Leaf Sensors

FIGURE 4.5 IoT equipped smart agriculture tools/products.

data related to various activities of farming. Wireless sensors may work independently and as part of other equipment and tools. A wide range of sensors are used to automate various agricultural activities. Among them, the tasks of seed classification and pest management are handled by acoustic sensors (Srivastava et al., 2013). FPGA-based sensors are used in humidity detection and irrigation (Millan-Almaraz et al., 2010). When it comes to the tasks of soil analysis and plant assessment, optical sensors are used (Povh et al., 2014). The process of weeding is automated with optoelectronic sensors. Soil nutrient-level analysis is performed with electrochemical sensors (Andújar et al., 2011). Mechanical resistance of the soil is measured with mechanical sensors (Schuster, Darr, & McNaull, 2017).

4.5.2 SMART TRACTORS

To meet the growing requirements of automated equipment, many tractor manufacturers have started to deliver driverless tractors with cloud-computing features. These self-driving tractors have many advanced features like no overlapping and the ability to make very precise turns. Until now, none of the manufacturers had provided fully automated tractors. But, anticipating the future demand of fully automated equipment, many manufacturers, in collaboration with researchers, are putting best possible efforts to provide fully functional high-tech tractors.

4.5.3 HARVESTING ROBOTS

To gain the maximum yield from crops, harvesting must be done at the right time. If it is done early or late, it may affect production severely. Keeping in view the role of this activity and the challenges related to manpower, domain experts suggest using automated equipment for this phase. So, harvesting robots are playing a very crucial role in managing the above discussed issues. Very productive work in the area of fruit harvesting is presented by Zhao et al. (2016) and Zujevs, Osadcuks, and Ahrendt (2015). For fruit harvesting, smart solutions are developed with the integration of IoT and image processing (Feng, Zeng, & He, 2019). Some dedicated robot-based harvesting solutions are available for peppers, apples, and strawberries.

4.5.4   COMMUNICATION

The most crucial factor in smart agriculture is the transfer of data in a timely manner. We can say that the availability of information acts as a backbone to precision agriculture. This target cannot be achieved without a reliable and secure communication system. To reap the true benefit of IoT platforms in the agricultural sector, a large communication network is always required. Poor communication networks cannot aid in transmitting data to the farm-management systems. A number of communication technologies, like cellular communication, Zigbee, Bluetooth, etc., are used to communicate the data between various smart devices and applications.

4.5.5   SMARTPHONES

Mobile phones are the most commonly used devices for transfering data in the agricultural domain. Thanks to technological revolution, smartphones have more features at a reduced cost, which increases their use in farming, especially in the rural areas. Very fast development in the communication networks provides better services to the farmers in the remote areas. Although only 8% of agricultural services are mobile phone based yet (Metcalfe, 2015), to improve this share, IT professionals are developing more attractive and efficient mobile apps for farmers (Alfian, Syafrudin, & Rhee, 2017). It has been further observed that from the last few years, researchers are investing additional efforts to enhance the share of smartphone agriculture applications.

4.6   ORGANIC FARMING

Organic farming is the process of cultivating crops without using synthetic agro-chemicals like fertilizers, pesticides, or genetically modified organisms, which can maintain the health of the soil, biodiversity, ecosystem, and people (Rigby & Cáceres, 2001). This farming approach focuses on cultivating the land and growing crops to keep the soil in good health by using organic wastes such as crops, animals and farm wastes, and aquatic wastes; other biological materials and bio-fertilizers are used to provide required nutrients to crops for better sustainable production in an eco-friendly pollution-free environment (Aghasafari et al., 2020), as shown in Figure 4.6.

The United States Department of Agriculture (USDA) defines "Organic farming as an agricultural technique where the usage of synthetic inputs is totally precluded and mainly depends upon organic waste, crop rotation, animal manure and eco friendly techniques to maintain nutrient level in the soil." Whereas the Food and Agriculture Organization (FAO) of the United Nations advised that "Organic

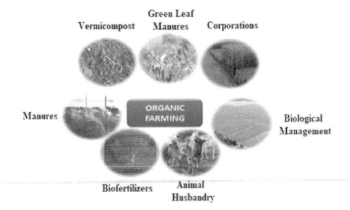

FIGURE 4.6   Organic farming.

agriculture as an exclusive production management system which encourages agro-ecosystem health, including biodiversity, biological cycles and soil biological activity, and is attained by making use of on-farm agronomic, biological and mechanical methods and strictly prohibiting the use of all synthetic off-farm inputs." Organic farming is not a new method of farming in India; it was used before the green revolution and was in practice from ancient times (Yadav et al., 2013). With a breakthrough in population, our concern would not be only to maintain production but to raise it further in a continual way. Researchers and scientists have analyzed that the "Green Revolution," with huge input and practice, has reached a stage where it is now sustained with a decreased recurrence of decreasing reward. So, an essential equity is required to be preserved in every possible way for the survival of life and property. The common available choice that would be more suitable in the current scenario is the diminishing of agrochemicals in availability. In 2016, Sikkim had drawn lots of attention by becoming the first "Organic State "in the whole world in the direction of organic farming. The farmers started following traditional, organic methods of farming. The harms arose from the extreme usage of synthetic chemicals on soil that took longer rehabilitation time, thereby increasing the time required to earn a certificate of organic farming.

   In 2011, the population of India was 1210 million and vegetable production was 146.55 million tons. Expected vegetable production in the same year was 230.40 g/person/day, whereas in 1952, expected consumption was 87.66 g/person/day. At present, the recommended level of dietary allowance (RDA) for vegetables is 300 g/person/day. But still, we are not self-sufficient to meet this requirement of vegetables and are experiencing a shortfall of approximately 30 million tones. Records show a 25% post-harvest depletion in vegetables. Only 5% of the total organic vegetable produce is exported. As the agricultural land is decreasing day by day because of urbanization, there is a big burden on the agricultural field to feed the regularly increasing population. There is a continuous bargain for quantity and quality of the agricultural produce. Although it's difficult, it would have been an achievement if both requirements could be targeted in one go.

The NSSO published in the 66th round that there is a transition in the consumption habits of the Indian population. Now, people have become health and diet conscious and are more interested in the consumption of fruits and vegetables rather than grains and cereals. The Indian population is more inclined toward quality instead of quantity of the food items. One of the important food products used each day are vegetables. The price of the vegetables mainly depends upon the factors like freshness, color, and supply, etc. One more important factor, on which the price of the vegetables depends, is the cultivation methods used to grow the vegetables. Awareness about the growing methods used for the agricultural products among the populations helps a lot in marketing and trade. Due to limitations on data availability, in this chapter we have focused on organic vegetables only.

4.7 SUSTAINABLE AGRICULTURE

Sustainable agriculture unifies economic equity, profitability, and environmental health into one main objective. A number of principles have accorded these objectives. Different stakeholders, farmers, researchers, manufacturers of agriculture equipment, and consumers all have aimed and contributed to achieve these goals (Yadav et al., 2013). Sustainability depends upon the concept to fulfill current requirements and assure that next generations will also be able to meet their needs without any impact. So, the management of all types of resources, i.e., human and natural, is really important in the domain of agriculture (Amutha, 2013). Under human resources, the various important traits like consumer needs, working conditions of the labor, health, and safety, etc., need to be given due importance, whereas in natural resources, fertility and the nutrient level of land need to be considered and maintained. Agricultural sustainability needs to be practiced across the complete system, ranging from small farms to the local agricultural system, along with the population touched at different geographical domains (Rigby & Cáceres, 2001). Consequently, the effect of these agricultural methods on population and environment can be studied and analyzed more accurately. A system aspect provides the techniques to analyze the relationship among agriculture and environmental aspects. A system approach allows the workers, farmers, consumers, experts, and scientists to contribute to the research and development in the stream. This shift in paradigm from traditional farming to sustainable agriculture is a complete process, where all the stakeholders need to play their expected roles to establish the concept of sustainable agriculture (Kerr, 1996).

4.7.1 ORGANIC FARMING FOR SUSTAINABLE AGRICULTURE

Modernization of agricultural processes, like the high use of synthetic inputs, advanced irrigation techniques, more produce-giving methods, etc., are the main contributors in achieving the astonishing growth in the agriculture sector in the last few decades. Although the high usage of synthetic inputs, like fertilizers and pesticides and water resources, has increased the yield of the crops to meet the current food needs, it has badly impacted human life and the ecosystem (Narayan, 2012). More and more adoption of pesticides in farming is the biggest risk to environmental sustainability.

There is also a huge danger for the survival of all types of life on Earth since, over time, these synthetic inputs become part of the food chain. So, there is a big trade-off between our food requirements and environmental sustainability issues (Rigby & Cáceres, 2001). Organic or eco-farming is the most viable solution to this challenge, which helps to meet all types of challenges. Under organic farming, the main objectives are to produce the yield with balanced nutrients under an eco-friendly environment. It accentuates the control of the application of synthetic inputs like fertilizers, pesticides, etc. Under organic farming, the use of organic manures is encouraged; it consists of all the required nutrients for plant growth. In the same way, in organic farming, natural methods for pest management are practiced, and these do not affect human life. A number of sustainability practices, like crop rotation, intercropping, etc., are inspired in organic farming.

4.7.2 Organic Manure in Organic Farming

Organic manure is a vast term. All types of crop residues, industrial biowaste, and rural and urban waste can be converted to organic manure after processing. Organic manure has many benefits in sustainable agriculture. It maintains the nutrient level of the soil required for crop growth, so it maintains soil health (Babar et al., 2011). It also increases the water-retaining capacity of the soil for sustainable and eco-friendly agriculture. Organic manure also acts as a soil stimulant.

4.7.3 Vermicomposting

Vermicomposting is a process in which organic wastes collected from different sources are decomposed with the help of earthworms. The level of nutrients in vermicomposting is much higher than any other manure (Narayan, 2012). It is rich in both macro and micro nutrients required for plant growth. The literature discusses that earthworms consume organic waste and transform it into nutrients. Vermicomposting enhances the fertility of the soil and also improves the water-retention capacity of the soil. It aids the process of nutrient management for the plants in sustainable agriculture.

4.7.4 Biological Nitrogen Fixation

As we all know, there is 78% nitrogen in the environment that can be used as an important soil nutrient. So, instead of using fertilizer nitrogen, eco-friendly agricultural processes to leverage environmental nitrogen should be practiced for sustainable agriculture. These processes have the potential to transform environmental nitrogen to ammonia. A number of bacteria like Azotobacter, Azospirillum, etc., are available; they have already been proven for best nitrogen fixing capabilities.

4.7.5 Crop Rotation and Intercropping

Crop rotation is one of the most important traits of organic farming and sustainable agriculture. When crops are grown on a rotation basis on a piece of land with due analysis, the rotation assists in the best possible utilization of the soil nutrients available at different levels (Wijnands, 1999). The process of intercropping is very useful to reduce the danger of crop damage due to pests and diseases. Intercropping also helps to get more yield from the crops.

4.8   SUSTAINABLE AGRICULTURE IN INDIA

A collection of agricultural methodologies that are economical, eco-friendly, and socially adequate is termed as sustainable agriculture. A farming method needs a huge amount of synthetic inputs and produces good yield. Although it may be sustainable economically, it will affect the environment and human life; therefore, it cannot be categorized as sustainable agriculture. Organic farming impresses upon "healthy soil" by minimizing the usage of synthetic inputs like pesticides and fertilizers. In the concept of sustainable organic agriculture, it is not possible to eradicate the use of chemicals, but their reduced usages is encouraged (Hodge, 1993). Another term, "alternative agriculture," is also becoming popular day by day in relation to sustainable agriculture. Any crop production that is more dependent on natural techniques, uses minimum synthetic inputs, is more dependent on biological processes, and makes the most optimized use of natural resources is defined as alternative agriculture. As in conventional methods of agriculture, there is a high use of chemicals like pesticides and fertilizers; it affects human life badly and also pollutes our water bodies. Consequently, this technique of farming is discouraged. As per the marks of sustainable agriculture, many factors like crop rotation, intercropping, environment sustainability, and ecosystem play a crucial role to impact human life and soil health.

4.8.1   GOVERNMENT POLICIES FOR SUSTAINABLE AGRICULTURE AND ORGANIC FARMING IN INDIA

Because of the huge population to be fed, agricultural policies of the Indian government always impress upon food self-sufficiency, which may not align with the concept of sustainable agriculture. In India, the yield of the imperishable agricultural products increased unquestionably during 1970 and 1980, but then decreased in 1990. Further, a downfall was observed in 2000, but a decline in 2001–2003 was very challenging. It affected both the livelihood as well as ecosystem. As a result, a thorough review of the agricultural policies for sustainable development was urgently needed (Narayan, 2012). These policies were not required to be reviewed by considering the local food requirements and environmental sustainability but also keeping in view of the international markets (Priyadarshini & Abhilash, 2020).

The fundamental targets of the National Program for Organic Production (NPOP) are focused on the export market. Various objectives of the NPOP include facilitation to run educational programs for organic agriculture, framing policies, and regulations for the certification of organic products, finalizing National Standards for Organic Products (NSOP), motivating farmers to practice organic farming, etc. A number of projects are initiated by many state governments to encourage the practice of organic agriculture among the farmers from time to time (Suradhaniwar et al., 2018). In 2000, an organic agriculture task force was constituted by the Ministry of Agriculture. This body plays an advisory role to the Ministry of Agriculture to establish organic farming models, to promote organic farming practices, and to advise financial aid and subsidies to the farmers for organic agriculture.

4.9 CONCLUSIONS

With a rapid change in the field of agriculture, there is a high need for a paradigm shift from production-focused agriculture to sustainable farming. To bring this shift into practice, Indian farmers need to adopt resource-conservation techniques and IoT-based agricultural solutions for the development of sustainable agriculture. In the country, the environment is becoming more congenial for green smart sustainable agriculture. Advanced techniques are showing a new roadmap to the farmers and all other stakeholders for the development of sustainable agriculture and ecological sustainability. As the environmental circumstances change continuously, so, too, our farmers should always be ready to adjust, experiment, and adapt. A suitable farm management system is required to look after the important parameters like productivity, profitability, ecosystem, and human health for practicing and managing green smart sustainable agriculture.

Technological advancement has caused a great revolution in the domain of agriculture. But there are a few challenges to addressfor the growth of green smart sustainable agriculture, as discussed below:

- To use smart farming tools and techniques efficiently, proper education and training of the farmers is required. Yet, Indian farmers are not trained enough to handle all high-end technologies.
- Integration of smart techniques enhances agricultural productivity; but, it is a costly approach to farming. It requires huge investment to swap old equipment with smart equipment. So, there is a high requirement of economical smart farming equipment and techniques to reap the maximum benefit.
- As in smart farming, most of the environmental data is captured by the sensors and predictions are made. If the methods associated with data gathering and processing are not accurate, it will badly affect the product. So, there is a need for more accurate data collection and analysis methods for more reliable predictions in terms of weather forecasting, fruit ripening, water and nutrient requirements, etc.

Along with these technology-centered challenges, there is also a need for more sophisticated and contemporary government policies for the development of green smart-sustainable agriculture. Additional attention is also needed to address pricing and marketing of the organic products.

REFERENCES

Aghasafari, H, Karbasi, A, Mohammadi, H, & Calisti, R (2020). Determination of the best strategies for development of organic farming: A SWOT – Fuzzy analytic network process approach. *Journal of Cleaner Production, 277,* 1–12.

Akkaş, MA, & Sokullu, R (2017). An IoT-based greenhouse monitoring system with Micaz motes. *Procedia Computer Science, 113,* 603–608.

Alfian, G, Syafrudin, M, & Rhee, J (2017). Real-time monitoring system using smartphone-based sensors & nosql database for perishable supply chain. *Sustainability, 9,* 2073.

Amutha, D (2013). Present status of Indian agriculture. https://papers.ssrn.com/sol3/papers.cfm?abstract_id=2739231

Andújar, D, Ribeiro, Á, Fernández-Quintanilla, C, & Dorado, J (2011). Accuracy and feasibility of optoelectronic sensors for weed mapping in wide row crops. *Sensors, 11,* 2304 –2318.

Ayaz, M, Ammad-Uddin, M, Sharif, Z, Mansour, A, & Aggoune, EHM (2019). Internet-of-Things (IoT)-based smart agriculture: Toward making the fields talk. *IEEE Access, 7,* 129551–129583.

Babar, S, Stango, A, Prasad, N, Sen, J, & Prasad, R (2011). Proposed embedded security framework for internet of things (IoT). *Second International Conference on Wireless Communication, Vehicular Technology, Information Theory and Aerospace & Electronic Systems Technology (Wireless VITAE),* 1–5.

Benincasa, P, Antognelli, S, Brunetti, L, Fabbri, CA, Natale, A, Sartoretti, V, et al. (2018). Reliability of NDVI derived by high resolution satellite and UAV compared to in-field methods for the evaluation of early crop N status and grain yield in wheat. *Experimental Agriculture, 54,* 604–622.

Berger, M, Pfister, S, & Motoshita, M (2016). Water footprinting in life cycle assessment: How to count the drops and assess the impacts?. *Special Types of Life Cycle Assessment.* Finkbeiner, Matthias (ed), Springer, 73–114.

Bruinsma, J (2003). World Agriculture: Towards 2015/2030: An FAO Perspective. Earthscan.

Charatsari, C, & Lioutas, ED (2020). Smart farming and short food supply chains: Are they compatible? *Land Use Policy, 94,* 1–8.

Chung, S-O, Choi, M-C, Lee, K-H, Kim, Y-J, Hong, S-J, & Li, M (2016). Sensing technologies for grain crop yield monitoring systems: A review. *Journal of Biosystems Engineering, 41,* 408–417.

Dinkins, CP, & Jones, C (2013). Interpretation of soil test reports for agriculture. *MT200702AG, Montana State University Extension,* Bozeman, MT, USA.

Elijah, O, Rahman, TA, Orikumhi, I, Leow, CY, & Hindia, MN (2018). An overview of Internet of Things (IoT) and data analytics in agriculture: Benefits and challenges. *IEEE Internet of Things Journal, 5,* 3758–3773.

Feng, J, Zeng, L, & He, L (2019). Apple fruit recognition algorithm based on multi-spectral dynamic image analysis. *Sensors, 19,* 949.

Glaroudis, D, Lossifieds, A, & Chatzimisios, P (2020). Survey, comparisons and research challenges of IoT application protocols for smart farming. *Computer Networks, 168.*

Greene, SA (2013). *Sittig's Handbook of Pesticides and Agricultural Chemicals*: William Andrew.

Hind, C (1988). Glass houses. A history of greenhouses, orangeries and conservatories. *JSTOR,* 44–49.

Hodge, I (1993). Sustainability: Putting principles into practice. An application to agricultural systems. *Rural Economy and Society Study Group,* Royal Holloway College: England.

Kerr, JM (1996). Sustainable development of rainfed agriculture in India, No: 581: 2016-39388.

Khanna, A, & Kaur, S (2019). Evolution of Internet of Things (IoT) and its significant impact in the field of precision agriculture. *Computers and Electronics in Agriculture, 157,* 218–231.

Kim, S, Lee, M, & Shin, C (2018). IoT-based strawberry disease prediction system for smart farming. *Sensors, 18*, 4051.

LaRue, J, & Fredrick, C (2012). Decision process for the application of variable rate irrigation. *2012 Dallas, Texas*, p. 1, American Society of Agricultural and Biological Engineers.

Lavanya, G, Rani, C, & Ganesh Kumar, P (2019). An automated low cost IoT based Fertilizer Intimation System for smart agriculture. *Sustainable Computing: Informatics and Systems, 28*, 100300.

Lin, J, Yu, W, Zhang, N, Yang, X, Zhang, H, & Zhao, W (2017). A survey on internet of things: Architecture, enabling technologies, security and privacy, and applications. *IEEE Internet of Things Journal, 4*, 1125–1142.

Liu, H, Wang, X, & Bing-Kun, J (2018). Study on NDVI optimization of corn variable fertilizer applicator. *INMATEH-Agricultural Engineering, 56*.

Mahbub, M (2020). A smart farming concept based on smart embeded electronics, internet of things and wireless sensor networks. *Internet of Things, 9*, 1–29.

Manfrini, L, Pierpaoli, E, Zibordi, M, Morandi, B, Muzzi, E, Losciale, P, et al. (2015). Monitoring strategies for precise production of high quality fruit and yield in Apple in Emilia-Romagna. *Chemical Engineering Transactions, 44*, 301–306.

Mcguire, D (2014). 2.1 Fao's Forest and landscape restoration mechanism. *Towards Productive Landscapes, 19*.

Metcalfe, H (2015). *Mobile for Development Impact Products and Services Landscape Annual Review*. GSM Association: London.

Meybeck, A, Gitz, V, Azzu, N, Batello, C, Chaya, M, De Young, C, et al. (2012). Greening the economy with climate-smart agriculture. *Second Global Conference on Agriculture, Food Security and Climate Change*, Hanoi, Vietnam, 3–7 September 2012.

Millan-Almaraz, JR, Romero-Troncoso, RdeJ, Guevara-Gonzalez, RG, Contreras-Medina, LM, Carrillo-Serrano, RV, Osornio-Rios, RA, et al. (2010). FPGA-based fused smart sensor for real-time plant-transpiration dynamic estimation. *Sensors, 10*, 8316–8331.

Narayan, BS (2012). Sustainable agricultural development and organic farming in India.

Navulur, S, & Prasad, MG (2017). Agricultural management through wireless sensors and internet of things. *International Journal of Electrical and Computer Engineering, 7*, 3492–3503.

Oliveira, K. V. de., Castelli, HME, Montebeller, SJ, & Avancini, TGP (2017). Wireless sensor network for smart agriculture using ZigBee protocol. *IEEE First Summer School on Smart Cities*, 1(S3C), 61–66.

Pimentel, D, & Burgess, M (2013). Soil erosion threatens food production. *Agriculture, 3*, 443–463.

Povh, FP, Anjos, W. d. P. G., dos Yasin, M, Harun, S, & Arof, H (2014). Optical sensors applied in agricultural crops. *Optical Sensors-New Developments and Practical Applications, 5*(4), 141–163.

Priyadarshini, P, & Abhilash, PC (2020). Policy recommendations for enabling transition towards sustainable agriculture in India. *Land Use Policy, 96*, 1–14.

Rigby, D, & Cáceres, D (2001). Organic farming and the sustainability of agricultural systems. *Agricultural Systems, 68*, 21–40.

Roy, TN (2015). Review on institutional agricultural credit facilities for growth of agriculture and related problems in India–A longitudinal analysis, (A report).

Schuster, JN, Darr, MJ, & McNaull, RP (2017). Performance benchmark of yield monitors for mechanical and environmental influences, *ASABE Annual International Meeting, 15*(7).

Shamshiri, RR, Kalantari, F, Ting, K, Thorp, KR, Hameed, IA, Weltzien, C, et al. (2018). Advances in greenhouse automation and controlled environment agriculture: A transition to plant factories and urban agriculture, *17*(7), 234–253.

Sisinni, E, Saifullah, A, Han, S, Jennehag, U, & Gidlund, M (2018). Industrial internet of things: Challenges, opportunities, and directions. *IEEE Transactions on Industrial Informatics*, *14*, 4724–4734.

Srivastava, N, Chopra, G, Jain, P, & Khatter, B (2013). Pest monitor and control system using wireless sensor network with special reference to acoustic device wireless sensor. *International Conference on Electrical and Electronics Engineering*.

Suradhaniwar, S, Kar, S, Nandan, R, Raj, R, & Jagarlapudi, A (2018). Geo-ICDTs: Principles and applications in agriculture. *Geospatial Technologies in Land Resources Mapping, Monitoring and Management*, Springer, 75–99.

Theopoulos, A, Boursianis, A, Koukounaras, A, & Samaras, T (2018). Prototype wireless sensor network for real-time measurements in hydroponics cultivation. *Seventh International Conference on Modern Circuits and Systems Technologies (MOCAST)*, 1–4.

Tripathi, AD, Mishra, R, Maurya, KK, Singh, RB, & Wilson, DW (2019). Estimates for world population and global food availability for global health. *The Role of Functional Food Security in Global Health*, Elsevier, 3–24.

Tripodi, P, Massa, D, Venezia, A, & Cardi, T (2018). Sensing technologies for precision phenotyping in vegetable crops: current status and future challenges. *Agronomy*, *8*, 57.

Tzounis, A, Katsoulas, N, Bartzanas, T, & Kittas, C (2017). Internet of Things in agriculture, recent advances and future challenges. *Biosystems Engineering*, *164*, 31–48.

Venkatesan, R, Kathrine, GJW, & Ramalakshmi, K (2018). Internet of Things based pest management using natural pesticides for small scale organic gardens. *Journal of Computational and Theoretical Nanoscience*, *15*, 2742–2747.

Walter, A, Finger, R, Huber, R, & Buchmann, N (2017). Opinion: Smart farming is key to developing sustainable agriculture. *Proceedings of the National Academy of Sciences*, *12*(2), 6148–6150.

Wietzke, A, Westphal, C, Gras, P, Kraft, M, Pfohl, K, Karlovsky, P, et al. (2018). Insect pollination as a key factor for strawberry physiology and marketable fruit quality. *Agriculture, Ecosystems & Environment*, *258*, 197–204.

Wijnands, F (1999). Crop rotation in organic farming: Theory and practice. *Designing and Testing Crop Rotations for Organic Farming. Proceedings from an international workshop. Danish Research Centre for Organic Farming*, 21–35.

Wolfert, S, Ge, L, Verdouw, C, & Bogaardt, M-J (2017). Big data in smart farming – A review. *Agricultural Systems*, *153*, 69–80.

Yadav, S, Babu, S, Yadav, M, Singh, K, Yadav, G, & Pal, S (2013). A review of organic farming for sustainable agriculture in Northern India. *International Journal of Agronomy*, *2013*.

Yousuf, O, Titikshya, S, & Singh, A (2018). Organic food production through green technology: An ideal way of sustainable development, *7*(6), 160–163.

Zhao, Y, Gong, L, Huang, Y, & Liu, C (2016). A review of key techniques of vision-based control for harvesting robot. *Computers and Electronics in Agriculture*, *127*, 311–323.

Zhang, C, Wohlhueter, R, & Zhang, H (2016). Genetically modified foods: A critical review of their promise and problems. *Food Science and Human Wellness*, *5*, 116–123.

Zhang, L, Dabipi, IK, & Brown Jr, WL (2018). Internet of Things applications for agriculture. *Internet of Things A to Z: Technologies and Applications*, *1*(5), 507–528.

Zhou, J, Reynolds, D, Websdale, D, Le Cornu, T, Gonzalez-Navarro, O, Lister, C, et al. (2017). CropQuant: An automated and scalable field phenotyping platform for crop monitoring and trait measurements to facilitate breeding and digital agriculture. *BioRxiv*, *161547*.

Zujevs, A, Osadcuks, V, & Ahrendt, P (2015). Trends in robotic sensor technologies for fruit harvesting: 2010–2015. *Procedia Computer Science*, *77*, 227–233.

5 Toward Circular Product Lifecycle Management through Industry 4.0 Technologies

Barbara Ocicka

Collegium of Business Administration, Department of Logistics, SGH Warsaw School of Economics, Warsaw, Poland

Grażyna Wieteska

Faculty of Management, Department of Logistics, University of Lodz, Lodz, Poland

Beata Wieteska-Rosiak

Faculty of Economics and Sociology, Department of Investment and Real Estate, University of Lodz, Lodz, Poland

CONTENTS

5.1   INTRODUCTION

The global challenge facing the modern world is to reduce human pressure on the environment while ensuring a high life quality and competitiveness of economies. The circular economy concept, which has recently become more popular, can be perceived as an approach that supports the achievement of sustainable development goals. The transition to a circular economy is an opportunity to create a low-carbon, resource-efficient, competitive economy and finally become climate-neutral (European Commission 2015; 2020). However, the dynamic growth of the world population, global consumption, and production are sources of the extremely high demand for natural resources, problems with waste growth, intensive greenhouse gas emissions, as well as the alarming climate change, which has resulted in the intensification of extreme climate phenomena (IPCC, 2013; Below and Wallemacq 2018).

Nowadays, when designing a product's lifecycle, one needs to consider the principles of the circular economy, which direct our attention toward the circular product lifecycle (European Environment Agency 2017). To successfully face the challenge of circular product lifecycle management, companies can exploit multiple diverse digital technologies in each of the stages, as well as throughout the entire lifecycle. The literature is filled with numerous theoretical studies dealing with this subject, although they approach the issue mostly from an overall perspective of the circular economy. It is broadly accepted that digital technologies support the transition toward a circular economy (Pagoropoulos et al. 2017; Antikainen et al. 2018; Okorie et al. 2018; Rosa et al. 2019). However, empirical studies that discuss the role of selected technologies are rare, and further studies on real company cases are desirable. Researchers underline the urgent need to conduct more empirical research on the adoption of technologies in case studies (Pagoropoulos et al. 2017).

This chapter explains how Industry 4.0 technologies can be integrated into the circular product life cycle within different industries. It starts with a presentation of the latest circular economy (CE) concept and its principles. This is the starting point for the next section, which introduces the definition and six phases of the circular product lifecycle (CPLC). The authors focus on Industry 4.0 technologies that they identified in the literature review and described regarding their functionality and role for CPLC development. Next comes an empirical study based on an analysis of multi-sector case studies that demonstrate the use of Industry 4.0 at different stages of product lifecycle management and their impact on the effective implementation of CE principles.

The reason behind the choice of sectors was twofold. On the one hand, the three selected industries (construction, furniture, and cosmetics) impact the quality of life of modern society by delivering products that meet the basic needs of consumers and households. On the other hand, however, they differ from each other regarding the complexity and value of these products. In the next section, a comparative analysis of the case studies is conducted based on CE principles and CPLC phases. It allows us to formulate the key conclusions and practical and theoretical implications outlined at the end of this chapter.

5.2 DEFINITIONS AND PRINCIPLES OF THE CIRCULAR ECONOMY

The global challenge of the modern world is to reduce human pressure on the natural environment while ensuring a high quality of life and competitiveness of economies. The current linear model is based on the principles of "take-make-dispose" (Ellen MacArthur Foundation 2013, p. 2; 2019a, p. 5), "take-make-consume-dispose" (European Environment Agency 2016, p. 5), or "take-make-consume and dispose" (European Commission 2014, p. 2); however, it has lost much of its pertinence. Shrinking natural resources, increasing amounts of waste, dynamic population growth, increasing consumption and production, the volatility of prices, and raw material supply risks, environmental pollution, high greenhouse gas emissions, as well as progressive climate change (IPCC, 2013; Deloitte 2016) have all shifted the attention toward sustainable development and a resource-efficient economy. The significant loss of resources, increasing resource extraction, and wastage that are inherent in the linear approach gave the impetus for the development of the circular economy model.

The assumption underpinning the CE concept is to keep resources in use. When a product reaches its "end-of-life" phase, the aim is to productively reuse materials and resources and create economic value in the economy, processes, or product life cycle. In this way, we achieve efficient resource management as well as waste minimization (European Commission 2014). The circular economy concept seeks to reduce the dependence of economic development on shrinking resources. At the same time, it works to achieve economic growth and create new jobs while minimizing waste and reducing economic pressure on the environment (Ellen MacArthur Foundation 2014; 2019a). This means that the transition of the economy to circularity brings with it environmental, economic, and social benefits. It is part of the design of sustainable development. The CE has become one of the tools that seek to support the shaping of sustainable development and enable the implementation of proposals in the document entitled Agenda for Sustainable Development 2030. The document was signed by 193 countries and is committed to achieving 17 global goals of sustainable development (United Nations 2015).

In the CE, raw materials, materials, and products must circulate in a closed loop, limiting waste generation and reducing pollution released to the environment. The CE is more than just a concept, it is a global strategy. As a framework that can be implemented in many sectors of the economy, it has also become an object of interest to the world of science and politics (Vermeulen et al. 2018).

One of the definitions presented in many publications is the following: The circular economy aims to maintain various products and raw materials as long as possible. The assumption is that they are useful and valuable in the economy (Ellen MacArthur Foundation 2015, p. 2). Numerous definitions presented in the literature help to identify the main features and specificity of the CE (Rizos et al. 2017, p. 6; Korhonen et al. 2018, p. 547).

The literature on the subject discusses numerous CE principles presented in a narrower or broader sense. The Ellen MacArthur Foundation points to three general principles: design out waste and pollution, keep products and materials in use, and regenerate natural systems (Ellen MacArthur Foundation 2019b). A wide variety of CE principles in the literature combine the "R-word." In the area of the circular economy, the literature most often refers to the 3R typology. The 3R principles (Reduce-Reuse-Recycle) are used the most often; an extended version, known as the 5R principles (Reduce-Reuse-Remanufacture-Recover-Recover), is, in turn, the most often suggested. These rules are related to the waste management hierarchy. There are also less frequently used classifications based on 4R, 6R and 7R, and even 10R (Reike et al. 2018; Campbell-Johnston et al. 2019). The most common rules are R0-R9. The literature contains two more principles: Re-mine and Rethink. The 10R principles are proposed in Table 5.1. The development of a circular economy model takes time. Achieving a lasting change in the economy does not happen overnight. It requires systemic thinking and the involvement of equal stakeholders: producers (new products, and business and market models), consumers (new consumer behavior and environmental awareness), public authorities (taking care of appropriate development policies, supporting the CE development, and supporting and financing circular initiatives), research and scientific institutions (technology development and innovation), and non-governmental organizations (promoting CE and circular activities). This means that it is necessary to involve public and private sector entities and resources.

The economy has evolved toward a digital economy. Research on the CE and technologies is developing rapidly, and the potential of digital technologies is increasingly more important for the development of the circular economy. The importance of technology is also emphasized by the European Commission in its development policy documents (European Commission 2020). Obviously, technology and the circular product lifecycle are positively interrelated. On the one hand, modern technologies support the design of a circular product lifecycle, and on the other hand, the development of a circular product lifecycle becomes an impulse for the creation of further digital innovations. The two stimulate each other for the benefit of the circular economy and sustainable development. Technologies offer opportunities for the CE principles implementation (Pagoropoulos et al. 2017; Sukhdev et al. 2018; Rosa et al. 2019).

TABLE 5.1
CE Principles

	Principle	Description
R0	Refuse	Refuse to use materials, resources, or products that are not indispensable, or replace them with other, more environment-friendly, digital ones. • consumers: refrain from buying; buy and use less; reject packaging waste. • producers/retailers/designers: refuse to use virgin and hazardous materials that pollute the environment; design processes to avoid waste.
R1	Rethink	Enhance the intensity of the use of a product by, e.g. making it available, re-using it. • consumers: share products – you do not have to own a product • producers/retailers/designers: sell products as services
R2	Reduce	Reduce demand for natural resources and materials, save resources, avoid waste in production, distribution, and consumption; generate less waste rather than think about how to dispose of it after it is generated, i.e. prevent rather than cure. • consumers: take good care of a product; use it carefully; use a product longer • producers/retailers/designers: design to minimize the demand for raw materials and resources in production
R3	Reuse	Reuse products, components, and resources in good technical condition, which are not waste and can fulfill their primary functions. • consumers: buy second-hand products, not or hardly used, which require minor repair or adaptation; use Internet auctions and offers of used products • producers/retailers/designers: sell or use unsold returns or products in damaged packaging; use recycled materials; re-use products, components, and resources in production; multiple uses of packaging (including for transport)
R4	Repair	Repair products, replace damaged parts, and continue using them in their primary functions; expand their useful life. • consumers: have a product repaired by a repair company or another consumer; repair it in a 'repair café' • producers/retailers/designers: send reclaimed products to service centers (own or third-party); treat repair as strategic; design easy to repair products; expand the useful life of products and avoid continuous repairs
R5	Refurbish	Renew your product and make it usable. • consumers: send a product to the service center • producers/retailers/designers: restore a product by replacing parts, modules, or components, leaving the structure of a complex product untouched; use technically advanced components (major repairs of buildings, machines).

(Continued)

TABLE 5.1 (Continued)
CE Principles

	Principle	Description
R6	Remanufacture	Use elements from a product that is no longer suitable for use. Use valuable parts in products that perform the same functions as the old one. • consumers: redirect the product to the service center • producers/retailers/designers: disassemble the structure of a multi-component product, replace or repair in an industrial process, parts to be recycled.
R7	Repurpose	Change the purpose for which a product or part of it is used. Give it value and new application goals. Use a discarded product, or part of a product, in a new product and in new applications.
R8	Recycle Materials	Recycle. The goal is to achieve the same or lower material characteristics. • consumers: segregate waste, buy products labeled as being made from recycled materials, • producers/retailers/designers: design products that can be made of recycled materials; design for recycling; process/remanufacture products and waste to recover materials; use recovered materials for primary or other purposes.
R9	Re-mine	Retrieve materials and components after the landfilling phase. • consumers: choose recovered/retrieved products • producers/retailers/designers: retrieve materials from landfills; reclaim valuable materials, products, and parts ("cannibalization")
R10	Recover (energy)	Recover energy from waste through incineration; use biomass.

Source: Vermeulen et al. 2018; Potting, et al. 2017; Reike et al. 2018; Morseletto 2020; European Investment Bank 2020; Ellen MacArthur Foundation 2017.

5.3   CIRCULAR PRODUCT LIFE CYCLE

Product lifecycle management (PLM) is an organized concept for designing, developing, and managing product and product-related information (Saaksvuori and Immonen 2008). It is implemented in various industries that design, manufacture, and support products (Stark 2016). The interest in PLM is especially driven by globalization, product complexity, shrinkage in product lifecycles, and environmental issues. Managing the product across the lifecycle connects various stakeholders who cooperate by jointly developing products and implementing lifecycle assessment (LCA) methodologies (Ameri and Dutta 2005).

The idea of product lifecycle has expanded significantly in recent decades, with examples including various lifecycle scenarios, approaches, and models (INCOSE UK 2017). As the circular economy concept directly relates to the product architecture, as well as designing, manufacturing, and distribution processes, we have decided to focus on engineering. The generally accepted definition of product lifecycle (PLC) in the ISO 14040 standard is the "consecutive and interlinked stages of a product system, from raw material acquisition or generation from natural resources to final disposal" (International Organization for Standardization 2006).

An engineering product lifecycle consists of several phases, which are usually based on a linear structure. The product lifecycle diagrams available in the literature on the subject differ in terms of the names of phases and the degree of detail, for example:

- Chen et al. (2009) depict the PLC using the following seven stages: product design, process development, product manufacturing, sales, the product in use, post-sales service, and retirement. In addition, each stage encompasses several activities.
- Terzi et al. (2010) illustrate a PLC in a four-level model. The main phases chronologically are: design and manufacturing (beginning of life), distribution, use and support (middle-of-life), and retirement (end-of-life). Each phase contains several elements, e.g. the design phase includes product design, process design, and plant design, while the support phase means repair and maintenance.
- Stark (2016) distinguishes only five phases: imagine, define, realize, support/use, retire/dispose.
- Ali et al. (2019), based on previous studies, indicate five PLC phases (design, manufacturing, usage, maintenance, end of life) and explain them through seven processes (needs recognition, design/development, production, distribution, usage, maintenance, disposal/recycling).

There is also clear evidence in the PLC descriptions that brings us to the topic of closing the loop, and it includes information and material flows that occur between the individual phases in the whole product lifecycle (Kiritsis 2011). The most thoroughly considered phase is the end of life, which aims to maintain the flows during or after a product is used. Scenarios such as reuse, recycling, remanufacturing, or disposal come up at this point, and implementing them connects the end-of-life phase with the previous phases (Ishii et al. 1994; Asiedu and Gu 1998; Kondoh et al. 2017). This kind of logic directly links with ensuring the circularity of raw materials as well as reducing waste. It is also a source of opportunities to continuously improve closed-loop flows. Namely, designers get feedback with detailed information from other PLC participants, and they can use it when reducing environmental impacts through designing improved products or processes (Young et al. 2007; Cao and Folan 2012; Lindkvist et al. 2017). There is a general agreement that the product design phase may determine all phases of its lifecycle, especially the end-of-life stage (Rose et al. 2002).

Discussions around the end-of-life phase led to other product-recovery scenarios being distinguished. Different reprocessing methods, depending on the product quality and the type of good, e.g. module and material, have been recognized (see, e.g. Parlikad et al. 2003; De Brito and Dekker 2004; Cao and Folan 2012). All these scenarios clearly overlap with the CE principles. Simultaneously, it is reflected in the publications on the circular economy, where there is a direct reference to the product lifecycle stages (Den Hollander et al. 2017; European Environment Agency 2017, pp. 7–11; Reike et al. 2018). This is not surprising since the CE is "the concept of closing material loops to preserve products, parts, and materials in the industrial system and extract their maximum utility" (Zink and Geyer 2017). Therefore, it is suggested to expand the circular product lifecycle term, which was presented in the literature (see, e.g. Kurilova-Palisaitiene et al. 2015), in the following definition:

> *"Circular product lifecycle (CPLC) is a product system consisting of several phases, from product design to product retirement, all of which are involved in the implementation of circular economy principles and are interlinked through closed loop material and information flows"*

The CPLC is presented as a circular graph, which consists of six PLC phases: design, raw material acquisition or generation, manufacturing, distribution, usage, and retirement (Figure 5.1). Both materials and information flow between these

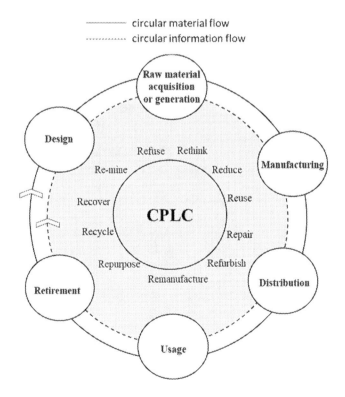

FIGURE 5.1   Circular product lifecycle.

Source: Own elaboration.

stages. The CPLC also includes all the 11 principles of the circular economy, which can be implemented at individual stages of the cycle and supported by various technologies.

The circular economy structure aims to retain the substances, resources, and final products for as long as possible and to eliminate waste at the same time. When a product's lifecycle is coming to an end, the challenge becomes maintaining resources in the loop and reprocessing them to create new value (European Commission 2014, p. 2). The traditional product lifecycle is an open-chain process rather than a closed-loop (Tóth Szita 2017). Growing production and consumption trigger the increase in demand for raw materials and waste generation, producing further environmental problems, which is why the design of the entire product lifecycle based on CE principles becomes crucial.

Since it is recommended that individual CE principles are simultaneously implemented by different product stakeholders (Potting et al. 2017; Reike et al. 2018; Vermeulen et al. 2018; European Investment Bank 2020), they generally concern more than one phase of the product lifecycle. All 11 principles lie at the center of design thinking and managing a product during the middle-of-life and end-of-life phases. To achieve the objectives of the CE, stakeholders within a sustainable product system should cooperate and exchange information and materials.

5.4 INTEGRATING INDUSTRY 4.0 INTO A CIRCULAR PRODUCT LIFECYCLE

The World Economic Forum, in cooperation with Accenture, highlights the significance of the 4th Industrial Revolution technologies (referred to as Industry 4.0 technologies) among the core pillars of the transformation toward a circular economy, including the digital, physical, and biological realms, and underlining their enormous impact on circularity (Lacy et al. 2020). Digitalization can boost the transformation from a linear to a circular product lifecycle, integrating the material and information flows regarding product characteristics, e.g. availability, location, or quality (Antikainen et al. 2018). Increasingly more advanced utilization of Industry 4.0 technologies creates novel ways to manage and optimize each phase of the CPLC, as well as the whole product-service system. These technologies can positively influence the development of CE capabilities and the circularity of resources within companies, increase efficiency in the use of resources and energy, improve maintainability, extend the lifecycle and value of the products, and enhance social and environmental performance (Dubey et al. 2019; Rajput and Singh 2019; Bag and Pretorius 2020; Piscitelli et al. 2020). Piscitelli et al. argued that the development of the CE is impossible without Industry 4.0 technologies, and it is impossible to develop a sustainable Industry 4.0 without the CE concept (Piscitelli et al. 2020). However, it should be noted that empirical research on the influence of Industry 4.0 technologies on implementing the CE is in its infancy (De Sousa Jabbour et al. 2018).

Note that, as yet, there is no consensus on the major types of Industry 4.0 technologies in the development of the CE. However, the following classifications can be mentioned as examples:

- Pagoropoulos et al. (2017) systemized digital technologies as follows: data collection, analysis and integration, RFID (Radio Frequency Identification), the Internet of Things (IoT), Relational Database Management Systems and database handling systems, PLM systems, machine learning, and Big Data analytics.
- Zhong et al. (2017) indicated cyber-physical systems, IoT, Big Data, and cloud manufacturing as key technologies.
- Kang et al. (2016) and De Sousa Jabbour et al. (2018) proposed an overview of the key Industry 4.0 technologies, including cyber-physical systems, manufacturing based on the cloud model, IoT, and additive manufacturing.
- According to the World Economic Forum's transformation map, the following technologies enable circularity: artificial intelligence (AI), robotics, aerospace, IoT, digital communication, mobile technologies, computing, advanced manufacturing, and production, as well as infrastructure (WEF 2020).

In the light of the literature review findings, the potential of the following Industry 4.0 technologies is presented in more detail: additive manufacturing, blockchain, cloud computing, data analytics, and the Internet of Things (Table 5.2).

Table 5.2 shows that Industry 4.0 can integrate information, improve decision-making, and enable the development and implementation of new strategies and processes in light of the CE concept. Their contribution to the phases of a product's lifecycle includes a range of 10R principles. Besides the fact that there are different ways to classify Industry 4.0 technologies and assess their significance for the development of the CE, academics and practitioners both emphasize the need for an interoperability approach to tap their full potential. At the same time, we should not ignore the various barriers that hinder the use of technologies that facilitate CPLC management. Rajput and Singh (2019) identified the following barriers: a lack of uniform standards and specifications, investment cost, the time required to implement and integrate technologies, challenges with compatibility and infrastructure standardization, difficulties in interfacing, networking, and upgrading technology.

5.5 CASE STUDIES ON IMPLEMENTING INDUSTRY 4.0 ACROSS INDUSTRIES

5.5.1 THE CONSTRUCTION INDUSTRY

The construction sector has great potential for implementing the CE principles. The reasons are twofold: the intensive use of primary resources and the large amount of waste it generates. Globally, buildings generate around 40% of total waste and use 40% of mineral resources (Becqué et al. 2016, p. 21). The number of studies and literature reviews of the CE in the construction sector is increasing. The studies focus on the construction process, materials, products, recycled materials, the end-of-life phase, and construction and demolition waste management (Jin et al. 2019; Ginga et al. 2020). Circular solutions can be implemented at each stage of a building's lifecycle. Buildings should be designed to minimize the use of natural

TABLE 5.2
An Overview of Major Industry 4.0 Technologies within the Circular Product Lifecycle

Technology	Brief Description of Functionality	Product Lifecycle Phase	CE Principles Integrated
Additive manufacturing (AM), also called 3D printing	• relies on digital designing in CAD/CAE software, enabling hyper-personalization (even the economy of one) • reduces raw material consumption through additive (not subtractive) manufacturing • makes it possible to print components and products without tools, minimizing waste • makes it possible to manufacture new value-added products by printing from recycled materials (Kellens et al. 2017; Nascimento et al. 2019; Colorado et al. 2020)	Design Raw material acquisition or generation Manufacturing Distribution Retirement	R0 R2 R3 R6 R7 R8
Blockchain	• enables P2P controlled and transparent transactions to close the vulnerabilities of the Industrial Internet of Things • increases sustainability through monitoring and reporting social and environmental metrics (including a product's carbon footprint) as well as enhancing visibility and trust among entities engaged in the PLC • secures the control and replacement of consumables • provides real-time data synchronization and aggregation on a global scale • develops a smart contract-enabled multi-agent model to negotiate with or exchange know-how between process participants in the pursuit of a precise objective • eliminates the need for manual work and control, minimizing the probability of human error	End-to-end PLC	R0 R1 R2 R6 R8 R9

(Continued)

TABLE 5.2 (Continued)
An Overview of Major Industry 4.0 Technologies within the Circular Product Lifecycle

Technology	Brief Description of Functionality	Product Lifecycle Phase	CE Principles Integrated
	• provides unchanging shared data across external links and internal departments, supporting business process planning or resource sharing (Seebacher and Schüritz 2017; Cole et al. 2019; Esmaeilian et al. 2020; Leng et al. 2020)		
Cloud computing	• creates virtual and global space for sharing manufacturing resources and capabilities via the internet • centralizes distributed manufacturing resources and enables their sharing and high use on demand • is service based, meaning that business counterparties sell and buy services as well as reused and refurbished components or products • facilitates data management regarding types and amounts of waste or collection sites (Benlian et al. 2018; De Sousa Jabbour et al. 2018; Nascimento et al. 2019; Rajput and Singh 2019)	Raw material acquisition or generation Manufacturing Distribution Usage Retirement	R1 R2 R3 R5 R7 R8
Data analytics	• it is useful to collect data from different sources (including mobile technologies, social media, or electronic trading platforms) and then process and interpret the data into valuable information. As a result, it ensures information processing capability to optimize planning and procurement processes • ensures high capabilities to monitor production and consumption processes • provides findings based on raw and embedded data on multiple machines, tools, or products	Raw material acquisition or generation Manufacturing Usage Distribution Retirement	R0–R8

	End-to-end PLC	R0–R10
Internet of Things (IoT), also referred to as cyber-physical systems	• enables the implementation of new strategies and techniques in maintenance, reuse, and remanufacturing (Ge and Jackson 2014; Moreno and Charnley 2016; Srai et al. 2016; Pagoropoulos et al. 2017; Bag and Pretorius 2020; Kristoffersen et al. 2020) • helps manage circular business models as dynamic and effective feedback control loops • provides current information about the activities of stakeholders throughout the entire PLC • increases efficiency in the use of tangible and intangible resources within PLM • improves data collection and sharing resource consumption and materials wastage, enhancing the effectiveness and precision of production systems • enables real-time product traceability, and improves the possibilities of collection at the end of lifecycle, refurbishment, remanufacturing, and recycling • supports predictive maintenance and service recovery while significantly reducing the probability of system failure (Manyika et al. 2015; Wortmann and Flüchter 2015; Reuter 2016; Antikainen et al. 2018; Bressanelli et al. 2018; Rajput and Singh 2019)	

Source: Authors' own elaboration.

resources, while the materials they are made of should be highly durable, environmentally friendly, and recyclable. Modular designs are preferred. At the construction site, the equipment should be used in an optimum way, and waste minimization needs to be ensured. For existing structures, revitalization is recommended. Buildings should be renovated and maintained regularly, using renewable energy sources and optimizing consumption. Resource sharing should be promoted by e.g. "sharing" office space after hours. In new investment projects, building materials should be recyclable or reusable (Deloitte 2016).

The use of technology in construction is recommended as it will ensure greater productivity, efficiency, and higher quality. Technologies are sources of the potential that can be used. They offer the opportunity to save costs and time, as well as reduce waste and pollutant emissions. Technologies that can be used in construction include robotics, drones, digital reality, IoT, AI, building information modelling (BIM), 3D printing, and blockchain (Deloitte 2019, p. 7). However, the intensity with which digital technologies are used in construction is still low, and the sector surely needs to advance in this area (Agarwal et al. 2016, p. 3). The literature identifies barriers to and drivers of the adoption of technology in construction. The barriers to using high technologies, such as AR or VR, include costs or the immaturity of the technologies, which are not developed enough to be comprehensively embraced by the sector. (Delgado et al. 2020a; 2020b). The same can be said about the awareness and use of BIM in construction (NBS 2020).

5.5.1.1   Cloud Computing and Mobile Technologies

Due to the role that the construction industry plays in the economy and its impact on the natural environment, the implementation of CE principles in the sector is extremely important. To maximize efficiency in the construction industry, it is important to ensure real-time access to the latest information from anywhere in the world since this facilitates cooperation and information exchange between employees and contractors. Cloud computing, in combination with mobile technologies, which are used by more and more employees, helps build a competitive advantage on the market.

These benefits have been recognized by Pico Volt LLC, which uses eSUB design software, which is based on mobile technologies and cloud computing. The company is engaged in consulting and electrical management services. The technology enables the proper planning of resources and materials when the company is working on several parallel projects, saving time and money. Ground Breakers Construction, thanks to eSUB, has control over reports, field work, and the flow of information between teams and office staff. All information is readily available in real time, and it is possible to view and log changes. As the company emphasizes, this solution has a clear advantage over paper document flows (eSUB 2020). Working in the cloud reduces errors and corrections, which, in turn, is reflected in project costs (Sage Construction and Real Estate 2015, p. 4). Cloud computing makes it possible to make and input changes from anywhere, not necessarily when in the office. It reduces the commuting burden on the environment and on employees who otherwise would have to travel to the office to deliver documents or make corrections in projects. The technology makes it possible to put the reduce principle in place by reducing design errors, demand for printing, and travel to the office.

5.5.1.2 Building Information Modeling

Building information modeling (BIM) is an increasingly widely known technology applied at the design stage because it can generate a digital model of a building. It supports cloud computing, and it is also designed to facilitate collaboration between different industries involved in the construction process by making the 3D model available to them. BIM is used throughout a building's entire lifecycle to ensure better cooperation in design, construction, and at the facility management stage.

A high-profile example of BIM in action was in the construction of the Shanghai Tower. By applying BIM in this investment project, the investor managed to reduce construction waste, errors made at different stages, and the resulting necessary technical corrections. The tool allows the user to determine the demand for resources necessary for the project, and in the Shanghai Tower construction, consumption of materials decreased by 32%. The software enables optimal management and maintenance of the building through appropriate renovation, building maintenance, and failure identification. The building uses renewable energy sources and solutions required by the LEED certificate (Autodesk 2020).

The use of BIM across the industry will hopefully grow with the development of sustainable construction and certification. BIM can also be used in the retirement stage. The technology makes it possible to optimize demand for materials, reduce their use in construction, minimize waste, and make decisions at the management stage about appropriate repairs (to extend the lifespan of buildings). It can also be used to recover construction and demolition waste (HISER project 2020). Thus, this technology contributes to the implementation of CE principles: Refuse, Reduce, Reuse, Recycling.

5.5.1.3 Augmented Reality and Virtual Reality

Design offices compete in the market to win potential clients' attention. AR and VR technologies are beneficial solutions helpful for designers. They make it possible to better adjust a project to the customer's needs and requirements. They are used by companies such as Design Group Latinoamerica, Design Hause Architecture, GASHU Arquitectos, and Darf Design. The technology lets customers choose optimal solutions, and as a result, minimize the number of errors and changes in the design. Used at the design stage, technology reduces the number of errors at the construction stage. It saves time and money, as well as resources and materials, including paper (Albornoz 2018; Darf Design 2020; Design Group Latinamerica 2020; GASHU Arquitectos 2020).

The potential of the technology is also seen at the construction stage, where it reduces possible clashes between different solutions occurring on the construction site. Smart helmets and glasses are examples of such technologies, while special applications can also be developed. The technology has several uses: it allows the user to make better decisions in the construction process, it can convert a 2D image into a 3D one, virtual measurements can be made, and the results can be compared with the design in real time. The technologies make it possible to get a 3D image of the constructed elements and installations, and they can help reduce the number of errors on the construction site through quick reactions and corrections made by the contractor. As a result, less waste is generated at the construction site, and demand for new materials is reduced (Suoza 2019). Under the CE, it is possible to implement the rules of Refuse and Reduce.

5.5.1.4   Artificial Intelligence and Big Data

Time-cost optimization is a major challenge facing the construction sector, which is why the manufacturing and construction stages require proper preparation. This was noticed by the Hawaiian Dredging Construction Company, Toda Corporation, Kajima Corporation, Build Group, Multiplex, Rio Tinto, Takenaka Corporation, and DPR Construction. Using ALICE technologies – a construction simulation and optimization platform – they focused on artificial intelligence and big data technologies. The tool lowers investment costs and reduces the time needed to complete the project. It helps develop the best investment scenario by analyzing several solutions and selecting the best options. Additionally, it optimizes investment projects when it comes to necessary resources (the number of employees), the location of machines on the construction site, the availability of materials and contractors, and planning the sequence of activities. As a result, it saves materials and resources, including fuel (ALICE Technologies 2020). Savings are subject to circular rules such as Refuse (abandoning resource-intensive solutions) and Reduce (reducing the demand for materials and raw materials).

5.5.1.5   Drones

The potential of drones is recognized in the design and manufacturing stages. An example of using this technology was its implementation in the T2 highway project in Estonia. Drones make it possible to take accurate geodetic measurements and accurately determine the demand for gravel and sand for road construction. The Estonian government was able to compare estimates obtained from drones with those from the contractor. The technology saves time and money (Wingtra 2020), allowing the user to save resources and put the Reduce principle into practice.

5.5.1.6   Blockchain

Designers and contractors are not the only stakeholders from the construction industry who should be involved in the CE. From the CE perspective, the materials from which a facility is built are crucial. How they are produced should fit into the principle of sustainable development and take place in a closed loop, and blockchain technology seems to be the best choice. It creates an opportunity to increase trust between companies and standardize and trace the origin of materials.

The Mining and Metals Blockchain Initiative is a consortium that has realized what this technology has to offer. Together with the World Economic Forum, in 2019, the consortium established seven mining and metallurgical companies. The goal was to design and implement blockchain solutions that would support the responsible extraction of raw materials (World Economic Forum 2019). One of the members of the consortium is Tata Steel, a steel manufacturer that operates in line with the CE principles (Hodgson and Brooks 2018). They used blockchain in a pilot project to determine the entire lifecycle of a steel beam from production to its reuse or recycling. The steel beam is tracked with a unique ID registered within the blockchain system. The material passport the company proposed allows the user to track the product (including transport) and use it in the BIM model. All information on steel beams is available to interested parties (Institution of Civil Engineers 2018, p. 31). Blockchain enables material identification and recovery, and it makes it possible to Reuse, Remanufacture, or Recycle.

5.5.1.7   3D Printing

3D printing technology has also demonstrated its high potential in the construction sector. It helps to quickly complete a building and save funds and resources. The largest building in the world made using 3D printing technology was completed in 2019 in Dubai. The building is 9.5 m tall, and its rentable area is 640 m^2. The building was constructed on-site with one printer using the local material supplier (Cheniuntai 2019). In 2018, the GAIA house in Italy was built in 3D. This is a passive building, built on-site in a few weeks from waste rice husks, raw earth, straw, and lime, all of which were sourced locally (Chiusoli 2018).

The technology helps to reduce possible errors on the construction site, and it reduces the demand for materials and the amount of waste. It can also be used to build permanent structures. The demand for equipment, machinery, and fuel decreases as only one printer is used, while the materials used for printing can be environmentally friendly and recyclable. 3D printing technology contributes to the implementation of the principles of Refuse, Reduce, Repurpose.

5.5.1.8   Internet of Things

The Internet of Things technology is used in smart buildings and is associated with using a Building Management System (BMS). As numerous sensors are installed in the buildings to help manage the facility, a BMS can control lighting, the HVAC system (heating, ventilation, air conditioning), and security systems. Additionally, it optimizes building conditions and reduces the demand for resources. A BMS identifies failures in installations and user preferences, seeks to meet them, and adjust ambient conditions to the user's needs. Examples of smart buildings are Google's New London HQ (Urie 2019, p. 7) or Cisco in Bedfont Lakes (Johnson Controls 2015).

The IoT enables the delivery of materials and components just in time using RFID (radio frequency identification). It allows the right number of orders for a given need, not more. The IoT and GPS data enable the tracking of machines and tools, improving their efficiency, extending their service life, and saving fuel. The IoT also makes it possible to track employees, which improves their efficiency and safety (Urie 2019).

RFID is a technology that supports construction companies. It is used to monitor materials, equipment, and workers on the construction site, which helps to better organize work and deliveries. The list of companies using this technology includes Kiewit, Twin Contractors, and HITT Contractor (Constructech 2012). The IoT offers the potential for the principles of Refuse, Reuse, Reduce, and Rethink. It takes care of machines, extends their useful life, and improves efficiency, while also saving resources and reducing waste.

5.5.1.9 Web Platforms

Excess materials-exchange platforms are becoming increasingly popular in the construction industry because they allow resources and building materials to be reused in new investment projects. For example, in 2018, VLA Architecture used the Opalis platform on a project to renovate an architectural office in Brussels (Belgium). Eighty percent of the materials that were used in the renovation project were recovered (Opalis 2020). Similarly, AgwA, another Belgian architectural firm, changed the function of an office building into a school, and the materials recovered from the office building were used in the school's design. The exchange of materials is a step forward in advancing the CE in construction, reducing demand for raw materials and giving materials a chance for a second life (Reduce, Reuse, and Repurpose).

5.5.2 THE FURNITURE INDUSTRY

The furniture sector faces many barriers that significantly hinder its transition toward the circular economy. These barriers include low-quality materials and poor design, which restrict the potential for a successful second life of products. They also have poor product design, low specification standards of the recycled content and design for the circular economy principles, poor consumer information, the availability of spare parts that could prolong and extend a product's lifespan, the limited, high cost of repair and refurbishment, low demand for second-hand furniture and recycled materials, and underinvestment in collection and reverse logistics and the reuse, repair and remanufacturing infrastructure (Forrest et al. 2017, pp. 15–16). These challenges confirm the validity of the research question: Is Industry 4.0 a driver of successful change towards circular furniture lifecycle management? In addressing this question, researchers focus on the innovative practices implemented by various furniture companies around the world.

5.5.2.1 Computer-Aided Design Programs and Generative Design AI-Based Algorithms

A breakthrough innovation for the furniture industry is the use of computer-aided design (CAD) programs that enable 3D design, the introduction of an infinite number of corrections, and the use of a wide database of digital components. In recent times, computer-aided design has been enriched with the possibilities offered by AI via generative design. This concept uses AI algorithms to create various design options, simultaneously considering real-life manufacturing limitations and specific product requirements (Love that Design 2020). Generative design and 3D printing open up the realm of new design opportunities that would be impossible using the human mind alone.

Herman Miller's COSM chair is given as an example because its design involved 3D technology. It is praised for its excellent ergonomic performance, and its suspension is tailored individually to each user's spine, providing total spinal support.

The company offers 3D models and planning tools that customers can use in space-planning applications, such as Revit, and 2D and 3D AutoCAD files (Herman Miller 2020). This practice helps to plan real needs, reduces demand, and prevents resource waste, reflecting the CE principle of Reduction.

5.5.2.2 Blockchain

The use of Industry 4.0 technologies in the phase of raw material acquisition or generation is the first stage of growth in the furniture industry, as managers increasingly pay attention to the applicability of blockchain technology, especially in global sourcing. One company that has declared an interest in implementing this technology is Bassett Furniture & Home Décor (Slaughter 2018). It relies on third parties for much of its global sourcing, especially in Asia and South America; thus, blockchain technology could significantly increase the security of furniture supply chains because it ensures product tracking through all phases and helps fulfill compliance requirements. The functionality of this technology allows companies to enhance the quality of their products and reduce the waste of resources.

5.5.2.3 The Internet of Things and 3D Printing

Furniture plants are increasingly more technologically advanced, and German furniture maker Goldbach-Kirchner is developing the concept of networked production. Electronic manufacturing services (EMS) systems transfer data to other systems, enabling the automatic organization of manufacturing processes. Thanks to automation, robotics, and machine-to-machine (M2M) communication in the IoT environment, many benefits are achieved, such as the optimization of resource consumption and equipment utilization, as well as preventive maintenance.

Another technology that is changing production systems in the furniture industry is additive manufacturing. The largest furniture manufacturer in the world, Ashley Furniture, serves as an example. The company benefits from the advantages of 3D printing, like increases in productivity and quality, as well as a reduction in waste. This technology ensures there are over 700 parts in service in its factories (Colyer 2019). 3D-printed parts are used to simplify adjustments between product changeovers, and they ensure quick replacement. In summary, using both the IoT and 3D printing integrates such CE principles as Refuse, Reduce, and Remanufacture.

Furniture companies individualize products and strive to ensure the highest possible levels of customer experience and furniture comfort by implementing the IoT. An example is the Navigo Smart chair offered by Polish company Nowy Styl. The chair is equipped with a microcomputer and ultra-low energy technology that connects it with a computer or a smartphone (Nowy Styl 2020). The sensors monitor the chair's adjustment and the user's behavior, and it measures the physical conditions of a workstation (e.g. temperature, humidity, and air pressure). The data are analyzed and used to determine whether the user is sitting in an ergonomic position and working in optimal conditions. This combination of product and service is in line with the CE principles of Rethink and Reduce.

Smart solutions increase the intensity with which the same piece of furniture can be used, extending its useful lifetime in the most suitable environmental conditions. Some furniture companies concentrate their efforts on integrating the CE principles in the retirement phase. An example of good practice is the *Circular Product Design Guide of IKEA*, which relates to principles such as Reuse, Refurbish, Remanufacture, and Recycle. One of the technologies used to extend the product lifecycle is the 3D printing of spare parts for furniture (Iles 2018).

5.5.2.4 Cloud Computing

Furniture companies such as Ahrend, Fernish, Gispen, Ikea, and Vepa, take advantage of the possibility to efficiently extend product lifecycles by offering a new service in distribution – Furniture-as-a-Service (FaaS) via cloud-computing applications. This distribution model is based on furniture leasing, for which customers pay a periodic fee and can return the furniture when it is no longer needed. The office furniture manufacturer Ahrend is one of the pioneers in this business model. The company reports numerous benefits of the FaaS model within the PLC, such as the reduced need for natural resources and energy related to material extraction and processing, lower carbon emissions and environmental impacts, closer relationships with customers, higher potential for profits, the possibility of recovering components and materials that can be used to make new furniture, reduced office set-up costs, and higher flexibility for customers (Ellen MacArthur Foundation 2020). The FaaS distribution model influences other phases of the PLC, such as raw material acquisition, generation, and use. Finally, this technology contributes to the implementation of the following CE principles: Rethink, Reuse, Repair, Refurbish, Remanufacture, and Recycle.

5.5.3 THE COSMETICS INDUSTRY

The global cosmetics market has grown rapidly in recent decades. It is especially driven by rising consumer income and changing lifestyles (Łopaciuk and Łoboda 2013), with clients increasingly looking for natural ingredients, green products, new packaging styles, and advanced beauty treatments (Rajput 2019). This forces companies to develop innovative products using the latest technologies and exerting less impact on the environment. Thus, strong players, in particular, increasingly start to focus on the principles of the circular economy (Fortunati et al. 2020). The literature contains publications that combine the latest research into innovative cosmetics ingredients and cosmetics packaging development with circular economy models (Cinelli et al. 2019; Fidelis et al. 2019; Lourenço-Lopes et al. 2020). There is a general recommendation and interest in both implementing LCA and tracking issues concerning the environmental, carbon, and water footprints in the cosmetics industry (Cosmeticseurope 2018, p. 4). Cosmetics companies also decide to deploy digital marketing and utilize information systems to enhance their market share (Kumar 2005). The latest reports show that Industry 4.0 technologies is an important determinant of the development and implementation of product and process innovations in the beauty sector.

5.5.3.1 Advanced Co-creation Platforms

Open innovations are becoming an important developing factor in the cosmetics industry. Procter & Gamble Co. (P&G) has launched a "Connect + Develop" co-creation platform to connect the company with companies and individuals who have innovative solutions for design, packaging, technologies, research, and engineering. The external collaborations allow P&G to improve its internal processes and design new products to meet the real needs of today's consumers, whose awareness of environmental problems is growing. They are involved in around 50% of P&G's designs (Procter & Gamble 2020). The result of the global company's open innovation strategy was about 2,000 formal contracts with partners from different countries (Ozkan 2015). P&G is actively involved in implementing the 3R (Reduce, Reuse, Recycle) principles through the PLC. The company plans to have 100% recyclable or reusable packaging by 2030, and it is also devoted to reducing energy use, greenhouse gas emissions, and truck transportation kilometers (Procter & Gamble 2019, pp. 5, 12). One of the last projects based on the company's extensive cooperation with NGOs and volunteers was the introduction of the recyclable shampoo bottle made from beach plastic (Head & Shoulders 2020).

5.5.3.2 Blockchain

Consumers increasingly care about reliable information concerning the origin and composition of cosmetics. However, the cosmetics industry is still not strictly regulated; thus, more and more companies have decided to implement blockchain in their supply chains. In particular, this technology ensures ingredient transparency, providing reliable information on the brand's story and products (CB Insights 2019, pp. 50–51). For example, Cult Beauty is a pioneer in bringing verified information into its online retail ecosystem. Thanks to Proof Points, the Provenance blockchain platform, a network of cooperating brands share information both on the supply chain links (from suppliers of raw materials to points of sale) and confirmed product characteristics (e.g. cruelty-free, vegan). This way, the company deals with social and environmental impacts across the whole cosmetics product lifecycle by being compatible with CE principles such as Refuse or Reuse.

5.5.3.3 3D Printing

In the beauty industry, many companies have already become interested in using 3D printing at different stages of the cosmetic lifecycle. For example, L'Oréal has implemented it for the fast prototyping of packaging, as well as for the rapid creation of spare parts or format parts in factories (L'Oréal 2020a). Thus, thanks to additive manufacturing, the use of raw materials and the generation of waste are reduced. L'Oréal also promotes the CE by preserving natural resources (e.g. water reuse in 'dry factories') and focusing on making 100% of packaging compostable, recyclable, reusable, or refillable by 2025 (L'Oréal 2018, pp. 6, 13).

Consumers can now also take advantage of the 3D makeup printer offered by New York-based beauty company Mink. This device transforms any image into wearable makeup (Mink 2020). Another example is Lancôme, which has introduced an innovative color-blending technology for its clients. At the sales point, the client's skin is scanned to determine its tone, and then, after some smart calculations are carried out, a unique customized foundation is printed (Lancôme 2020). This allows for personalized shopping, with each client receiving an individual complexion ID and the name on the foundation bottle for easy refill. Consumers not only take less time to choose the right product but, above all, they avoid buying an inadequate cosmetic product that would quickly cause dissatisfaction and then be left unused and turn into waste. As a result, waste is reduced, and the glass packaging can be reused.

5.5.3.4 Cloud Computing

Polish cosmetics manufacturer Farmona decided to change its IT model to ensure flexibility in responding to the current needs of online customers. The cosmetics industry is characterized by relatively high seasonality, e.g. in the holiday months, there is a significant increase in the number of page hits and transitions between the product websites, the company website, and the online store. Clients also stay longer on individual subpages than in other months (Polcom 2020).

Using the computing power provided by the Polcom Data Center guarantees that the servers are constantly monitored and covered by a comprehensive backup system. Cloud computing also allows for the maximum use of e-commerce sales channels and ensures production volumes appropriate for the size of the real demand. Thus, excess production is avoided, reducing inventories and the waste of products that have exceeded their shelf life.

5.5.3.5 The Internet of Things, Augmented Reality, Artificial Intelligence, and Big Data Analytics

In the beauty industry, the Internet of Things is also exploited at the distribution and use stage. Kérastase, which specializes in hair care products, offers a smart hairbrush, the Kérastase Hair Coach, which uses technology developed by Withings, the French consumer electronics company. The product has several built-in sensors, a vibrator, and a microphone. It connects to a dedicated smartphone app, which collects data (e.g. the number of brush strokes), provides tips based on hair type and brushing method, and suggests other products from the Kérastase line (Allure 2017).

HiMirror, offered by Taiwan's New Kinpo Group, is another popular smart product in the beauty industry. It is a technologically advanced device that supports the user during make-up and skin care routines. It analyzes skin through photos and stores measurement data allowing users to track their skincare results over time. An additional function is that it recommends cosmetics or reminds the user when products are about to expire. Apart from being connected to the internet, HiMirror uses AI technology. It scans the face and helps to explore and select the best make-up products by showing the results of their application virtually (HiMirror 2020).

The Makeup Genius smartphone app, which incorporates thousands of L'Oreal products and over a hundred unique facial expressions, lets users test and shop for color cosmetics online (L'Oréal 2020b). A similar beauty maker tool for consumers was launched by Coty. This beauty company also developed a smart Magic Mirror in collaboration with Holition (a creative innovation studio) and PERCH (an in-store retail marketing platform) (Coty 2020). The smart products are used not only to choose color cosmetics, but also to advise on the most appropriate skin care cosmetics.

LuluLab is a spin-off company created under the Samsung C-Lab program. It has launched LUMINI, an innovative device that digitalizes facial skin using AI and a fast algorithm for face scanning. It then recommends the best cosmetics (e.g. creams) following big data analytics (LuluLab 2020).

Selecting cosmetics is still one of the biggest problems facing consumers today. Not only is it time-consuming, but most of all, it often leads to unnecessary purchases, which then become unnecessary waste in the consumer's home. The examples of phone apps and digital mirrors discussed above not only solve the consumer's decision-making problems, but they also reduce the use of raw materials and promote waste minimization in the cosmetics industry.

5.6 THE COMPARATIVE ANALYSIS OF INDUSTRIES

Table 5.3 presents a comparative analysis of case studies from the selected industries. It identifies the similarities and differences in the implementation of technologies at different stages of the PLC, and it is an opening for a discussion on the potential of technologies in supporting the implementation of particular principles of the circular economy.

In this study, the CPLC consists of several phases (Figure 5.1) that integrate various CE principles (Table 5.1). The case study analysis shows that for each stage of a product's lifecycle, regardless of the type of industry, one or more Industry 4.0 technologies can be recognized. Companies are interested in implementing advanced approaches for different products with varying degrees of complexity, demand uncertainty, or lifecycle length. The comparative analysis shows that the use of individual technologies in the construction industry affects a much greater number of PLC phases compared to other industries. Depending on the industry, technologies also dominate in different phases. In the cosmetics and furniture industries, the distribution and use phases are characterized by greater exploitation of technologies, suggesting that the companies are focused mainly on the consumer market. In the construction industry, the technologies are implemented in the design, manufacturing, and use phases. Generally, it can be noted that the potential of Industry 4.0 is not fully exploited, especially at the raw material acquisition or generation phase. Furthermore, the more complex the architecture, the more technically advanced the product, and the higher the unit price, the more companies are likely to implement Industry 4.0, especially at the first phases of the PLC.

Technologies clearly integrate the adjacent phases of the PLC. Thus, the same technology usually determines more than one phase. This should encourage participants in the lifecycle of individual products to foster collaboration and joint investment in technology.

The key closing phase in the engineering product lifecycle is retirement. Unfortunately, while it is very important for closing the product lifecycle, it is not found in every industry. In the case of cosmetics, after the product has been used, only the packaging remains for recovery scenarios. In the furniture industry, packaging has mainly a logistical function, and at the retirement stage, there are products that either no longer satisfy the customer or require replacing for technical and safety reasons. The situation is completely different for long-lasting and complex construction structures. Another aspect is that, in the case of fast-moving consumer goods, the product moves faster to the retirement stage. Therefore, in this

TABLE 5.3

Industry-Specific Analysis of Industry 4.0 Technologies that Support the Implementation of CE Principles

Case	Company Name	Implemented Technology	Phase of the PLC that the Technology Relates to	The CE Principle that is Implemented through the Technology or Supported by the Implementation of the Technology
Construction	Pico Volt LLC Ground Breakers Construction	Cloud computing Mobile technologies	Design, Manufacturing	Reduce
	Gensler (project: Shanghai Tower)	Building information modeling (3D-7D)	Design, Manufacturing, Use, Retirement	Refuse Reduce Reuse Recycle
	Design Group Latinoamerica, Designhaaus, Gashu Arquitectos, Darf Design, construction companies	Augmented Reality, Virtual Reality	Design, Manufacturing	Refuse Reduce
	Hawaiian Dredging Construction Company, Build Group, Toda Corporation, Kajima Corporation, Multiplex, Rio Tinto, Takenaka	Artificial intelligence, Big Data	Design, Manufacturing	Refuse Reduce

	Company / Project	Technology	Lifecycle Stage	R-Strategy
	Corporation, DPR Construction, ALICE Technologies 2020 Hades Geodeesia	Drones	Design, Manufacturing	Reduce
	Tata Steel, The Mining and Metals Blockchain Initiative	Blockchain	Raw material acquisition or generation, Manufacturing	Reuse Remanufacture Recycle
	Apis Cor, WASP in association with Rice House	3D printing	Manufacturing, Retirement	Refuse Reduce Repurpose
	Heatherwick Studio and Bjarke Ingels Group (project: Google's New London HQ). Cisco, Johnson Control (project: building in Bedfont Lakes), Kiewit, Twin Contractors, HITT Contractor	Internet of Things (smart building, Building Management Systems) RFID	Use, Distribution, Manufacturing	Refuse Reuse Reduce Rethink
	VLA Architecture, AgwA	Web platform	Retirement	Reduce Reuse Repurpose
Furniture	Herman Miller	Computer-Aided Design programs and generative design AI-based algorithms	Design	Reduce
	Basset Furniture & Home Décor	Blockchain	Raw material acquisition or generation	Reduce

(Continued)

TABLE 5.3 (Continued)
Industry-Specific Analysis of Industry 4.0 Technologies that Support the Implementation of CE Principles

Case	Company Name	Implemented Technology	Phase of the PLC that the Technology Relates to	The CE Principle that is Implemented through the Technology or Supported by the Implementation of the Technology
	Goldbach-Kirchner, Ashley Furniture	Internet of Things, 3D printing	Manufacturing	Refuse Reduce Remanufacture
	Ahrend	Cloud computing	Distribution Use	Rethink Reuse Repair Refurbish Remanufacture Recycle
	Nowy Styl	Internet of Things	Use	Rethink Reduce
	IKEA	3D printing	Retirement	Refurbish
	Procter & Gamble	Advanced co-creation platform	Design	Reduce Reuse Recycle
Cosmetics industry	L'Oréal, Mink	3D printing	Design, Manufacturing	Refuse Reduce Reuse Recycle Repurpose

Cult Beauty	Blockchain	Raw material acquisition or generation, Manufacturing, Distribution	Refuse Reduce
Lancôme	3D printing	Manufacturing, Retirement	Reduce Reuse
Farmona	Cloud computing	Manufacturing, Distribution	Reduce
LuluLab	Artificial intelligence, big data analytics, Internet of Things	Distribution, Use	Refuse Reduce
Kérastase, New Kinpo Group, Coty	Internet of Things, augmented reality, artificial intelligence	Distribution, Use, Retirement	Refuse Reduce

Source: Authors' own elaboration.

industry, it is particularly important to develop efficient ways to collect and re-process waste.

Product price is important to the consumer. In each industry, the end-product has a different price and is produced on a different scale. In construction, each investment implies extremely high costs, while furniture and cosmetics have a different economic dimension. The profitability of using the technology will depend on economies of scale at each stage of the product lifecycle. In the age of sustainable development and dwindling resources that are increasingly more expensive, every technology and every industry should strengthen its resilience and competitiveness to be able to survive. This is true of all industries because they all face a common problem of environmental degradation and fewer available resources. The solution lies in adopting the circular economy principles, and technologies that are now becoming the key to building enterprises' and economies' competitive advantage are helpful in this.

The specifics of the industries are reflected in the use of technology and its significance for the implementation of CE principles in a product's lifecycle. Complying with these principles or achieving sustainable benefits are rarely mentioned among the main reasons behind Industry 4.0. The key factors that lead to their implementation are economic, and include costs, flexibility, quality, and customer service. This is a surprising observation because, as the analysis of the case studies shows, fulfilling each CE principle can be supported by Industry 4.0 technologies, which significantly determine whether the sustainable goals are achieved. However, only a few companies (for example, IKEA) have so far developed strategies and policies that respond to the needs and requirements of managing the circular product lifecycle.

It should be clearly highlighted that industries have considerable potential – although not fully tapped – to implement all CE principles through the use of technology. The comparative analysis of industries demonstrated that Industry 4.0 technologies are at different levels of application maturity, depending on the product, stage of its lifecycle, and industry. The most common principles implemented by technologies are 3R, in particular, Reuse, Reduce, Recycle, with the Reduce principle being applied most often. As indicated in the literature on the subject, these principles are most popular in works on the CE. They are associated with directly reducing the demand for resources, raw materials, and materials, as well as reducing waste.

Companies integrate different Industry 4.0 technologies and benefit from their interoperability, and the synergy of using different technologies allows them to achieve greater sustainable effects. The same technologies lead to similar results, regardless of the industry. This empirical observation allows us to recognize significant opportunities for benchmarking and transferring technology projects not only between companies from the same industry but also from different industries.

5.7 CONCLUSIONS AND IMPLICATIONS

The circular economy is defined as a term, an economic strategy, a strategy, an alternative to the linear economy, an industrial system (restorative and re-generative), and an approach to resources. The flows of materials and resources in production and consumption are being arranged in a new, closed cycle – the opposite of the linear approach. Circularity is achieved through resource efficiency and effectively designing business models, and a product's lifecycle. Resources, materials, and products are maintained in the economy for as long as possible by giving them a circular flow. The aim is to use them to the maximum in the economy. Resources are recovered and pumped into the next circulation, extending their economic value. Waste is considered even at the design stage, as the goal is to minimize waste and the wastage of resources. At the end of the lifecycle, waste can become a resource with new value in the economy. A sustainable and circular economy is based on the use of renewable energy sources. The CE ensures economic growth (while reducing dependence on primary resources), the competitiveness of economies, new jobs, enterprises with new specializations, and it brings forward the development of innovation and technology. The underlying assumptions are the protection of the environment and the improvement of life quality of the present and future generations. Many definitions of CE indicate these features (Rizos et al. 2017, p. 6; Korhonen et al. 2018, p. 547).

The implementation of CE principles in the economy is already taking place. The significant role of technologies is noted in this process, although their implementation in the presented industries is still low on a global scale. This is certainly related to the continuous development, gradual implementation, and low availability of technologies. When introducing Industry 4.0, companies focus mainly on fulfilling the needs and expectations of their future users. Consumer and business clients' demand, which is mainly influenced by the cost and availability of the technologies, stimulates the design and implementation of new solutions in all three industries. Companies most often implement a selected Industry 4.0 technology at one stage of the product lifecycle. However, it is recognized that the influence and benefits of a single technology spread to other stages, integrating them, and causing a looping effect.

To meet the targets of the UN's Agenda 2030 for Sustainable Development, it is important to put more emphasis on achieving goals and implementing the CE principles by using technology in each industry. The process is difficult, complicated, time-consuming, and capital-intensive. It requires high technological maturity, compatibility between technologies, and adapting them to specific industries and specific needs, and this needs greater expenditure on R&D. The impulses for disseminating different technologies and putting the CE principles into practice should come from the government's development policy, backed up by financial support schemes that promote the use of new technologies. The role of bottom-up and local initiatives cannot be overestimated in this area, as customer demand is the main driver that stimulates supply and production. To direct the manufacturing sector toward the CE solutions, we need to promote and strengthen environmental awareness across the consumer society. Thus, the Refuse principle, which is at the heart of circular rules, is crucial. It is also frequently confirmed by practical examples. Refuse can be practiced at the very beginning of the product lifecycle. The product designer's knowledge of sustainable development and the CE is critical here, as is the consumer's environmental awareness.

The implementation of CE principles is becoming a common objective in the analyzed industries. It can integrate suppliers and users of Industry 4.0 around their efforts to ensure social, environmental, and financial benefits. Noticeably, however, in the light of current conditions, enterprises should focus on more than merely observing the most popular 3R principles through the use of new technologies. In the face of shrinking raw material resources and increases in their prices, companies should take the next step and start implementing the other rules, even though they seem more difficult and complex. To ensure circularity, they require the conscious involvement of the user and close cooperation with enterprises. This means that the incentives that direct companies towards a full transition to the CE framework and the implementation of Industry 4.0 for this purpose should simultaneously target business and society.

Industry 4.0 has recently become one of the vital pillars of the circular economy, contributing to the development of new management concepts such as the CPLC. Industry 4.0 has had a positive impact on the reduction of demand for resources and waste generation, and it also plays an increasingly important role in the development of industries. There is great potential in integrating the activities of PLC participants to achieve synergy effects that are satisfactory for the users. More and more advanced projects lead not only to continuous improvement, but also to further technological innovations. For example, 4D printing, which refers to how products change shape over time, meeting consumer needs, is an evolution of 3D printing technology. Industry 4.0 technologies can contribute to the development of new enterprises emerging based on business models, such as start-ups and spin-off companies across industries.

Moreover, they positively influence the integration of various industries, e.g. cosmetics producers with manufacturers of care devices. In addition, their use in one sector affects sustainability in other sectors. For instance, companies turning to the cloud reduce demand for paper and printing services and the transport needs of their employees. We are convinced that technological projects will increase the integration of different industries in the circular economy of the 21st century. Industries such as cosmetics, medicine, astronautics, construction, and agriculture provide interesting examples today. Currently, however, the use of Industry 4.0 in various industries does not come close to being a universal trend. Therefore, the exchange of knowledge and experience between companies that use Industry 4.0 technologies in accordance with the CE principles, and spreading them across different industries, should remain at the top of the agenda.

When referring to the implications of the presented research results, several issues should be noted. Since the principles of sustainable development and the CE concept increasingly determine the economic development policies of many countries, including restrictive legal regulations based on environmental protection, enterprises should start to intensively prepare for these changes. Today, CE principles in the PLC are a must. Enterprises designing products must consider not only their lifecycle but also circularity. In this regard, there is an urgent need to educate market participants on how to transform linear product lifecycles into circular ones. However, the key challenge seems to be managing a circular product so as to not only obtain an environmental effect, but also to reduce costs in the long term and increase strategic and operational flexibility. It seems that the development and implementation of new technologies have become particularly appropriate in this regard.

Companies offering Industry 4.0 should consider and promote the benefits of their use for the development of the circular economy. Integrating technologies is extremely beneficial because it can foster sustainable effects across the entire CPLC. Companies should more intensively exploit the potential of Industry 4.0 to close product lifecycles. For example, blockchain technology can effectively support the cycle closure opportunities. The greater the scale of the implementation and dissemination of technologies, the lower the total cost of their acquisition and ownership, and the greater their availability, the greater their ability to generate sustainable effects.

ACKNOWLEDGEMENTS

Sections 5.3 and 5.5.3 are the result of the project "Flexibility in relationships with suppliers in terms of supplier-purchaser models of cooperation on product development in the B2B market", no. 2016/21/B/HS4/00665, financed by the National Science Centre (NCN) in Poland.

REFERENCES

Agarwal, R, Chandrasekaran, S, & Sridhar, M (2016). *Imagining construction's digital future*. McKinsey & Company.

Albornoz, M (2018). https://www.archdaily.com/899599/5-architecture-offices-using-vr-to-present-their-designs (accessed September 14, 2020).

Ali, MM, Rai, R, Otte, JN, & Smith, B (2019). A product life cycle ontology for additive manufacturing. *Computers in Industry*, *105*, 191–203. DOI: 10.1016/j.compind.2018.12.007.

ALICE Technologies. (2020). https://www.alicetechnologies.com/case-studies/ (accessed September 11, 2020).

Allure. (2017). https://www.allure.com/story/kerastase-hair-coach (accessed September 14, 2020).

Ameri, F, & Dutta, D (2005). Product lifecycle management: Closing the knowledge loops. *Computer-Aided Design and Applications*, *2*(5), 577–590. DOI: 10.1080/16864360.2005.10738322

Antikainen, M, Uusitalo, T, & Kivikytö-Reponen, P (2018). Digitalisation as an enabler of circular economy. *Procedia CIRP*, *73*, 45–49. DOI: 10.1016/j.procir.2018.04.027

Asiedu, Y, & Gu, P (1998). Product life cycle cost analysis: State of the art review. *International Journal of Production Research*, *36*(4), 883–908. DOI: 10.1080/002075498193444

Autodesk. (2020). https://damassets.autodesk.net/content/dam/autodesk/www/case-studies/shanghai-tower/shanghai-tower-customer-story.pdf (accessed September 10, 2020).

Bag, S, & Pretorius, JHC (2020). Relationships between technologies 4.0, sustainable manufacturing and circular economy: Proposal of a research framework. *International Journal of Organizational Analysis*. ISSN: 1934-8835. Vol. ahead-of-print No. ahead-of-print. https://doi.org/10.1108/IJOA-04-2020-2120

Becqué, R, Mackres, E, Layke, J, Aden, N, Liu, S, Managan, K, & Graham, P (2016). Accelerating building efficiency: Eight actions for urban leaders. *World Resources Institute, Washington, DC*. ISBN-13: 978-1-56973-887-0.

Below, R, & Wallemacq, P (2018). Annual disaster statistical review 2017. *CRED, Centre for Research on the Epidemiology of Disasters: Brussels, Belgium.*

Benlian, A, Kettinger, WJ, Sunyaev, A, Winkler, TJ Guest Editors. (2018). The transformative value of cloud computing: A decoupling, platformization, and recombination theoretical framework. *Journal of Management Information Systems, 35*(3), 719–739. DOI: 10.1080/07421222.2018.1481634

Bressanelli, G, Adrodegari, F, Perona, M, & Saccani, N (2018). Exploring how usage-focused business models enable circular economy through digital technologies. *Sustainability, 10*(3), 639. DOI: 10.3390/su10030639

Cao, H, & Folan, P (2012). Product life cycle: The evolution of a paradigm and literature review from 1950–2009. *Production Planning & Control, 23*(8), 641–662. DOI: https://10.1080/09537287.2011.577460

Campbell-Johnston, K, ten Cate, J, Elfering-Petrovic, M, & Gupta, J (2019). City level circular transitions: Barriers and limits in Amsterdam, Utrecht and The Hague. *Journal of Cleaner Production, 235*, 1232–1239. DOI: 10.1016/j.jclepro.2019.06.106

CB Insights. (2019). 15 trends changing the face of the beauty industry in 2020. https://www.cbinsights.com/research/report/beauty-trends-2019/ (accessed September 14, 2020).

Chen, YJ, Chen, YM, & Chu, HC (2009). Development of a mechanism for ontology-based product lifecycle knowledge integration. *Expert Systems with Applications, 36*(2), 2759–2779. DOI: 10.1016/j.eswa.2008.01.049

Cheniuntai, N (2019). https://www.apis-cor.com/dubai-project (accessed September 9, 2020).

Chiusoli, A (2018). https://www.3dwasp.com/en/3d-printed-house-gaia/ (accessed September 9, 2020).

Cinelli, P, Coltelli, MB, Signori, F, Morganti, P, & Lazzeri, A (2019). Cosmetic packaging to save the environment: Future perspectives. *Cosmetics, 6*(2), 26. DOI: 10.3390/cosmetics6020026

Cole, R, Stevenson, M, & Aitken, J (2019). Blockchain technology: Implications for operations and supply chain management. *Supply Chain Management: An International Journal, 24*(4), 469–483. ISSN: 1359-8546.

Colorado, HA, Velásquez, EIG, & Monteiro, SN (2020). Sustainability of additive manufacturing: The circular economy of materials and environmental perspectives. *Journal of Materials Research and Technology, 9*(4), 8221–8234. DOI: 10.1016/j.jmrt.2020.04.062

Colyer, J (2019). Ashley Furniture transforms factory floor with 3D printing. https://3dprintingindustry.com/news/ashley-furniture-transforms-factory-floor-with-3d-printing-156330/ (accessed September 14, 2020).

Constructech. (2012). https://constructech.com/contractors-adopt-rfid/ (accessed September 14, 2020).

Cosmeticseurope. (2018). Environmental Sustainability: The European Cosmetics Industry's Contribution. https://www.cosmeticseurope.eu/files/9615/2872/3399/CE_Environmental_Sustainability_Report_2018.pdf (accessed September 14, 2020).

Coty (2020). https://www.coty.com/in-the-news/press-release/coty-introduces-blended-reality-beauty-magic-mirror (accessed September 14, 2020).

Darf Design. (2020). https://www.darfdesign.com/mixedrealityvisualisation.html (accessed September 28, 2020).

De Brito, MP, & Dekker, R (2004). A framework for reverse logistics. In *Reverse logistics.* Dekker, R., Fleischmann, M., Inderfurth, K. & Van Wassenhove, L.N. (Eds.) (pp. 3–27). Springer, Berlin, Heidelberg. DOI: 10.1007/978-3-540-24803-3_1

De Sousa Jabbour, ABL, Jabbour, CJC, Godinho Filho, M, & Roubaud, D (2018). Technologies 4.0 and the circular economy: A proposed research agenda and original roadmap for sustainable operations. *Annals of Operations Research, 270*(1–2), 273–286. DOI: 10.1007/s10479-018-2772-8

Delgado, JMD, Oyedele, L, Beach, T, & Demian, P (2020a). Augmented and virtual reality in construction: Drivers and limitations for industry adoption. *Journal of Construction Engineering and Management*, *146*(7), 04020079. DOI: 10.1061/(ASCE)CO.1943-7862.0001844

Delgado, JMD, Oyedele, L, Demian, P, & Beach, T (2020b). A research agenda for augmented and virtual reality in architecture, engineering and construction. *Advanced Engineering Informatics*, *45*, 101122. DOI: 10.1016/j.aei.2020.101122

Deloitte. (2016). Deloitte sustainability. Circular economy potential for climate change mitigation. https://www2.deloitte.com/content/dam/Deloitte/fi/Documents/risk/Deloitte%20-%20Circular%20economy%20and%20Global%20Warming.pdf (accessed September 15, 2020).

Deloitte. (2019). Point of view on digital construction. The business case of incorporating digital technologies into the construction industry. https://www2.deloitte.com/content/dam/Deloitte/nl/Documents/energy-resources/deloitte-nl-eri-point-of-view-digital-construction.pdf (accessed September 15, 2020).

Den Hollander, MC, Bakker, CA, & Hultink, EJ (2017). Product design in a circular economy: Development of a typology of key concepts and terms. *Journal of Industrial Ecology*, *21*(3), 517–525. DOI: 10.1111/jiec.12610

Design Group Latinamerica. (2020). https://www.dg-la.com/en/home/ (accessed September 16, 2020).

Dubey, R, Gunasekaran, A, Childe, SJ, Papadopoulos, T, Luo, Z, Wamba, SF, & Roubaud, D (2019). Can big data and predictive analytics improve social and environmental sustainability?. *Technological Forecasting and Social Change*, *144*, 534–545. DOI: 10.1016/j.techfore.2017.06.020

Ellen MacArthur Foundation. (2013). Towards the circular economy. *Journal of Industrial Ecology*, *2*, 23–44.

Ellen MacArthur Foundation. (2014). Towards the circular economy. Vol. 3: Accelerating the scale-up across global supply chains.

Ellen MacArthur Foundation. (2015). Towards a circular economy: Business rationale for an accelerated transition.

Ellen MacArthur Foundation. (2017) https://www.ellenmacarthurfoundation.org/assets/galleries/CEinaction-_Activity06-nine-Rs-6R3_from-graham-081217.pdf (accessed September 6, 2020).

Ellen MacArthur Foundation. (2019a). *Circular economy in cities*. Project Guide.

Ellen MacArthur Foundation. (2019b). Introduction to the circular economy.

Ellen MacArthur Foundation. (2020). Ahrend bringing office furniture full circle.

Esmaeilian, B, Sarkis, J, Lewis, K, & Behdad, S (2020). Blockchain for the future of sustainable supply chain management in Technologies 4.0. *Resources, Conservation & Recycling*, *163*, 105064. DOI: 10.1016/j.resconrec.2020.105064

eSUB. (2020). eSUB construction software. https://esub.com/case-study-pico-volt/, https://esub.com/case-study-ground-breakers-construction/ (accessed September 14, 2020).

European Commission. (2014). COM(2014) 398 final/2. *Towards a Circular Economy: A Zero Waste Programme for Europe*.

European Commission. (2015). COM(2015) 614 final. *Final—Closing the Loop—An EU Action Plan for the Circular Economy*.

European Commission. (2020). COM(2020) 98 final. *A New Circular Economy Action Plan for a Cleaner and More Competitive Europe*.

European Environment Agency (EEA). (2016). Circular economy in Europe developing the knowledge base. *EEA*. ISBN-13: 978-92-9213-719-9.

European Environment Agency (EEA). (2017). Circular by design – Products in the circular economy. *EEA*. Report No 6/2017. ISBN-13: 978-92-9213-857-8.

European Investment Bank. (2020). The EIB circular economy guide supporting the circular transition. *EIB*. DOI: 10.2867/578286. ISBN-13: 978-92-861-4672-5.

Fidelis, M, de Moura, C, Kabbas Junior, T, Pap, N, Mattila, P, Mäkinen, S, Putnik P, Granato, D, Bursac Kovacevic D, Tian Y, and Yang B (2019). Fruit seeds as sources of bioactive compounds: Sustainable production of high value-added ingredients from by-products within circular economy. *Molecules*, *24*(21), 3854. DOI: 10.3390/molecules24213854

Forrest, A, Hilton, M, Ballinger, A and Whittaker, D (2017). Circular economy opportunities in the furniture sector. *European Environmental Bureau: Brussels, Belgium.* https://eeb.org/library/circular-economy-opportunities-in-the-furniture-sector/ (accessed September 13, 2020).

Fortunati, S, Martiniello, & L. Morea, D (2020). The strategic role of the corporate social responsibility and circular economy in the cosmetic industry. *Sustainability, 12*(12), 5120. DOI: 10.3390/su12125120

GASHU Arquitectos. (2020). https://gashuarquitectos.com/servicios/ (accessed September 28, 2020).

Ge, X, & Jackson, J (2014). The big data application strategy for cost reduction in automotive industry. *SAE International Journal of Commercial Vehicles*, *7*(2014-01-2410), 588–598. DOI: 10.4271/2014-01-2410

Ginga, CP, Ongpeng, JMC, Daly, M, & Klarissa, M (2020). Circular economy on construction and demolition waste: A literature review on material recovery and production. *Materials*, *13*(13), 2970. DOI: 10.3390/ma13132970

Herman Miller. (2020). 3D models and planning tools. https://www.hermanmiller.com/en_apc/resources/3d-models-and-planning-tools/ (accessed September 12, 2020).

Head & Shoulders. (2020). https://www.headandshoulders.ca/en-ca/whats-new/new-head-shoulders-bottle-to-be-made-with-recycled-beach-plastic (accessed September 14, 2020).

HiMirror. (2020). https://www.himirror.com/en-US/eshop/product/HiMirrorSlide (accessed September 14, 2020).

HISER project. (2020). http://hiserproject.eu/index.php/our-activities/smart-demolitions-and-refurbishments (accessed September 9, 2020).

Hodgson, P, & Brooks, P (2018). Tata steel Europe's approach to the circular economy in a life cycle perspective. *Asian Steel Watch*, *5*, 24–33.

Iles, J (2018). Brands team up to see how 3D printing can revolutionise repair. https://medium.com/circulatenews/brands-team-up-to-see-how-3d-printing-can-revolutionise-repair-86d882d2a95d (accessed September 8, 2020).

INCOSE UK. (2017). Guide to lifecycles and lifecycle models issue 1.1 https://webcache.googleusercontent.com/search?q=cache:qrR1SN98Z-0J:https://www.apm.org.uk/media/13835/guide-to-lifecycle-models.pdf+&cd=1&hl=pl&ct=clnk&gl=pl&client=firefox-b-d (accessed September 10, 2020).

Institution of Civil Engineers. (2018). Blockchain technology in the construction industry. *Digital Transformation for High Productivity.* https://www.ice.org.uk/ICEDevelopmentWebPortal/media/Documents/News/Blog/Blockchain-technology-in-Construction-2018-12-17.pdf (accessed September 11, 2020).

International Organization for Standardization. (2006). *Environmental management: Life cycle assessment; Principles and framework* (No. 2006). ISO.

IPCC Working Group I. (2013). *Climate Change 2013: The Physical Science Basis: Summary for Policymakers.* Intergovernmental Panel on Climate Change. https://www.ipcc.ch/site/assets/uploads/2018/03/WG1AR5_SummaryVolume_FINAL.pdf (accessed August 20, 2020). ISBN-13: ISBN 978-92-9169-138-8. DOI: 10.1017/CBO9781107415324

Ishii, K, Eubanks, CF, & Di Marco, P (1994). Design for product retirement and material life-cycle. *Materials & Design*, *15*(4), 225–233. DOI: 10.1016/0261-3069(94)90007-8

Jin, R, Yuan, H, & Chen, Q (2019). Science mapping approach to assisting the review of construction and demolition waste management research published between 2009 and 2018. *Resources, Conservation and Recycling*, *140*, 175–188. DOI: 10.1016/j.resconrec.2018.09.029

Johnson Controls. (2015). https://g2electrical.co.uk/wp-content/uploads/2018/03/be_case_study_cisco.pdf (accessed September 14, 2020).

Kang, HS, Lee, JY, Choi, S, Kim, H, Park, JH, Son, JY, ... & Do Noh, S (2016). Smart manufacturing: Past research, present findings, and future directions. *International Journal of Precision Engineering and Manufacturing-green Technology*, *3*(1), 111–128. DOI: 10.1007/s40684-016-0015-5

Kellens, K, Baumers, M, Gutowski, TG, Flanagan, W, Lifset, R, & Duflou, JR (2017). Environmental dimensions of additive manufacturing: Mapping application domains and their environmental implications. *Journal of Industrial Ecology*, *21*(S1), S49–S68. DOI: 10.1111/jiec.12629

Kiritsis, D (2011). Closed-loop PLM for intelligent products in the era of the Internet of things. *Computer-Aided Design*, *43*(5), 479–501. DOI: 10.1016/j.cad.2010.03.002

Kondoh, S, Tateno, T, Kishita, Y, Komoto, H, & Fukushige, S (2017). The potential of additive manufacturing technology for realizing a sustainable society. In *Sustainability through innovation in product life cycle design*. Matsumoto, Mitsutaka, Masui, Keijiro, Fukushige, Shinichi, & Kondoh, Shinsuke (Eds.) (pp. 475–486). Springer, Singapore.

Korhonen, J, Nuur, C, Feldmann, A, & Birkie, SE (2018). Circular economy as an essentially contested concept. *Journal of Cleaner Production*, *175*, 544–552. DOI: 10.1016/j.jclepro.2017.12.111

Kristoffersen, E, Blomsma, F, Mikalef, P, & Li, J (2020). The smart circular economy: A digital-enabled circular strategies framework for manufacturing companies. *Journal of Business Research*, *120*, 241–261. DOI: 10.1016/j.jbusres.2020.07.044

Kumar, S (2005). Exploratory analysis of global cosmetic industry: Major players, technology and market trends. *Technovation*, *25*(11), 1263–1272. DOI: 10.1016/j.technovation.2004.07.003

Kurilova-Palisaitiene, J, Lindkvist, L, & Sundin, E (2015). Towards facilitating circular product life-cycle information flow via remanufacturing. *Procedia Cirp*, *29*, 780–785. DOI: 10.1016/j.procir.2015.02.162

Lacy, P, Long, J and Spindler, W (2020). How can businesses accelerate the transition to a circular economy? https://www.weforum.org/agenda/2020/01/how-can-we-accelerate-the-transition-to-a-circular-economy/ (accessed September 10, 2020).

Lancôme. (2020). https://www.lancome-usa.com/le-teint-particulier/LAN233.html (accessed September 14, 2020).

Leng, J, Ruan, G, Jiang, P, Xu, K, Liu, Q, Zhou, X, & Liu, C (2020). Blockchain-empowered sustainable manufacturing and product lifecycle management in technologies 4.0: A survey. *Renewable and Sustainable Energy Reviews*, *132*, 110112. DOI: 10.1016/j.rser.2020.110112

Lindkvist, L, Movilla, NA, Sundin, E, & Zwolinski, P (2017). Investigating types of information from WEEE take-back systems in order to promote Design for Recovery. In *Sustainability through innovation in product life cycle design*. Matsumoto, Mitsutaka, Masui, Keijiro, Fukushige, Shinichi, & Kondoh, Shinsuke (Eds.) (pp. 3–19). Springer, Singapore. DOI: 10.1007/978-981-10-0471-1_1

Łopaciuk, A, & Łoboda, M (2013, June). Global beauty industry trends in the 21st century. In *Management, Knowledge and Learning International Conference* (pp. 19–21).

L'Oréal. (2018). Progress report, Sharing beauty with all. https://www.loreal-finance.com/en/annual-report-2018/responsibility-1-5/ (accessed September 14, 2020).

L'Oréal. (2020a). https://www.loreal.com/en/news/group/how-is-digital-transformation-helping-loreal-work-wonders/ (accessed September 14, 2020).

L'Oréal. (2020b). https://www.lorealparisusa.com/beauty-magazine/makeup/makeup-looks/makeup-genius-from-day-to-night.aspx (accessed September 14, 2020).

Lourenço-Lopes, C, Fraga-Corral, M, Jimenez-Lopez, C, Pereira, AG, Garcia-Oliveira, P, Carpena, M, … & Simal-Gandara, J (2020). Metabolites from macroalgae and its applications in the cosmetic industry: A circular economy approach. *Resources*, *9*(9), 101. DOI: 10.3390/resources9090101

Love that Design. (2020). Technology disrupts the furniture manufacturing industry. https://www.lovethatdesign.com/article/technology-disrupts-the-furniture-manufacturing-industry/ (accessed September 10, 2020).

LuluLab. (2020). http://www.lulu-lab.com/en/ (accessed September 14, 2020).

Manyika, J, Chui, M, Bisson, P, Woetzel, J, Dobbs, R, Bughin, J, & Aharon, D (2015). *Unlocking the potential of the Internet of Things*. McKinsey & Company. https://www.mckinsey.com/business-functions/mckinsey-digital/our-insights/the-internet-of-things-the-value-of-digitizing-the-physical-world (accessed September 15, 2020).

Mink. (2020). https://www.minkbeauty.com/products/mink-makeup-printer (accessed September 14, 2020).

Moreno, M, & Charnley, F (2016). Can redistributed manufacturing and digital intelligence enable a regenerative economy? An integrative literature review. In *Sustainable design and manufacturing*. Scholz, Steffen G., Howlett, Robert J., & Setchi, Rossi (Eds.) (pp. 563–575). Springer, Cham. DOI: 10.1007/978-3-319-32098-4_48

Morseletto, P (2020). Targets for a circular economy. *Resources, Conservation and Recycling*, *153*, 104553. DOI: 10.1016/j.resconrec.2019.104553

Nascimento, DLM, Alencastro, V, Quelhas, OLG, Caiado, RGG, Garza-Reyes, JA, Rocha-Lona, L, & Tortorella, G (2019). Exploring Technologies 4.0 to enable circular economy practices in a manufacturing context. A business model proposal. *Journal of Manufacturing Technology Management*, *30*, 607–627. DOI: 10.1108/JMTM-03-2018-0071

NBS. 2020. 10th Annual BIM Report (2020). https://www.thenbs.com/knowledge/national-bim-report-2020 (accessed September 21, 2020).

Nowy Styl. (2020). https://nowystyl.com/en/competences/technology/ (accessed September 15, 2020).

Okorie, O, Salonitis, K, Charnley, F, Moreno, M, Turner, Ch., & Tiwari, A (2018). Digitalisation and the circular economy: A review of current research and future trends. *Energies*, *11*, 1–33. DOI: 10.3390/en11113009.

Opalis. (2020). https://opalis.eu/fr/projets/vla-architecture (accessed September 10, 2020).

Ozkan, NN (2015). An example of open innovation: PG. *Procedia-Social and Behavioral Sciences*, *195*,1496–1502. DOI: 10.1016/j.sbspro.2015.06.450

Pagoropoulos, A, Pigosso, DC, & McAloone, TC (2017). The emergent role of digital technologies in the circular economy: A review. *Procedia CIRP*, *64*, 19–24. DOI: 10.1016/j.procir.2017.02.047.

Parlikad, AK, McFarlane, D, Fleisch, E, & Gross, S (2003). The role of product identity in end-of-life decision making, White paper. *Auto-ID Center*, Institute for Manufacturing, Cambridge, 1–26.

Piscitelli, G, Ferazzoli, A, Petrillo, A, Cioffi, R, Parmentola, A & Travaglioni, M (2020). Circular economy models in the technologies 4.0 era: a review of the last decade. *Procedia Manufacturing*, *42*, 227–234. DOI: 10.1016/j.promfg.2020.02.074

Polcom. (2020). Farmona stawia na cloud computing w Polcom Data Center, case study. https://polcom.com.pl/newsroom/laboratorium-kosmetykow-naturalnych-farmona-stawia-na-chmure (accessed September 12, 2020).

Potting, J, Hekkert, MP, Worrell, E, & Hanemaaijer, A (2017). *Circular economy: Measuring innovation in the product chain* (p. 2544). PBL Publishers, The Hague.

Procter & Gamble. (2019). Environmental sustainability 2019. https://us.pg.com/sustainability-reports/ (accessed September 12, 2020).

Procter & Gamble. (2020). https://www.pgconnectdevelop.com/ (accessed September 14, 2020).

Rajput, N (2019). Cosmetics market by category (skin sun care products, hair care products, deodorants, makeup color cosmetics, fragrances) and by distribution channel (general departmental store, supermarkets, drug stores, brand outlets) – Global opportunity analysis and industry forecast, 2014–2022. [online] Allied Market Research. https://www.alliedmarketresearch.com/cosmetics-market (accessed September 14, 2020).

Rajput, S & Singh, SP (2019). Connecting circular economy and technologies 4.0. *International Journal of Information Management*, 49, 98–113. DOI: 10.1016/j.ijinfomgt.2019.03.002

Reike, D, Vermeulen, WJ, & Witjes, S (2018). The circular economy: New or refurbished as CE 3.0?—Exploring controversies in the conceptualization of the circular economy through a focus on history and resource value retention options. *Resources, Conservation and Recycling*, 135, 246–264. DOI: 10.1016/j.resconrec.2017.08.027

Reuter, MA (2016). Digitalizing the circular economy: Circular economy engineering defined by the metallurgical internet of things. *Metallurgical and Materials Transactions B*, 47, 3194–3220. DOI: 10.1007/s11663-016-0735-5

Rizos, V, Tuokko, K, & Behrens, A (2017). *The circular economy: A review of definitions, processes and impacts (No. 12440)*. Centre for European Policy Studies. https://www.researchgate.net/publication/315837092_The_Circular_Economy_A_review_of_definitions_processes_and_impacts (accessed September 8, 2020). ISBN-13: 978-94-6138-597-0.

Rose, CM, Ishii, K, & Stevels, A (2002). Influencing design to improve product end-of-life stage. *Research in Engineering Design*, 132, 83–93.

Rosa, P, Sassanelli, C, Urbinati, A, Chiaroni, D, & Terzi, S (2019). Assessing relations between circular economy and technologies 4.0: A systematic literature review. *International Journal of Production Research*, 58(6), 1662–1687. DOI: 10.1080/00207543.2019.1680896

Saaksvuori, A, & Immonen, A (2008). *Product lifecycle management*. Springer Science Business Media, Berlin, Heidelberg. ISBN-13: 978-3-540-78172-1.

Sage Construction and Real Estate. (2015). Cloud computing and the construction industry. https://www.teamtag.net/wp-content/uploads/2015/07/Sage-CRE-Whitepaper-Cloud-Computing-and-the-Construction-Industry.pdf (accessed September 15, 2020).

Seebacher, S & Schüritz, R (2017). Blockchain technology as an enabler of service systems: A structured literature review. *Eighth International Conference on Exploring Service Science, IESS 1.7*. https://www.researchgate.net/publication/315858662_Blockchain_Technology_as_an_Enabler_of_Service_Systems_A_Structured_Literature_Review (accessed September 12, 2020). DOI: 10.1007/978-3-319-56925-3_2

Slaughter, P (2018). Blockchain: Could it be supply chain game changer? https://www.furnituretoday.com/business-news/blockchain-could-it-be-supply-chain-game-changer/ (accessed September 14, 2020).

Srai, JS, Kumar, M, Graham, G, Philips, W, Tooze, J, Ford, S, Beecher, P, Raj, B, Gregory, M, Tiwari, MK, Ravi, B, Neely, A, Shankar, R, Charnley & Tiwari, A (2016). Distributed manufacturing: Scope, challenges and opportunities. *International Journal of Production Research*, 54, 6917–6935. DOI: 10.1080/00207543.2016.1192302

Stark, J (2016). Product lifecycle management. In *Product lifecycle management volume 2*. Stark, John (Eds.) (pp. 1–35). Springer, Cham. ISBN-13: 978-3-319-24436-5.

Sukhdev, A, Vol, J, Brandt, K, & Yeoman, R (2018). *Cities in the circular economy: the role of digital technology*. Ellen MacArthur Foundation, Cowes, UK. https://www.ellenmacarthurfoundation.org/assets/downloads/Cities-in-the-Circular-Economy-The-Role-of-Digital-Tech.pdf (accessed September 14, 2020).

Suoza, E (2019). https://www.archdaily.com/914501/9-augmented-reality-technologies-for-architecture-and-construction (accessed September 11, 2020).

Terzi, S, Bouras, A, Dutta, D, Garetti, M, & Kiritsis, D (2010). Product lifecycle management–from its history to its new role. *International Journal of Product Lifecycle Management*, 44, 360–389. DOI: 10.1504/IJPLM.2010.036489

Tóth Szita, K (2017). The application of lifecycle assessment in circular economy. *Hungarian Agricultural Engineering*, *31*, 5–9. DOI: 10.17676/HAE.2017.31.5

United Nations. (2015). Transforming our world: The 2030 agenda for sustainable development. https://sustainabledevelopment.un.org/content/documents/21252030%20Agenda%20for%20Sustainable%20Development%20web.pdf (accessed September 20, 2020).

Urie, M (2019). The Internet of things in construction. https://marketintel.gardiner.com/uploads/1901_IoT-in-Construction.pdf (accessed September 15, 2020).

Vermeulen, WJV, Reike, D, & Witjes, S (2018). Circular economy 3.0: Getting beyond the messy conceptualization of circularity and the 3R's, 4R's and more …. https://www.cec4europe.eu/wp-content/uploads/2018/09/Chapter-1.4._W.J.V.-Vermeulen-et-al._Circular-Economy-3.0-getting-beyond-the-messy-conceptualization-of-circularity-and-the-3Rs-4-Rs-and-more.pdf (accessed September 8, 2020).

Wingtra. (2020). https://wingtra.com/case_studies/wingtraone-estonian-road-survey/ (accessed September 14, 2020).

World Economic Forum. (2019). https://www.weforum.org/press/2019/10/seven-mining-metals-companies-partner-on-responsible-sourcing-with-world-economic-forum (accessed September 21, 2020).

World Economic Forum. (2020). https://intelligence.weforum.org/topics/a1Gb0000000pTDMEA2?tab=publications (accessed September 12, 2020).

Wortmann, F, & Flüchter, K (2015). Internet of things: Technology and value added. *Business & Information Systems Engineering*, *57*(3), 221–224. DOI: 10.1007/s12599-015-0383-3

Young, RI, Gunendran, AG, Cutting-Decelle, AF, & Gruninger, M (2007). Manufacturing knowledge sharing in PLM: A progression towards the use of heavy weight ontologies. *International Journal of Production Research*, *457*,1505–1519. DOI: 10.1080/00207540600942268

Zhong, RY, Xu, X, Klotz, E & Newman, ST (2017). Intelligent manufacturing in the context of Technologies 4.0: A review. *Engineering*, *3*, 616–630. DOI: 10.1016/J.ENG.2017.05.015

Zink, T, & Geyer, R (2017). Circular economy rebound. *Journal of Industrial Ecology*, *213*, 593–602. DOI: 10.1111/jiec.12545

6 A Taxonomy on Smart Grid Technology

Anurag Jain
Virtualization Department, School of Computer Sciences,
University of Petroleum and Energy Studies, Dehradun,
Uttarakhand, India

Rajneesh Kumar
Department of Computer Science and Engineering,
Maharishi Markandeshwar (Deemed to be University),
Mullana, Ambala, Haryana, India

Sunil Kumar Chawla
Department of Computer Science & Engineering,
Chandigarh University, Gharuan, Mohali, Punjab, India

CONTENTS

6.1   INTRODUCTION

The conventional system had evolved when energy was low cost to satisfy the need of the world's electric power transmission. The distribution system in earlier times was the centralized, unidirectional, and demand-driven control system. To meet the increasing demand for electricity consumption, more power stations came into existence. This scenario of generation, distribution, and transmission is shown in Figure 6.1 (Farhangi, 2009).

DOI: 10.1201/9781003127819-6

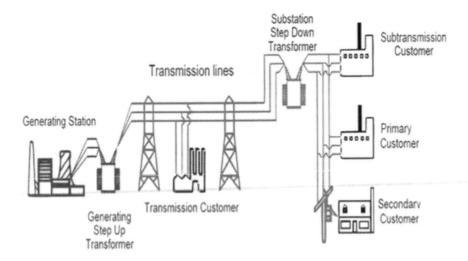

FIGURE 6.1 The conventional electricity grid system.

Over time, local grids expanded and had become interrelated. The load centers receive the power from vital power-generating stations via high-capacity transmission lines, which is further divided to meet the power requirements over the entire supply area. The user had to pay bills according to the single tariff because of the system's limited data collection and processing capabilities. The lower night demand motivated for dual-tariff arrangements means using low-cost power at night. However, having extensive load during the daytime means that the supply of electricity gets interrupted in some areas, resulting in power cuts and even blackouts. To handle this situation, electricity demand patterns were recognized. The daily domestic demand peaks were controlled by an arrangement of high-power generators that were switched on for a short span per day. However, their low utilization could not meet their initial cost of integration, resulting in high costs to the electricity companies and high electric bills for consumers.

Highly polluting power stations were situated possibly far away from populated areas, but they were located near fossil fuel reserves. The structure of the emerging grid was influenced by hydroelectric dams in mountain areas and nuclear power plants. The conventional grid has upgraded to meet up with the expanding energy demand and to overcome the limitations. The lousy impact of fossils on the environment has motivated the use of renewable energy instead. Because of the variable nature of wind and solar power, more refined control systems are needed to assist the connectivity of renewable sources to the grid. The conventional grids were considered soft targets for potential attacks. The need for a robust power grid was felt in some countries for this concern (Fang et al., 2011).

Smart grid (SG) evolution has taken place because it is a key that integrates a vast collection of IT resources that allow the primitive and smart grids to lower electric wastage and cost. Smart grid technology has been implemented in many developed countries, and still, many are moving slower to accept and implement this technology (Gungor et al., 2011).

The chapter highlights the emergence of the smart grid as a call to update the existing grid. Multiple research papers have been studied to extract accurate knowledge of the basics of the smart grid. Several of the various components of smart grid technologies and the application and the role of different tools in implementing the smart grid, its functions, needs, characteristics, and opportunities in this domain have also been highlighted.

6.2 SMART GRID

A smart grid is a network or an electrical grid that integrates the performance and operation of smart devices and energy-efficient resources, especially renewable sources, in a cost-effective way. A smart grid is capable of controlling the generation and distribution of electronic power. High standards and supply with security and safety of power with low losses are key smart grid features. The schematic representation of the smart grid is shown in Figure 6.2 (Gungor et al., 2011).

6.2.1 CHARACTERISTICS OF SMART GRID

The smart grid is the evolution of the primitive energy system to a new period of a reliable and more efficient system. The characteristics of the smart grid have to be

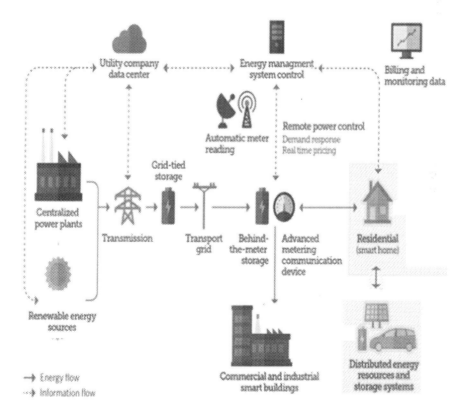

FIGURE 6.2 Conceptual view of smart grid.

defined for proper communication between the projects. Accordingly, the smart grid is designed for the generation of cleaner energy and is characterized by the following parameters:

Consumer Participation: Consumers are the key to any power system because they help maintain a balance between power generation and consumption. Their behavior to modify the usage of electricity may increase system reliability. The choices include the implemented technology and new tariffs.

Distributed Generation: Assimilation of various renewable-energy resources ultimately into the grid. A smart grid integrates large, centralized power plants and the distributed-energy generating sources, which rapidly increases the value chain from suppliers to customers via marketers.

Power System Efficiency: An efficient power system is attained by working with smart technologies that use its resources to the fullest. When the assets are used dynamically at higher loads by incessantly sensing and ranking the capacities, the maintenance effectiveness can be enhanced.

Prediction of Disturbances: The smart grid smartly isolates the challenging elements and the normal working of the remaining system to attain resiliency. The self-healing characteristic of the smart grid facilitates the provider to manage the delivery infrastructure in a better way, thereby providing the consumer with disruption-free service.

New Products and Services: Smart grid enables the markets for the consumers where they can select from the competing services. Markets play an important role in managing the variables like location, time, capacity, changing rate, and quality of power. The markets should be flexible enough to alter businesses' rules and operating conditions to attract regulators and consumers.

Power Quality for Various Requirements: The power requirement is different for different consumers. SG enables the consumer to use varying power as per their needs. The cost of multiple grades of power is also different and allows the consumer to select the use and pay option. The automatic monitoring components are deployed to diagnose and solve power consumption and issues that affect the quality of power for natural or artificial disasters (Cecilia & Sudarsanan, 2016; Edris & D'Andrade, 2017).

6.2.2 COMPONENTS OF SMART GRID

A smart grid includes subsystems with layers of technology together with intelligent monitoring, innovations, and new tools. It involves the production, distribution of sustainable power. An essential component of the smart grid are as follows:

Smart Device Interface: Smart devices are the ones that connect with other independently operating devices via various wireless methods. Smart cars, doorbells, meters, phones, etc., are capable of monitoring and controlling real-time information processes.

Storage Component: The disparity between energy consumption and availability leads to finding energy storage methods for future use to improve reliability and grid resiliency.

Transmission System: The adaptive transmission lines provide the interconnection between the primary substations and the load centers. Their adaptive nature allows them to tolerate eventual and vibrant changes in the load without interrupting the service, making them the backbone of the integrated power system. Being efficient, reliable, and cost-effective is the main aim of the transmission subsystem.

Monitoring and Control Technology: The smart monitoring devices are attack-resistant, capable of self-healing, predictable, adaptable, and can smartly handle reliability issues and instability issues to avoid congestion. These devices are self-aware and can take independent decisions according to real-time changes.

Demand-Side Management: These components are deployed for reliable generation and distribution of power and to reduce greenhouse gas emissions. Such components reduce the use of expensive generators, thereby lowering the operation cost as demanded by the consumer.

Intelligent Grid Distribution Subsystem: Distribution of power to the consumers is the final step of the grid network, which is accomplished by distribution feeders that deliver power to residential areas, small industry and business areas. The distribution subsystem should be capable of monitoring and utilizing the communication links between utility control and consumers. Furthermore, they must detect the faults and restore, optimize the voltage, transfer the load, and produce the automatic bills in real-time (Ghosal & Conti, 2019; Kabalci & Kabalci, 2019).

6.2.3 FUNCTIONS OF SMART GRID

The integration of innovative products and services with intellectual monitoring, managing, communication, and self-healing properties make the smart grid able to:

- Connect the generators of various sizes operating on different technologies.
- Allow the consumers to participate in optimizing the working of the system.
- Make the consumers aware of various use and pay policies.
- Relax the consumers by lowering the electricity prices paid by them, which made the technology highly affordable.
- Employ renewable DERs leading to the reduction in emission of CO_2 and other harmful gases and carbon particles that reduce the pressure on the environment, thereby providing cleaner power.
- Maintain and improve the reliability and security of the system together with upgrading the existing services, which in turn reduces the occurrence of natural and artificial attacks.
- Lower the power quality disturbances and probability of blackouts that improve the reliability of the grid.
- Reduce the injuries and loss of life, thereby enhancing safety concerns.
- Revolutionize the transport sector by the deployment of electric vehicles for generation and storing purposes.
- Improve efficiency by reducing electricity losses and energy wastage (Bayindir et al., 2016; Gonzalez-Longatt & Torres, 2018).

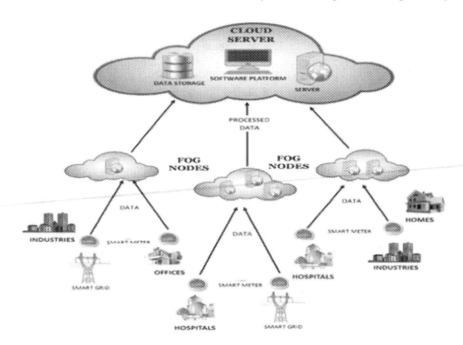

FIGURE 6.3   Usage of different technology in smart grid.

6.3   ROLE OF TECHNOLOGY IN SMART GRID

Many technologies are used in the realization of the smart grid concept. Figure 6.3 shows the usage of other technologies in the implementation of the smart grid concept. Some of them are listed below with their role:

Integrated Communications: Such communications will permit real-time management, exchange of information, resource utilization, and theft prevention to make the system more consistent.

Sensing and Measurement: They sense the stability of the grid by monitoring congestion, the health of the equipment by making use of various technologies like smart meters, large-area monitoring systems, EM signature analysis, real-time measurement and pricing tools, RF technology, sophisticated switches, cables, and defensive relays.

Smart Sensor: Smart sensor integrates sensors and the processing elements on a single chip that can record, process, compute, and communicate the analog signal and the digital representation of derived signals via a bidirectional digital bus. The smart sensors are required to be integrated with the system to support smart grid monitoring applications so that they can integrate and support the distributed sensor system, and monitor the main substation and line equipment.

Sensor and Actuator Networks (SANET): SANET in smart grid are used for managing and optimizing the flow of energy by using the flow of information. SANETs to work effectively require the sensing of parameters, device control, and decision-making powers. The information flow is used to transmit the sensed data to controllers for decision making, where the controller issues control commands to the actuators and execute the task accordingly.

Smart Meters: The meters that can record the readings and can communicate are termed a smart meter. The meter can be operated remotely from a distant location to read and measure energy consumption units, and the same is safely displayed on a device. It is also capable of two-way communication and can receive information regarding payment modes or tariff updates. Smart meters help lower the cost and consumption of energy units by making the consumers aware of the same. It also sends data to calculate the peak-load demands, take necessary measures to control the load, and form pricing policies based on consumption data. Smart meters are deployed for measuring gas, water, or electricity consumption. Smart meters inform the consumers about the best pricing strategies, and more accurate bills will be generated.

Distributed Power Flow Control: For controlling the power flow in the existing transmission lines, the control devices fasten on them. Such transmission lines sustain the use of renewable sources by facilitating the real-time control of power flow in the grid, thereby enabling the grid for efficient storage of irregular renewable energy for future use.

Smart Power Generation Using Advanced Components: Load balancing refers to the concept of a match between supply and demand of power generation. Any mismatch would result in frequency variations causing blackouts. The load balancing has become a challenging task as more variable generators are incorporated into the grid, compelling other producers to adjust their output more often than in the past. In smart power generation, balancing is achieved by using numerous similar generators that can operate automatically and efficiently at the chosen load.

Power System Automation: The automation system provides diagnosis and solutions to explicit grid interruptions. The technologies that rely on and contribute to each other for highly developed control methods are distributed intelligent agents, analytical tools, operational applications, and artificial intelligence.

Vehicle to Grid (V2G): V2G utilizes electricity-driven vehicles to supply power to specific electric markets. The prime advantage of this technology is that it can store renewable energy. It also sustains the generation of large-scale wind energy via direction. To store power in a vehicle's fuel cells, a battery or hybrid of both is used.

Plug-in Hybrid Electric Vehicle (PHEV): PHEV is a hybrid electric vehicle with a larger pack of batteries charged by a plug during off hours. The vehicle uses electric power if the battery is fully charged; otherwise, it uses the alternative petroleum-based source. PHEV helps cut down the emissions of gases because of its ability to use hybrid sources of energy. PHEV is equipped with a device to establish a connection with an external electrical source for charging. These vehicles are beneficial in emergencies when fuel prices spike or supplies decline, as they then use the stored energy preserved by the utilities during off-peak hours.

Automated Meter Reading (AMR): AMR devices in smart grid perform communication in full-duplex mode and let the utilities read meters from distant locations. Their home display allows the consumer to either lower their consumption or reschedules it in off-peak hours to reduce the bills. AMR collects the data of consumer's energy consumption from electric meters or smart meters and processes it for bill generation (Ma et al., 2016; Tuballa & Abundo, 2016).

6.4   APPLICATIONS OF SMART GRID

The new technology permits the efficient use of smart devices and detects and isolates the faulty devices for the smooth functioning of the rest of the system. Applications of smart grid in various areas are described below:

Automation of Building & Home: A smart home is an automated home in which the energy sources and equipment are controlled and coordinated using smart controllers and meters to fulfill the objectives of the smart grid.

Smart Substation: The workstations and the protection devices, and the small transducers carry out communication with fiber optical networks that have emerged with digital technology advancements. The communication process in the substation system takes place in three stages:

  i. Operations and reporting takes place at the station level
 ii. Bay level deals with the control and protection functions of the system
iii. Third process level deals with signal transmission from VTs, CTs, and other transducers.

Feeder Automation (FA): Distantly control and monitor the distribution of power is the critical aspect of the feeder automation system. It also facilitates consumers to collect and provide the data efficiently. It manages the voltage and reactive power for automatic switching in transmission lines, monitoring the health equipment. It uses switches and digital sensors, which are equipped with communication and control technologies. FA products can be easily induced in the existing grid as they are interoperable and designed to reduce the functioning costs, thereby satisfying the consumers (Dileep, 2020).

6.5   ADVANTAGES, OPPORTUNITIES, AND CHALLENGES IN SMART GRID

6.5.1   Benefits of Smart Grid

Self-Healing: The smart grid detects and responds to usual issues and quickly recovers from any short fault in transmission lines or other components, thereby reducing downtime and economic losses.

Inspire and Invite the Consumers by Cost Reduction: Visibility into synchronized pricing allows consumers to select the level of consumption and pricing option according to their needs.

Distributes High-Quality Power to Compete with the Needs of the New Era: The smart grid provides quality power free from any turbulence and interruptions.

Boost Up Assets and Efficient Operation: Enable cost reduction by transmitting power through existing infrastructure to operate and maintain the grid efficiently.

Grid Resiliency: Enhance grid reliability and shrink the occurrence of power blackouts.

The Decline of Peak Demand: Share and reduce the load on a large scale by integrating it with renewable energy resources.

It facilitates the smart devices to charge during low load hours to reduce the bills. It also assimilates imperishable resources like solar and winds energy entirely into the grid (Taha, 2020).

6.5.2   OPPORTUNITIES IN SMART GRID

Smart Grid Technologies Help in:

- Enlarge and advance the infrastructure to improve connectivity and communications among multiple parties.
- Assemble smart tools to utilize DR, power efficiency demand load control.
- Educate consumers through training sessions.
- Promote the investments in the smart grid, and create regulatory framework models.
- Ensure the cybersecurity and grid resilience infrastructure must be built up (Yoldaş et al., 2017).

6.5.3   CRITICAL CHALLENGES IN SMART GRID

Strengthening the Grid: The utility grid must be capable enough to have room for additional energy resources, especially renewable resources.

Offshore Movement: To reduce pollution and utilize offshore energies' random behavior (tidal and wave), effective and efficient connections should be developed. Development of a decentralized structure to promote the smooth functioning of small power generating systems with the larger systems.

Communications: Development of a strong structure for communication to allow millions of parties to operate and trade in a single market.

Involvement of Consumers: Enabling the consumers to play a dynamic role in the system's operation to be aware of all new tariffs with or without their generation.

Stability Factors: Assimilation of distributed generation from renewable sources and micro-grids on a massive scale.

Preparing for Electric Vehicles: Electric vehicles are pretty prominent because of their mobility and dispersive characteristic that will help them to deploy in bulk in the coming years.

Inadequacies in Grid Infrastructure: The grid network should provide clean and distributed power generation, which may pose several challenges in terms of architecture design, formation, operation, and maintenance.

Cybersecurity: The development of an advanced tool for handling the growing cyber threats is a significant issue for smooth functioning.

Storage: Proper storage of renewable and power generation from renewable sources must be ensured.

Data Management: Voluminous data from smart devices like sensors, controllers, cameras, meters, and from sources like weather forecast systems has to be appropriately managed to avoid breakdown or damage before occurrence.

Stakeholder's Engagement and Capital Investment: Awareness programs should be carried out for organizations and individuals; to make them know the smart grid benefits and utilities to induce the faith of stakeholders in the smart grid.

Trained Workforce: The success of a smart grid depends upon the empowered human resource of various sectors that may be prepared through suitable training sessions (Piricz, 2020; Rathor & Saxena, 2020).

6.6   CONCLUSION

Limitations of the conventional grid, increasing & fluctuating power demand, the need for a more transparent & robust system, more pollution due to usage of fossil fuels, and high cost & less availability of fossil fuels are some of the motivating factors behind the emergence of the smart grid system. The emergence of high-speed internet technology, sensor technology, machine learning, big data, and cloud computing has helped realize this concept. Its integration with renewable energy resources will help to prevent climate change conditions. Many developed and developing countries are adopting a smart grid system. An initial investment in the upgrade of the conventional grid and security of user data are critical challenges in the widespread adaption of the smart grid system. Using renewable energy resources available at the diversified location may reduce the upgrade cost. Also, installing different cybersecurity and big data tools will help manage consumer and grid information more securely. In the future, the authors have planned to work for the usage of machine learning in the estimation of the consumer demand and assessment of power generation from renewable energy resources.

REFERENCES

Bayindir, R, Colak, I, Fulli, G, & Demirtas, K (2016). Smart grid technologies and applications. *Renewable and Sustainable Energy Reviews*, 66, 499–516.

Cecilia, AA, & Sudarsanan, K (2016, February). A survey on smart grid. In 2016 International Conference on Emerging Trends in Engineering, Technology and Science (ICETETS), 1–7. IEEE.

Dileep, G (2020). A survey on smart grid technologies and applications. *Renewable Energy*, 146, 2589–2625.

Edris, AA, & D'Andrade, BW (2017). Transmission grid smart technologies. In *The Power Grid*. D'Andrade, B.W. (Ed.), 37–55. Academic Press.

Fang, X, Misra, S, Xue, G, & Yang, D (2011). Smart grid—The new and improved power grid: A survey. *IEEE Communications Surveys & Tutorials*, 14(4), 944–980.

Farhangi, H (2009). The path of the smart grid. *IEEE Power and Energy Magazine*, 8(1), 18–28.

Ghosal, A, & Conti, M (2019). Key management systems for smart grid advanced metering infrastructure: A survey. *IEEE Communications Surveys & Tutorials*, 21(3), 2831–2848.

Gonzalez-Longatt, F, & Torres, JLR (2018). Introduction to smart grid functionalities. In *Advanced Smart Grid Functionalities Based on PowerFactory*. Gonzalez-Longatt, Francisco, & Torres, José Luis Rueda (Eds.), 1–18. Springer, Cham.

Gungor, VC, Sahin, D, Kocak, T, Ergut, S, Buccella, C, Cecati, C, & Hancke, GP (2011). Smart grid technologies: Communication technologies and standards. *IEEE Transactions on Industrial Informatics*, 7(4), 529–539.

Kabalci, E, & Kabalci, Y (2019). Introduction to smart grid architecture. In *Smart Grids and Their Communication Systems.* Kabalci, Ersan, & Kabalci, Yasin (Eds.), 3–45. Springer, Singapore.

Ma, Y, Liu, F, Zhou, X, & Gao, Z (2016, August). Key technologies of smart distribution grid. In 2016 IEEE International Conference on Mechatronics and Automation, 2639–2643. IEEE.

Piricz, N (2020). Management Challenges of Smart Grids. In *Integration of Information Flow for Greening Supply Chain Management.* Kolinski, Adam, Dujak, Davor, & Golinska-Dawson, Paulina (Eds.), 393–415. Springer, Cham.

Rathor, SK, & Saxena, D (2020). Energy management system for smart grid: An overview and key issues. *International Journal of Energy Research,* 44(6), 4067–4109.

Taha, MQ (2020). Advantages and advances of smart energy grid: A technology review. *Bulletin of Electrical Engineering and Informatics,* 9(5), 1739–1746.

Tuballa, ML, & Abundo, ML (2016). A review of the development of smart grid technologies. *Renewable and Sustainable Energy Reviews,* 59, 710–725.

Yoldaş, Y, Önen, A, Muyeen, SM, Vasilakos, AV, & Alan, İ (2017). Enhancing smart grid with microgrids: Challenges and opportunities. *Renewable and Sustainable Energy Reviews,* 72, 205–214.

7 Towards Urban Sustainability: Impact of Blue and Green Infrastructure on Building Smart, Climate Resilient and Livable Cities

Prabha Roy
Ernest & Young, Global Delivery Services - IA,
New Delhi, India

CONTENTS

DOI: 10.1201/9781003127819-7

7.1   BACKGROUND CONTEXT

Urban population has increased drastically; and, by 2050, it will cover around 70% of the total global share. Current urbanization trends, inadequate access to basic services, and outdated urban-planning models have impacted the environmental footprints of cities and led to climate change, depletion of natural resources, degradation of the quality of urban environment, etc. Globally, rapid urban sprawl has created a significant impact on the natural landscape, thus affecting the environmental, social, and economic aspects of an urban ecosystem. The land cover is changing rapidly, leading to farmland displacement, encroachment of floodplains, loss of water bodies, deforestation, loss of arable land, etc. A few of the major impacts of urbanization on the urban ecosystem are a reduction in urban green cover, habitat destruction of wild flora and fauna, encroachment of catchment area and flood plains, reduced infiltration leading to decline in ground water level, urban flooding, urban-heat island, and numerous impacts due to climate change, etc. These impacts are exacerbated due to lack of awareness about the importance of integrating Blue-Green Infrastructure (BGI) within the development process and understanding its importance for mitigating the drastic impacts on the urban ecosystem. There is a need to devise adaptation and mitigation strategies through a collaborative and participatory effort of the stakeholders.

Because of the rising concern of rapid urbanization, countries across the globe are working with an ecosystem of stakeholders to create inclusive, resilient, and sustainable cities/communities. The New Urban Agenda and United Nations Sustainable Development Goals (UN SDGs) have set the standards for sustainable urban development and provide a recommendation framework for sustainable development. The United Nations Environment Program (UNEP) also stresses the integration of nature-based solutions as a key strategy for sustainable development. UN SDGs and the World Bank's shift toward the Sustainable Infrastructure Action Plan (SIAP) is a major paradigm shift to address the global agenda of climatic and environmental well-being. There is a need to align the national objectives with the global agenda and to focus on adopting an integrated strategy for transforming cities as thriving centres of resilience, innovation, and sustainability that would further mitigate the impact of disaster and climate change. There is a need to revise the urban-planning strategy and adapt building codes to the changing environment. Doing so would help stakeholders to enable evidence-based planning and informed decision making to adapt to the changing urban environment and morphology. The role of compact cities, mixed land use, transit-oriented development approaches, and nature-based solutions are being adopted across Indian cities to attain sustainable urban development while creating a scope for the integration of blue and green infrastructure in the urban-planning process. This research study elaborates on various types of BGI interventions, approaches ways to integrate it with the urban environment, and reflects upon the relevant policy framework and regulations necessary to counteract the impact of urbanization and its consequences.

7.1.1 INTRODUCTION TO BLUE-GREEN INFRASTRUCTURE (BGI)

The concept of green cities already existed as early as 1900 with a focus on preserving the natural ecosystem and realizing its importance in the planning and development process. Over the years, it has been further evolved as Blue-Green Infrastructure (BGI) to plan, design, and manage the water and green resources of the environment. The term BGI was coined around the tenth century because of the necessity to create an sustainable urban ecosystem. BGI is an integration of natural elements as an essential component of the urban-planning and design process for grey infrastructure such as roads, settlements, buildings, etc. It connects urban hydrological functions with the planning and design approach of the urban areas and could lead to a healthier, socially cohesive, connected city ecosystem and a liveable urban environment. The BGI-led planning approach acts as an effective tool in tackling the barriers of unplanned development and paving the way for future sustainable cities. The cohesive layer of vegetation and the water resources could effectively manage the drastic impacts of rapid urbanization. In turn, this enhances the quality of the urban environment while preserving the natural ecosystem.

BGI components may include manmade and natural green spaces, urban forests, streets and pathways, swales, parks, drainage systems, rivers, canals, and waterways, including building-level interventions such as green roofs, rain-water harvesting systems, etc. BGI components can be planned, designed, and implemented as small-scale urban-design interventions, such as green boulevards and gardens/parks, to more technical and large-scale interventions, such as vegetated depressions, Integrated Urban Water Management, Water Sensitive Urban Design Solutions, etc. The interventions could be efficient in managing the interconnected set of urban issues, e.g water-resources management, to improve sanitation system, solid waste management, urban flooding, and conservation of natural ecosystems. The benefits could be demonstrated in the form of improved water quality, reduced urban flooding, rain-water retention, improved groundwater quality, reduced pressure on grey infrastructure, enhanced biodiversity, and improved the quality of urban and natural environment. Globally, the approach is being applied as an overarching infrastructure component for all scales of urban-planning and design interventions, and thus, creating a multifunctional impact on the urban ecosystem.

7.2 STRUCTURE OF THE RESEARCH STUDY

BGI as one of the sustainable engineering solutions lacks comprehensive knowledge and understanding among the ecosystem of stakeholders responsible for promoting sustainable development. The research study aims to build strategies for mainstreaming the integration of Blue-Green infrastructure (BGI) as resilient and adaptive measures across urban-development processes to achieve environmental sustainability. It would focus on developing a framework for urban practitioners/ policy makers to integrate layers of blue (water-sensitive design components) and green (forest, parks, open spaces, etc.) zones for cities to be climate resilient, enhance livability, and achieve sustainability in the long run. The chapter highlights various concepts of the BGI and its application on various scales of urban development. Multiple benefits of the BGI on the urban environment are mapped as a part of the study. The literature part elaborates on various global and national case studies to understand the concept of BGI from planning, designing, and implementation perspectives. A detailed analysis of the urban-development pattern of the New Delhi and National Capital Territory region has been illustrated to understand the impact of rapid urban growth on the natural ecosystem and assess the mitigation measures adopted by the diverse stakeholders.

The study also highlights the role of relevant policies and programs the government of India (GoI) adopted to focus on sustainable development while adapting to BGI practices and approaches. The review process of the relevant policy and regulatory framework has assisted in understanding the extent up to which sustainability approach is being adopted within the urban-planning and design measures at macro and micro levels. The outcomes of the assessment process have been utilized in mapping the gap in the adaptation process, devising strategies to incorporate the components of BGI across urban sectors to achieve sustainable land-use management, ensure water-resource management, and enhance livability and quality of life, improve the public realm and spaces, promote biodiversity conservation, and reduce the impacts of climate change.

The last section of the study identifies the key challenges and barriers to implement BGI, framing of the relevant policy/strategies and recommendations necessary to mainstream the adoption of BGI approach within the the urban-development process.

7.2.1 AIM OF THE RESEARCH STUDY

The purpose of this research study is to create a knowledge base on the understanding of the BGI approach from the perspective of urban practitioners and policy makers. The paper illustrates the knowledge outcomes of the research as a beginner's guide for urban practitioners, city leaders, and decision makers to apprehend the role of BGI and implement the concepts in the urban development and planning agenda. The study highlights various examples of integrating nature-based solutions within the urban-development process and the environmental, ecological, socio-cultural, and economic benefits that could play a pivotal role to improve the overall quality of an urban environment.

7.2.2 OBJECTIVES OF THE RESEARCH STUDY

The study is developed and based on the following broad objectives:

- Explore the concepts of BGI and its multiple benefits for sustainable urban development.
- Map the key learnings from the evolving BGI practices at global/national level.
- Identify the approach to integrate BGI interventions within the urban planning process through understanding of the current policy and regulatory framework.
- Recommendations for adopting the BGI strategies at macro/micro scale of an urban environment.

7.3 CONCEPT OF BLUE-GREEN INFRASTRUCTURE (BGI)

BGI approach aims to integrate the network of natural and artificial components of the water system and green spaces within the urban-planning and design process. It balances the function of the natural ecosystem and provides enormous benefits in terms of improving the quality of the urban environment, as well as promoting the socio-economic and mental well-being of the users. The design principles are based on the hydrological and ecological aspects of an urban environment, which create scope for an integrated multifunctional infrastructure system. It can exist at various geographical levels (e.g. regional, city, urban & rural, peri-urban, coastal area, hills, river and catchment, project specific interventions, etc.) and can function across boundaries, functionalities, and jurisdictions. The last few decades have seen the increasing demand, innovative trends, and approaches to urban development and recognize its benefits in terms of reducing heat-island effect, flood control, reducing pollution, increasing land value, conducting urban renewal/redevelopment of the areas, improving the well-being of the communities, and enhancing the overall quality of the urban environment.

7.3.1 HISTORY OF BGI CONCEPT IN INDIA

In India, the concept of integrated water-management systems and similar sustainable approaches can be dated back to the prehistoric era of Indus Valley Civilization (3000 BC). The BGI elements were a very integral part of the human settlements, and the means of livelihood were based on the natural ecosystem. The City of Mohenjo-Daro had an excellent water-management system planned at the city scale. Cities like Udaipur (1559 AD) had a system of designed artificial lakes to fulfil the water demand of the city dwellers, as well as create recreational spaces along the lakes. The planning of Jaipur City (1726 AD) was based on the grid network of green and open spaces. The plan of Bengaluru City (1799 AD) was based on an integrated network of water bodies. New Delhi was designed during the British era, and the planning was based on an integrated hierarchy of open/green spaces. Post-British era cities like Chandigarh and Gandhinagar were designed linking hierarchical layers of green spaces with natural topographical features to maintain natural runoff for storm water/flood water. Architectural monuments from Mughal eras e.g. Humayun's Tomb, at New Delhi, Taj mahal in Agra, Shalimar Gardens in Kashmir, etc., are few of the living examples that had recognized the importance of integrating natural elements within the built environment.

7.4   APPLICATION OF BGI IN AN URBAN ENVIRONMENT

The water and green planning approach are interconnected in nature, and the scale of application varies from buildings to city and regional level. Various forms of applications are based on different attributes, e.g. function, scale, and local urban context (space, topography and climate).

The water-related infrastructure is different from the conventional water-management approaches and integrates the discipline of urban/landscape planning. It involves the urban water cycle and includes solutions such as rainwater use, wastewater treatment, detention, and infiltration, etc. At the building level, rain-water harvesting mechanisms with focus on the reuse of wastewater would play an important role. At the macro level, it includes large-scale solutions to prevent urban flooding, stormwater management, runoff and groundwater recharge, etc., while relieving the existing grey infrastructure from collapse/overburden of sewer system. The green infrastructure focuses on resolving issues related to micro-climatic conditions, enhancing capabilities of ecological systems, enhancing social and health values by integrating vegetation within the urban areas, e.g. parks/open spaces/river landscapes, green spaces, vertical greening, etc.

7.4.1   APPROACH TO INTEGRATE BGI WITH URBAN ENVIRONMENT

The integration of BGI approaches into the urban-planning process and strategy en-ables the effective management of water and green resources while preserving the natural ecosystem nd creating a sustainable and resilient urban environment. However, there is a need for systematic integration for an effective, flexible, adaptable solution. Figure 7.1 illustrates the various applications of BGI components at an urban scale. The applications of green belt, green wall, and green roof at a neighbourhood scale act as sound barrier and pollutant absorbents. They also assist in reducing the urban heat-island effect. Installation of rainwater harvesting systems and the use of permeable pavements could reduce the rate of stormwater runoff, improve the quality of ground water, as well as increase the water-storage capacity.

Application of BGI in the urban-planning and design process can be further enhanced through contextualizing the BGI strategy and frameworks and colla-borative and coordinated efforts of the stakeholders to maintain, restore, improve, and connect existing natural ecosystems within the man-made urban environment. It would act as a feasible solution to curb the impacts of rapid urbanization and cli-mate change. Bringing policy and spatial planning-level changes, preparation of

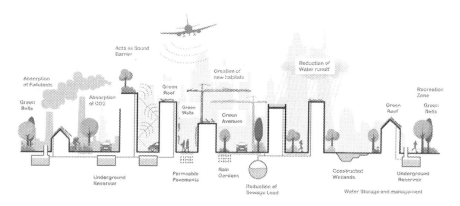

FIGURE 7.1   Integrating BGI components in an urban environment.

strategic action plans, understanding of local/regional need and context, along with a multi-stakeholder ecosystem would play a major role in integrating BGI components within the urban environment. It needs to be integrated within different context and scale of urban-planning and design measures while coherent with the urban-development sectors, e.g. water, agriculture, forestry & biodiversity, tourism, planning and policy, transport, built environment, etc. Table 7.1 highlights the important enabling factors for implementation of BGI.

7.4.2   CLASSIFICATION OF BGI COMPONENTS

The BGI components have evolved in the past several years from small-scale interventions to more engineered and designed versions for large-scale implementation. The components can be broadly categorized according to the functions, local

TABLE 7.1
Enabling Factors for Implementation of Blue Green Infrastructure

Major Drivers	Process	Success Factors
Enabling Policy and Legislation Framework	Need assessment Scoping of the project	Evidence-based planning
Physical Planning	Context setting Objectives to achieve Planning and design measures	Cross-sectorial & stakeholders participation
Strategies and Action Plans	Revenue, funding, and implementation measures	Awareness and community participation
Local/Regional Needs	Monitoring and evaluation framework	Stakeholders engagement
Stakeholders Intent	Dissemination/mainstreaming across sectors and departments	Advocacy efforts policy changes

context, and scale of implementation. The scale of application may vary in an urban environment and can range from a building level to urban-scale and from a city to a regional-level planning process. Few examples of the large-scale application of BGI interventions can be illustrated as follows:

- **Integrated Urban Water Management (IUWM):** IUWM acts as an effective planning tool and framework to manage the urban water systems and promote water conservation through alternative use of water sources e.g. greywater, stormwater, rainwater, etc. It is an effective tool to meet the escalating urban water demands, as well as transform the urban water systems as resilient and sustainable in nature. Through adaptation to effective policy framework and strategies along with stakeholder's participation, the tool can be implemented at various scales of an urban environment. The following tools are a few of the examples from the IUWM model.
- **Waste Water Management**: Urban areas produce large quantities of wastewater in the form of sewer discharge, industrial pollutants, etc., which can be harmful to the environment if not treated properly. Recycling wastewater can provide an alternative source of water to be fit for secondary/ tertiary uses, e.g. irrigation/landscaping, industrial usage, non-potable uses in buildings, such as toilet-flushing, cleaning, etc. The higher nutrient loads of recycled water can be beneficial for improving the natural landscape and promote the growth of urban forests.
- **Storm Water Management:** Urbanisation process has been replacing the natural ecosystem with grey infrastructure. The impervious surface (streets, concrete roads, man-made drains, etc.) results in increased runoff rates and volumes, causing changes to water quality and pollutant loads in the stormwater runoff. Stormwater management aims to reduce pollutant loads entering into the natural waterways by filtering and absorbing nutrients and trapping/removing litter and sediments from the source. Runoff, rainwater, and stormwater can be captured, treated, and stored for secondary/tertiary uses. Some commonly used BGI infrastructure includes bioretention systems, wetlands, tanks/ponds, permeable pavements, rainwater harvesting systems, etc.
- **Restoration of Waterways:** It is the process of preserving the natural drainage channels. It can be beneficial in improving the groundwater quality, restoring the natural ecosystem, mitigating urban flood, and heat-island effect, creating scope for green and open spaces, and adding aesthetic value to the urban environment.
- **Aquifer Recharge:** It is the practice of recharging aquifers with recycled water under human-controlled means. Canals, infiltration basins, ponds, etc., are few of the artificial sources of recharge. It can also be recharged by injecting of treated stormwater. The process can assist in maintaining the groundwater level, while diverting the flow of excess stormwater from entering and damaging the natural water sources.
- **Water Sensitive Urban Design (WSUD):** WSUD is the process of integrating water management, urban design, and landscape techniques within the urban-planning and development process. It ensures an effective

use of stormwater/rainwater/wastewater and reduces the impact of urbanization on the natural water system. It has a wide range of applications from a street, precinct, neighbourhood to a regional scale. Few of the examples can be described as follows:

- **Water Sensitive Urban Precincts:** The development of new precincts provides an opportunity for urban renewal at various scales (e.g. housing locality, neighbourhood, etc.). An integrated network of green spaces, water networks, streets, etc., at a precinct scale can provide valuable recreation and amenity assets for the users, as well as improving the aesthetic value of the urban environment.
- **Water Sensitive Parks and Open Spaces:** Urban landscaping typically requires a significant amount of water for sustenance. Parks and greenbelts act as sinks for carbon dioxide and counteract the urban heat-island effect because of the dense built-up areas. Designing open spaces/parks through integrating water-sensitive techniques can reduce the water demand for landscaping purposes, provide space to build urban amenities, enhance walkability factor, scope for integrated stormwater treatment, while mitigating urban flood and head-island effect.
- **Water Sensitive Streets:** Water-sensitive solutions across streets/roads may include permeable pavements, swales, etc. Streetscapes have ample scope to integrate such solutions given that streets occupy a significant portion of land use. It is also efficient for stormwater treatment, improves the natural ecosystem, and adds aesthetic values to the urban environment.
- **Building Level Interventions (Green Roofs, Walls/Facades, etc.):** Transitioning to water-sensitive cities requires interventions at all scales of an urban ecosystem. At a building scale, using water-efficient fixtures, reusing waste water and alternative water supplies, employing landscaping techniques, and reducing the impervious footprint of buildings can contribute significantly toward water-sensitive development. This creates a scope for integrating natural elements within the built environment, thereby increasing the exposure of users to nature in a highly urbanised surrounding while providing additional benefits, e.g. improvement in the natural ecosystem, stormwater runoff reduction, and improving the micro climatic condition of the urban areas/zones.

7.4.3   TYPES OF BGI INTERVENTIONS

Different types of BGI interventions can be applicable at various scales of urban plans and designs. The interventions can be implemented in silos or as an integrated network within an urban environment.

Interventions at the city level can be in the form of parks and open spaces, tot lots, playgrounds, green belts, streetscapes, ponds/lakes and wetlands, rivers, and floodplain management. For regional levels, the interventions can be in the form of canal beautification, urban forest, water reservoirs, road and street networks, greenbelt on highways/roadways, shorelines and embankment protection measures, etc. Table 7.2 illustrates the various major types of BGI Interventions.

TABLE 7.2

Types of BGI Interventions

BGI Solution		Definition
	**Bioswale**	Vegetated channel designed to convey stormwater runoff while removing the sediments and pollutants. It can be beneficial for groundwater recharge.
	**Constructed Wetland**	An artificial wetland designed to treat greywater or stormwater runoff. It uses natural vegetation and microbial organisms, algae, aquatic plants, etc. to improve the water quality.
	**Retention Basins**	It acts as an artificial pond to manage stormwater runoff, prevent flooding, reduce erosion, and improve water quality.
	**Ecosystem based Planning**	An approach to integrate natural landscape (including blue and green elements) within the urban-planning and development process.
	**Filter Strip**	Gently sloped planted strip of grass or densely planted vegetation. It is designed to filter runoff, reduce sediment load and other contaminants for improving the water quality.
	**Green Roof**	Roof with layer of vegetation to manage stormwater runoff and reduces energy requirement of the buildings. It also acts as an effective tool to reduce the urban heat island effect in urban dense areas.
	**Green Wall**	Vegetated wall surface designed to absorb air pollutants. It acts as an effective mechanism for controlling temperature within the buildings, acts as a sound barrier, and adds aesthetic value to the urban environment.

TABLE 7.2 (Continued)
Types of BGI Interventions

BGI Solution	Definition
 **Hedgerow**	Planted strips of shrubs and trees that act as an important landscaped habitat for flora and fauna. It also acts as a visual screen, sound barrier, and can be effective to reduce air pollution.
 **Permeable Pavement**	It acts as a porous urban surface that is effective for absorbing surface runoff and slowly allowing it to infiltrate into the ground. It can be used in the parking lots, sidewalks, driveways, etc.
 **Parian Buffer**	It is a vegetative buffer-strip along water streams that diverts runoff into streams, thus improving the water quality and creating a varied wildlife habitat.
 **Soakaways/Infiltration Trenches**	It acts as a sub-surface reservoir to store, infiltrates stormwater runoff, and allows clean water to percolate into the ground. It can be used along the streets, roads, parks, green spaces, etc.
 **Tree Canopy**	Trees help to clean air, absorbs pollutants, filter stormwater runoff and mitigates the impact of urban heat island. It can also improve the physical and mental well-being of the citizens while improving the overall quality of the urban environment.
 **Xeriscaping**	It is a landscape-design practice with certain groups of vegetation to reduce the watering requirements.

7.5   BENEFITS OF BGI INTERVENTIONS

The integrated network of blue/green infrastructure acts as a multifunctional system. It strengthens the urban ecosystems by introducing sustainable elements in the urban-planning and design process. BGI creates immense potential for

TABLE 7.3

Benefits of BGI Interventions

Urban Sectors	Process
Ecological	Preserves the natural ecosystem, improves the habitat for biodiversity proliferation, and mitigates the drastic impacts of climate change
Social	Improves overall quality of the urban environment, connects people/communities with the natural environment, creates scope for physical activities and thus leads to physical and mental well-being of the users, increases real estate values, and enhances aesthetic and amenity values of the surrounding areas
Environmental	Assists in pollution reduction, mitigates flood risk, controls the overflow of stormwater, improves groundwater quality, mitigates the adverse climate impacts e.g. drought, flood, urban-heat island, land cover changes, etc.

hydrological, socio-economic, and environmental benefits for an urban environment. Table 7.3 illustrates the major benefits of BGI.

The benefits are cost effective and can assist the cities to achieve sustainable urban development. Few of the tangible benefits can be in the form of land regeneration, which acts as a tool to revive underutilised land and improvise the urban centres. Small-scale interventions like drainage/canal improvement, trees, vegetation growth, clean water bodies, etc. can bring improvement in the natural ecosystems' health. Creating water-sensitive parks/open spaces can improve the performance of urban drainage systems through intercept and store runoff to minimise downstream flood risk, and also to adapt urban spaces to control flooding. Water and vegetation play a very important role in balancing temperature of dense built-up areas and can mitigate the urban heat-island effect. Raising awareness, driving behaviour change, and adopting sustainable and innovative approaches can help in sustaining the BGI interventions in the long run.

7.6   CASE STUDIES

In alignment with the mandates of the New Urban Agenda 2020, UN SDGs, countries across the globe are adapting the sustainable way of development while focusing on integrating BGI interventions within a variety of urban context, scale, and function. The case studies have been selected based on certain factors /indicators e.g. locality, type of infrastructure, purpose, etc., to highlight the key impacts and success stories from each of the studies as well as gain insights from relevant city/country wide policy, regulatory, and implementation framework.

7.6.1   GLOBAL CASE STUDIES

The implementation of BGI elements and techniques to create a sustainable urban ecosystem has yet to be recognised in India, as well as by the global counterparts. However, few countries such as the Netherlands, Singapore, etc. have been integrating the BGI components as an integral part of their urban development agenda. The reasons cited include a general lack of awareness and understanding of the BGI aspects among the decision makers, urban practitioners, policy makers, etc. The transition from conventional approach to sustainable infrastructure design would imply change in the social, legal, policy, and political setting at a national, regional, city, and grassroots level. Despite the multi-functional benefits, there are notable challenges in wide-scale adaptation of the BGI concept. While it is costly to implement BGI, the expense is quickly recovered and is justifiable because of the multiple benefits it offers. Few of the case studies of successful interventions across the globe are listed as below:

7.6.1.1 Name of the Project: Green Tokyo (2007)

Project Background: Tokyo, the largest city of Japan, has a population of 13.5 million (2015 data) and is also considered as one of the most densely populated cities of the world. The city has seen rapid growth and urbanization in the 20th century, leading to urban issues such as lack of green spaces, ageing society, dense urban areas, change in land use, and land cover, etc. To monitor the urban development process, Tokyo Metropolitan Government (TMG) has formulated a ten-year policy known as "Green Tokyo" in 2007. The green policy, aimed at increasing the green coverage of the city, develops green road networks and integrated open spaces/parks with size of 1000 ha, creates urban forests and green islands, and revives water bodies to control urban flooding, countering the urban heat-island effect, etc.

    Strategy Adopted: TMG had issued strict guidance, regulatory frameworks, enforced green standards, and followed the strict adherence and implementation measures in support from the key stakeholders and engaging community and citizens in the process. The policy also promoted green campaigns to encourage participation from private entities, pioneering lucrative models/schemes to invite participation from business opportunities, introducing leading-edge greening technologies, raising funds and inviting Public Private Partnerships, through innovative finance and revenue generation mechanisms, etc. Awareness and sensitization of the community had played an important role in achieving the objectives of the plan. Convergence mechanisms between schemes and projects had also been an important tool of the policy framework. Under the policy, TMG had enforced implementation of multiple interventions at various scales e.g. reinforcement of greening plans for buildings being constructed or renovated, formulating guidelines for creating green spaces in unused public land, advocating the need for implementing greening projects across schools, etc.

    Figure 7.2 illustrates few of the implemented interventions (open and green spaces across streets, green boulevards, etc.) under the Green Tokyo Policy.

7.6.1.2 Name of the Project: Water Square Benthemplein – Rotterdam, The Netherlands (2013)

Project Background: Referred to as the "City of the Future," Rotterdam is the second largest city in the Netherlands. It is home to one of the world's biggest ports. Being a coastal city, Rotterdam is vulnerable to extreme climatic conditions and other challenges such as housing demand, declining natural resources, etc. Considering

FIGURE 7.2   Successful implementation of green initiatives under the Green Tokyo plan.

resilience and sustainability as key strategies for development, the city authorities have taken numerous initiatives to mitigate the impacts of rising sea level, extreme rainfall, etc.

Strategy Adopted: One of such initiatives is the concept of Water Plazas (public open space). This recreational space acts as a buffering and water-storage infrastructure and is specially designed to collect the overflowing flood water, improve water quality, and discharge the excess water to nearest water bodies at a slower rate. Alternatively, the plazas can be used as a recreational space and consist of sports areas, sloped terrain, landscaping and green zones, etc. The plazas add value to the neighbourhood, enhance aesthetics, and act as a recreational space most times of the year. During heavy rains, the collected rainwater diverts toward the square (because of the fluent slope) and retains the water to prevent overflow and urban flooding while reducing the stress on the city's sewerage system. It also helps in minimizing the financial and cost implications on building additional grey infrastructure for the city in the future. Figure 7.3 illustrates the pictorial representation of the Water Square in Rotterdam. It demonstrates an unique example of BGI intervention where play and

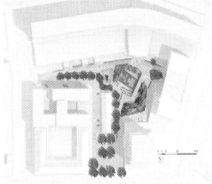

FIGURE 7.3   Water Square, Rotterdam.

sports areas with permeable surfaces have been designed as water reservoir and act as stormwater catch basins during flood and heavy rains.

7.6.1.3   Name of the Project: Sponge City Concept – China (2015)

Project Background: Sponge cities are based on the concept of absorbing stormwater and slowly infiltrating it into the ground. Rapid urbanization and increasing population have created havoc on China's overall infrastructure due to urban flood, inundation, and water-scarcity issues. To curb the menace of rapid urban growth, the Chinese government had launched the Sponge City Program (SCP) in 2015. The SPC program is based on the concept of comprehensive urban water-management strategies. The program aims to promote water security, environmental protection, and ecological restoration. The program emphasizes the urban areas should be designed to absorb and reuse rainwater/stormwater, create more greener urban spaces, reduce the use of impervious surfaces, and thus, reduce flood risks and lower the burden on drainage systems.

Strategy Adopted: The sponge city approach aims to control urban flood, prevents water pollution, and thus improves the microclimatic condition and the degrading urban environment. It can include an integrated network of green and open spaces, interconnected waterways/channels/ponds that can naturally detain and filter water, and create space for cultural and recreational opportunities. Small-scale and porous design interventions, e.g. bioswales, bio-retention systems, and porous roads have been an important part of the project, which allows water to be absorbed and recharge ground water. Interventions such as green roofs and rainwater harvesting systems assist in water saving and recycling techniques. Some of the positive outcomes of the program are cleaner groundwater, flood-risk reduction, lower burden on drainage systems, enriched biodiversity, and a green, healthier urban environment. Figure 7.4 illustrates the pictorial representation of the few of the application of BGI components, e.g. permeable pathways, retention basins,

FIGURE 7.4 Implementation of innovative green interventions under the sponge city program – Lingang, Shanghai.

etc. at Lingang, Shanghai. It acts as a renovated "sponge park" covering an area of 14,200 square meters near a major residential community in the Lingang area.

7.6.2 National Case Studies

In India, the existing urban-planning and design approach widely lacks integration of natural and built environment. Over the past few years, Indian cities have faced the brunt of climate change issues in the form of urban flooding, environmental degradation, biodiversity loss, etc. Cities are realizing the value of a sustainable urban environment and are paving the way to adopt innovative urban solutions. Few of the successful examples have been captured in the following section of this chapter.

7.6.2.1   Name of the Project: Restoration of Urban Ponds – Gurugram (2018)

Project Background: The project aims to rejuvenate and preserve the existing water bodies in Gurugram through adaptation of sustainable and low-cost measures of water conservation. It also aims to achieve urban renewal of degraded zones and develop active urban spaces for recreational activities while transforming Gurugram as a socio-cultural capital. The approach is based on the model of identifying contextual sustainable solutions based on community participation and active stakeholder's engagement process. The project sets forward a certain set of policy/guidelines for the relevant stakeholders, e.g. communities and citizens, city leaders and decision makers, urban practitioners, policy makers, etc., to adopt the approach as pilot projects and replicate the same across the city.

   Strategy Adopted: The principal objective of the project is eco-restoration of the surrounding watershed and biodiversity habitats in an ecologically sustainable manner, preservation of green areas as accessible public spaces, and the arrestment of the deteriorating groundwater hydrology and restoration of the water level of the ponds. The biodiversity conservation of the severely endangered Aravalli biotype and adaptation to climate change mitigation measures for the city also forms the larger part of the objective that the project aims to address during the future course of action. Figure 7.5 illustrates few of the related ongoing interventions implemented by Municipal Corporation of Gurugram (MCG). Citizen and stakeholders' engagement have been an important component of the revival process. The local communities have been involved at each stage of the planning, designing, and implementation process.

7.6.2.2   Name of the Project: Restoration of Kaikondrahalli Lake – Bengaluru, Karnataka (2016)

Project Background: Bengaluru has experienced rapid urbanization and sprawl during the last decades because of concentrated industrial and economic development of the region. This has resulted in environmental degradation leading to encroachment of catchment areas, decline in green spaces, pollution and drying up of existing water bodies. The consequences can be seen in the form of drastic changes within the city's land-cover pattern. The satellite images of the city depicts that 93%

FIGURE 7.5   Successful implementation of revival of lakes in Gurugram.

of the city's landscape is paved surface with a drastic decline in the urban green cover (ENVIS technical report – IISc, 2017IISc, Bengaluru 2017). Kaikondrahalli Lake at the southeastern periphery of the city is one such lake that has suffered degradation over the past decade. The lake has been revived because of the collaborative efforts of the stakeholders and active participation from the communities. Figure 7.6 illustrates the comparative change of Kaikondrahalli Lake before and after the rejuvenation process from 2007 to 2016. Through community efforts, involvement of local NGOs and relevant stakeholders has led to drastically reduce the level of water pollution, improve the existing surrounding area, and develop the lake as an important social and cohesive hub.

Strategy Adopted: The lake has been severely polluted because of illegal dumping of solid waste and untreated sewer, and rise in real estate values, thus damaging the ecological and environmental values of the lake. It has been ignored in this condition over the past several years. Local communities along with the (Bruhat Bengaluru Mahanagara Palike) BBMP officials have worked collaboratively between 2009–2011 to map the major issues and identify contextual and low-cost solutions for controlling the degradation. The lake has been developed as a cohesive urban space while maintaining the social and cultural fabric of the

FIGURE 7.6   A pictorial illustration of before and after image of the Kaikondrahalli lake.

surroundings. The strategies adopted by the city stakeholders to revive the city lakes can be summarised as follows.

- Enhance water quality
- Managing illegal dumping of waste
- Restriction on discharge of untreated wastewater
- Strengthen legal and governance mechanism
- Removal of encroachments
- Mapping of water bodies and buffer zones
- Provision for constructed wetlands and STPs
- Enhance recharge of ground water aquifers

7.6.2.3 Name of the Project: Kahn Riverfront Development Project – Indore (2015)

Project Background: Indore Municipal Corporation (IMC) under the government of India's Smart City Mission has taken the initiative for the riverfront development of Kahn River. It covers a stretch of 21 km in the city of Indore, and this stretch has been taken up for rejuvenation under the area-based development plan of Indore Smart City. The river, once an urban heritage, is now in a degraded state because of illegal encroachment, discharge of untreated sewage, municipal solid waste, and construction waste leading to groundwater contamination.

Strategy Adopted: The riverfront development project has been planned based on various innovative solutions such as check dams for water retention, rainwater harvesting, stormwater management, groundwater recharge, etc. It also aims to restrict discharge of untreated water to reduce pollution of the river and improve the water quality. The approach adopted has led to an overall improvement of the urban environment, developing the riverfront as a public realm with scope for recreational and social activities. The urban-rejuvenation process also helped to promote the tourism potential of the city. Figure 7.7 illustrates the existing and proposed plan for the riverfront development. The project focuses on reviving the degraded water

FIGURE 7.7 Existing and proposed image of the Kahn riverfront development.

quality of the river, creating green spaces across the stretch, and developing the riverfront as an important public realm for the city.

7.7 SUSTAINABILITY ASSESSMENT OF URBAN DEVELOPMENT IN NCT REGION

The assessment study of urban expansion in Delhi and National Capital Territory (NCT) region aims to highlight the major impact in terms of degrading urban environment and identify major initiatives taken by the concerned authorities as remedial measures. The assessment has been done based on various attributes such as:

- Impact of urbanization
- Population expansion
- Land use and land cover changes
- Water scarcity
- Reduction in green/open spaces
- Innovative and sustainable approaches to mitigate the impact

The assessment highlights the importance of adopting BGI approach and integrating it with the urban-planning and development strategies while preserving and managing the natural resources in the NCT region to create a sustainable environment for all.

7.7.1 RAPID URBANIZATION IN NCT REGION

The drastic consequences of urbanization in the NCT region can be seen in terms of degrading urban environment, pollution, flood, draught, decreasing green/open spaces, decline in groundwater, etc. The geographical size of the capital city has taken a two-fold increase almost from 1991 to 2011, with an increase in the number of urban households and decline of rural households. Cities along the periphery of the capital city, e.g. Bahadurgarh, Ghaziabad, Noida, Faridabad, Gurugram, etc. have also witnessed tremendous growth over the last few decades. Figure 7.8 illustrates the extent of rapid urbanization in NCT Region from 1989 to 2018. The images show the expansion of NCT boundary with formation of new towns/cities at the outskirts of capital city.

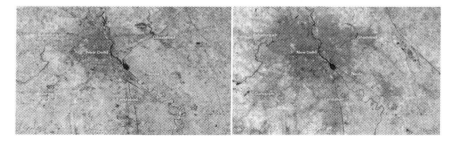

FIGURE 7.8 Landsat image showing the extent of urban sprawl in the NCT region from 1989 to 2018.

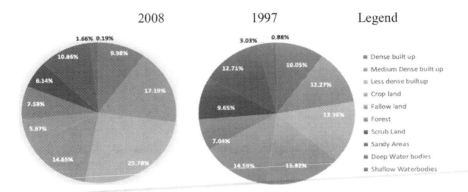

2008   1997   Legend

FIGURE 7.9   Changes in land-use of Delhi from 1997 to 2008.

As evident from Figure 7.9, the built-up area between 1997 and 2008 has increased by 16.87%. There has been a decrease in the agricultural land and significant reduction in the area covered under water bodies. The net decrease in agricultural land, wasteland, and waterbodies together accounts for a total decrease of 258.20 sq.km. Total area under waterbodies has reduced from 58.26 sq.km in 1997 to 27.43 sq.km in 2008. Many freshwater wetland/ponds are in a degrading state, and others are already extinct due to the repercussions of rapid urbanization, population growth, illegal encroachment along the catchment's areas, etc. It has identified that, out of 629 demarcated water bodies in NCT Region, around 232 cannot be revived because of large-scale encroachment.

The urban green spaces of the NCT Region have also seen drastic implications. Delhi had 300 sq.km of green area in 2009 (Forest Dept of NCT, Delhi), which accounts for 20% of the region with 22 sq.m per capita availability of green spaces. However, as per Urban & Regional Development Plans Formulation & Implementation (URDPFI) guidelines, 2014 around 25%–35% of a city's area should be earmarked as recreational and open spaces. There is a meagre increase of 0.5% in the forest cover as compared to around 17% increase in the built-up area of the region. Rapid pace of urban growth has equally impacted the ecology of River Yamuna, degrading the river water quality and the overall ecosystem. The high demand of water supply leading to extraction of river water and resultant release of wastewater from the neighbouring cities and towns has had an adverse impact on the Yamuna River ecosystem. Currently, the river acts like a drain where the catchment areas are being encroached illegally. Several action plans have been prepared to protect the river, even though none of them have been successfully implemented at its full capacity and the restoration plans are proving inadequate for reclaiming the river strength and improving the degraded water quality. There is a need to focus on developing integrated urban water-management systems toward an effective revival and implementation process, revisiting the vision and planning strategy. Measures such as check dams to increasing water demand from the neighbouring cities/towns ensure zero flow of untreated sewage and waste water into the river, restriction in encroachment of the floodplains, focus on recharging of aquifers, etc., would be considered as an immediate component of the river-restoration project. Role of awareness and sensitization campaigns, public participation, collaborative approach, improved policy regulations, effective implementation, as well monitoring process, need to be considered to restore the River Yamuna from further degradation.

7.7.2 INITIATIVES UNDER SMART CITY MISSION

The New Delhi Municipal Corporation (NDMC) zone in New Delhi has been selected for development under the Smart City Mission. Several area-based and pan city initiatives have been taken under the mission to develop the zone. Integration of blue and green layers of infrastructure within the existing city planning and design approach has been a major point of focus of the NDMC Smart City Proposal. Few of the major Blue-Green initiatives can be highlighted as follows:

- Availability of 20 sq.m of green spaces per 100,000 population
- Recharging aquifers along green areas/parks with the help of mini/micro STPs
- Mandatory use of rainwater harvesting techniques in major public buildings
- Improved access to public amenities (public toilets, parks, open spaces, recreational facilities, etc.)
- Improving aesthetics values of the city
- Creating more open and green spaces
- Green infrastructure to manage storm water runoff
- Improved water supply, wastewater, and solid waste management system

7.7.3 GREEN INITIATIVES FOR MASTER PLAN FOR DELHI – 2041

Delhi Development Authority (DDA) is currently preparing MPD – 2041. The proposed master plan would focus on a sustainable approach through adopting a blue-green policy. The policy stresses upon restoring the area along the major canals/drains (blue areas) and the adjacent land (green areas) in the urban-planning and development process. The master plan would focus on reviving 50 such drains and develop them as an integrated corridor, whereas the surrounding zones can consist of parks, recreational spaces, wetlands, ponds, pedestrian walkways, green-mobility circuits, etc. DDA, along with other stakeholders, would work through a collaborative process and focus on checking pollution-related aspects, and restrict flow of untreated wastes to the water bodies/canals/drains, etc., to revive and develop the zone as an integrated network of green and open spaces.

7.8 ONGOING POLICIES/SCHEMES IN INDIAN CONTEXT

Infrastructure development in India has mostly focused on the conventional approach of urban planning and design where grey infrastructure acts as a predominant component of the infrastructure plans. Many piecemeal approaches have been adopted across programs/missions to adhere to the needs of sustainable development goals and agenda. Several policies, acts, and guidelines have been enacted from time to time to protect the urban environment and mainstream the blue-green development approach across urban plans and policies.

7.8.1 MAJOR INITIATIVES OF GoI TO PROMOTE SUSTAINABLE DEVELOPMENT

India has enacted many acts/guidelines and has launched missions/schemes e.g. Smart City Mission (SCM), Atal Mission for Urban Rejuvenation and Transformation (AMRUT), Jal Shakti Abhiyan (JAL), etc., to assist the cities/towns in understanding the efficiency and effectiveness of sustainable urban solutions, administering and enforcing their implementation in a paced manner at national/regional/state/city level.

The earliest discussion on environmental policy was initiated by the Planning Commission of India during the fourth five-year plan (plan period 1964–1969). To interlink urban plans with green and blue components of the urban environment, Town and Country Planning organization (TCPO) Act, 1962 was formulated in India. Its major objective was to enact the preparation of master plans for cities with focused importance to be given to create green and open spaces. However, the response has not been very active, and only metropolitan cities have tried to enact it as per the TCPO Act. A review of the timelines of the major initiatives and the objectives has been done as part of this research study. Table 7.4 illustrates the

TABLE 7.4

Launch of Related Administrative/ministerial Bodies by GoI

Name of the Ministerial Bodies	Year of Formulation	Objectives
1. Federal Department of Environment	1980	Focused on protection of natural resources in India. It was renamed as the Ministry of Environment and Forests in 1985. Restructured as the Ministry of Environment, Forest & Climate Change (MoEFCC) in 2014.
2. Ministry of Water and Irrigation (MoWR)	1985	Established after the bifurcation of Ministry of Irrigation & Power with an objective to manage nation-wide water resources.
3. National Action Plan on Climate Change (NAPCC)	2002	Formulated in response to United Nations Framework & Convention on Climate Change (UNFCCC). The state level NAPCC was enforced in 2008 with an aim to mitigate the impacts of climate change.
4. Ministry of Water Resources, River Development & Ganga Rejuvenation (MoWRRD & GR)	2015–2018	Responsible for development and management of water resources in India through enactment of policies/guidelines. Implementing and monitoring river and water specific infrastructure programs.

launch of major administrative/ministerial bodies by the GoI under the vision of promoting sustainable urban development.

There are eight sub-missions launched under NAPCC in 2008 to adopt the sustainable approach for urban development with a vision to protect the natural environment. Table 7.5 highlights the sub-missions launched under NAPCC in favour of the objectives of the concerned Governing/Administrative Bodies.

TABLE 7.5

Sub-missions Launched under NAPCC

Name of the Ministerial Bodies	Year of Formulation	Objective
1. National Water mission under MoWRRD & GR	2018	Focus on integrated water-resource management. Promote water conservation, minimize wastage and ensure more equitable distribution of water resources amongst states.
2. National mission for a Green India Mission under MoEFCC	2008	Protect, restore and expand forest cover, adaptation of mitigation measures in response to climate change.
3. National Solar Mission, under the Ministry of New and Renewable Energy	2008	Framing of enabling policy environment and implementation strategies for mainstreaming the adaptation of Solar Energy across sectors in India.
4. National mission for enhanced energy efficiency, under the Ministry of Power, Government of India	2010	Fostering innovative policies and effective market instruments to promote the use and business market of energy efficient measures.
5. National mission for sustaining the Himalayan ecosystem, under the Ministry of Science and Technology	2014	Protecting the fragile Himalayan ecosystem through addressing climate change impacts and providing technical assistance for sustainable development in the Himalayan region.
6. National mission on Sustainable habitat under Ministry of Housing & Urban Affairs (MoHUA)	2010	Build a dynamic knowledge system, support national policy and actions relevant from the point of sustainable development.
7. National mission for sustainable agriculture under Ministry of Agriculture & Family Welfare	2010	Promote adoption of sustainable agriculture practices (e.g. integrated farming, soil health management, synergizing resource conservation, etc.) especially in rainfed areas.
8. National Mission for Sustainable Habitat under Ministry of Housing & Urban Affairs	2014	Build sustainable cities through enforcing adoption of energy efficiency measures in buildings, waste management, bringing modal shift to public transport, etc.

7.8.2 MAJOR URBAN REFORMS OF GoI

The Ministry of Housing and Urban Affairs (MoHUA) is the apex authority of the government of India. It formulates policies and programmes, funds, supports, and co-ordinates the management and implementation of urban development activities across central ministries, state/city governments and other nodal authorities. It also monitors the programmes concerning all the issues of urban development across the country.

It has a vision to transform urban India based on a smart, inclusive, and sustainable approach. Numerous acts, policies, and reforms have been launched to bring a paradigm shift in the approach toward synergetic urban development and tackle the implications of rapid urbanization. Urban green guidelines were launched in 2014 by MoHUA in collaboration with TCPO to provide the states/cities, particularly state town planning departments, urban development authorities, urban local bodies, etc., with a framework for increasing the urban green coverage and preserving the urban green areas. URDPFI guidelines launched in 2014 were conceptualised to provide framework and guidelines to the stakeholders in planning and formulating urban plans through integrating the concept of sustainable development, focusing on integrating green spaces as a major part of the urban plans for both macro- and micro-level planning and development process.

On-going schemes/missions of MoHUA, such as Smart Cities Mission, focus on targeting smart and sustainable urban development. Integration of BGI components has been a priority area for 100 cities selected under this mission. Other schemes, such as AMRUT, also focus on sustainable water and waste management, creating a hierarchy of green spaces in an urban area ranging from neighbourhood to city level. The scheme also emphasises on the availability of 10–12 sq.m of green spaces per capita at city level.

Jal Skahti Abhiyan (JSA) has been launched by MoWRRD & GR in 2019, which is an intensive water-conservation campaign built on active citizen participation approach. It aims to accelerate water-conservation measures across the country. Under JSA, five targeted interventions have been identified to be implemented across water-stressed districts in India. The interventions are focused on adapting innovative approaches to preserve and conserve groundwater, and promote integrated water-management systems through a collaborative and participative manner.

7.9 RECOMMENDATIONS TO PROMOTE USE OF BGI

Despite numerous benefits, the wider implementation of the BGI approach faces challenges due to lack of awareness, rigid regulatory framework, and policies that support grey infrastructure and traditional approaches for infrastructure development. The approach to urban planning and design should strongly be aligned to create a clean and healthy environment and create opportunities for the protection of the natural resources. Also, the urban plans/infrastructure need to be more adaptive and multifunction in nature. Few of the recommendations are as follows:

- **Governance Mechanism:** Efficient governance mechanism would play a major role in devising effective policies, legal, regulatory framework which is necessary for improving performance delivery, implementation, and monitoring the outcomes of BGI. Enforcement of existing regulations, adaptation of cross-sectoral and collaborative approach between national/ state government bodies and ULBs would assist in the wide scale implementation of BGI interventions.

- **Policy and Plans:** The process should focus on conducting urban diagnostics of existing urbanizations status, review of existing plans/policies to assess the issues, and build in climate change consideration in the planning process. Existing legal, policy framework should consider the scope to allow innovative approaches for pilot demonstration of such interventions. The policy makers need to work in a collaborative manner to create scope for cross-sectorial integration of BGI interventions across relevant schemes/programs.
- **Shift from Conventional Approach:** The shift from the conventional approach to BGI is based upon the technical, policy, and governance model, setting standards and regulatory requirements, bringing change at the local, city or regional scale. Incentives, resources and leadership support from government, funders, and communities would help in shifting the focus toward the approach of sustainable development. It should focus on achieving short-term interventions within the ongoing development process, whereas it should aim to integrate the long-term interventions within the project cycles for phase-wise implementation.
- **Multi-stakeholders Engagement:** Collaborative efforts of stakeholders would play a major role in planning and implementation of BGI. It creates a greater impact when a regulatory framework, effective governance and institutional structures, capacity, working practices and platforms are in place to promote such interventions. Involvement of an ecosystem of stakeholder, e.g. government bodies, NGOs, academic and research institutes, sector experts, private companies, communities, etc., can act as an effective and collaborative approach with defined shared priorities and goals for designing and implementing BGI.
- **Community Engagement:** Sensitization and awareness drives through involving NGOs and involving the local community would play a major role to ensure community driven actions. Engaging with the community creates a sense of ownership, improvises the decision-making process, and can deliver more effective outcomes to build sustainable, resilient, productive, and liveable communities.
- **Innovative Finance Mechanism:** Exploring opportunities to tap revenue-generation mechanisms and achieve participatory budgeting responses. Securing international grants and funds for running design innovations, accelerated programs, and support innovators to and scaling pilot solutions would act as an effective tool for the decision makers/leaders. There is ample scope to leverage support from global/national missions on adopting and adhering to a sustainable development approach.
- **Data and Technology:** Lack of disaggregated and local data on climate impact act as a barrier to devise strategies of BGI approach. There is a need to identify data gaps, including identifying key priorities and

opportunities to tackle the issues. Collaborative and data-sharing platforms amongst key stakeholders would be a great step to support and refine the ecologic, economic, and social arguments for BGI.

- **Sustainable Water Planning:** A water-planning approach is being adopted globally. Many countries are ensuring to generate water plans, which tend to be localized in nature, while considering the voices of all sections of the society. Sharing of similar methods and tools through engagement with global actors could play a major role in disseminating learning and knowledge to all for effective decision making process.

- **Social Inclusion of Water Models:** To ensure equitable access to water resources, policies should incorporate gender and social equality aspects. The models should be built upon sustainable strategies, aim toward poverty reduction, and promote equal access to water sources.

- **Integrated Urban Planning Approach:** City development plan/master plans and relevant urban-planning and strategy documents should focus on the long-term land use and environmental challenges, and integrate BGI interventions as an important part of the plans and to be made applicable across various scale of development.

- **MEL Mechanism:** Exploring the use of monitoring and evaluation (M&E) tools and robust decisions support system would assist in assessing the outcomes of the project in terms of benefits and cost effectiveness. It would help to identify lacunas in terms of funding modalities, technical implications, stakeholder interests, etc.

7.10 CONCLUSION AND FINDINGS

Cities are in imperative need to cope with the unprecedented challenges of rapid urbanization and mitigate the impacts of climate change. Urban practitioners would need to recognise innovative ways to integrate the natural ecosystem as an integral part of the urban plans at various scale, context and functions. Adapting to BGI practice would augment change in planning and policy framework, which favours the wide-scale adoption and implementation of pilot interventions to create a resilient, flexible, and adaptable infrastructure. City leaders and decision makers should adopt a collaborative approach model (ensure interdepartmental and disciplinary cooperation) to break the challenges of working in silos and solve the issues of fragmented or piecemeal approach for urban development.

A comparative analysis of the listed case studies reveals that approach is widely accepted globally. Examples from China, Netherlands, and Japan reveal that these countries have realised the importance of a sustainable approach for development to curb the drastic impacts of climate change. Alternatively, in India, where the ancient planning and design approach was based on integrating the natural elements, the stakeholders still need to realise the value of BGI interventions and consider its wider scale implementation. Cities like Indore, Gurugram, Bengaluru, etc. have taken up pilot interventions to restore the degraded water bodies.

The Delhi government is taking up few innovative interventions at policy and planning level to curb the menace of rising air pollution level in the city. It has launched a "Green War Room" which aims to monitor the pollution crisis through an app. The Blue-Green Policy enforced by DDA would focus on creating an interconnected network of open spaces across the drains/canals. Visionary missions of GoI such as SCM, AMRUT, etc. also focus on the sustainable approach for urban development. The missions have taken innovative measures to empower the city stakeholders to make use of the natural systems and develop a robust and resilient infrastructure rather than adopting the conventional approach of investing on grey infrastructure. Enabling policy framework and data-driven and evidence-based approach would act as an enabler for the stakeholders to plan, design, and implement such pilot interventions at a scale of street/neighbourhood/ward/zone level and identify opportunities to replicate them on a city wide scale.

There is a need to build technical capacities of the stakeholders, create a repository of knowledge about the subject, create awareness, partnerships, and collaboration, and explore innovative means of funding and revenue generation through a participatory and decentralised approach of implementation. This would also improve the quality of life in the urban environment and yield cross-sectorial benefits especially creating access to basic services and infrastructure such as water supply, sanitation facilities, wastewater management, etc. The vision and the strategies highlighted in this paper aim to act as guiding principles for the target audience and stakeholders to grasp, understand, and enable the implementation of BGI at various scales of the urban-planning and development process.

REFERENCES

Adriana, A, & Andrea, K 2020. Challenges of mainstreaming green infrastructure in built environment professions. *Journal of Environmental Planning and Management*, 63(4), pp. 710–732. doi: 10.1080/09640568.2019.1605890

Akanksha, G 2019. Pond rejuvenation project gets underway in Gurugram. https://www.cityspidey.com/news/10517/pond-rejuvenation-project-gets-underway-in-gurugram

Arup. 2014. Cities Alive – Rethinking Green Infrastructure.

Atkins in Partnership with University College of London, Indian Institute for Human Settlements and DHAN Foundation. 2015. Future Proofing of Indian Cities.

3 Big Ideas to Achieve Sustainable Cities and Communities. 2018. https://www.worldbank.org/en/topic/inclusive-cities

Centre for Science & Environment. 2017. Green Infrastructure – A Practitioner's Guide.

CIRIA London. 2013. Water Sensitive Urban Design in the UK – Ideas for Built Environment Practitioners.

Delhi Development Authority. 2007. Master Plan for Delhi 2021.

Delhi Jal Board. 2016. Action Plan for Cleaning the River Yamuna.

Engineering and Physical Science Research Council. 2013. Delivering and Evaluating Multiple Flood Risk Benefits in Blue-Green Cities. EPSRC Project EP/K013661/1.

Five Water Solutions for an Uncertain Climate Future. 2020. https://www.sei.org/featured/fivewater-solutions-for-an-uncertain-climate-future

Ghofrani, Z, Sposito, V, & Faggian, R 2017. A comprehensive review of blue-green infrastructure concepts. *International Journal of Environment and Sustainability (IJES)*, ISSN 1927-9566. 6(1), pp. 15–36.

Grellier, J, Mishra, HS, & Elliott, LR 2020. The Blue Health Toolbox – Guidance for Urban Planners and Designers.

IISc, Bengaluru. 2017. Bellandur and Varthurlakes rejuvenation blueprint. *ENVIS Technical Report*, pp. 15&16.

Indore Municipal Corporation. 2015. Smart City Proposal.

Jian, Y 2019. Lingang sponge city nears completion. https://www.shine.cn/news/metro/1908220642/

Ministry of Housing and Urban Affairs. 2014. Urban and Regional Development Plans Formulation & Implementation (URDPFI) Guidelines.

Mohan, M, Kandya, A & Kolli, N 2011. Dynamics of Urbanization and Its Impact on Land-Use/Land-Cover – A Case Study of Megacity Delhi. *Journal of Environmental Protection*. ISSN: 1274-1283, 2(9), November 2011, pp. 2–5.

Nagendra, H 2016. *Restoration of the Kaikondrahalli lake in Bangalore: Forging a new urban common*. Pune, Maharashtra: Kalpavriksh.

New Delhi Municipal Corporation. 2015. Smart City Proposal.

Patel, K 2018. Urban Growth of New Delhi. https://earthobservatory.nasa.gov/images/92813/urban-growth-of-new-delhi

Ramachandra, TV, Sincy, & Asulabha, KS 2020. Efficacy of rejuvenation of lakes in Bengaluru, India. *Green Chemistry & Technology Letters. eISSN: 2455-3611*, 6(1), pp. 14–26.

Ramboll. 2016. Strengthening Blue – Green Infrastructure in Our Cities.

Roxburgh, H 2017. China's 'sponge cities' are turning streets green to combat flooding. https://www.theguardian.com/world/2017/dec/28/chinas-sponge-cities-are-turning-streets-green-to-combat-flooding

Singh, G, Deb, M, & Ghosh, C 2016. Urban Metabolism of River Yamuna in the National Capital Territory of Delhi, India. *International Journal of Advanced Research*. ISSN: 2320-5407, 4(8), pp. 1240–1248; pp. 9&10.

The Indian Express. 2020. In Master Plan Delhi 2041, Focus on City's Water Bodies, Green Lungs and Unauthorised Colonies. https://indianexpress.com/article/cities/delhi/in-master-plan-delhi-2041-focus-on-citys-water-bodies-green-lungs-and-unauthorised-colonies-6585895

Tokyo Metropolitan Government. 2007. Basic Policies for the 10-Year Project for Green Tokyo.

Town and Country Planning Organisation, Government of India. 2014. Urban Greening Guidelines.

Urbanisten, DE 2013. Water square Benthemplein in Rotterdam, the Netherlands. *Landscape Architecture Frontiers*, 1(4), pp. 136–143.

Vernon, B, & Tiwari, R 2009. Place-making through water sensitive urban design. *Sustainability*, 1, pp. 789–814. doi: 10.3390/su1040789

Well, F, & Ludwig, F 2020. Blue-green architecture – A case study analysis considering the synergetic effects of water and vegetation. *Frontiers of Architectural Research*, 9, pp. 191–202.

Whelans, Maunsell, HG, Palmer, T, & Institute for Science and Technology Policy. 1993. *Water Sensitive Urban Design Guidelines for Perth Metropolitan Region*. Murdoch University.

8 Smart Factories: A Green Engineering Perspective

Amit Chopra
Guru Nanak Dev University, Amritsar, Punjab, India

CONTENTS

DOI: 10.1201/9781003127819-8

8.1 INTRODUCTION

The technological change has impacted the manufacturing industry in a considerable way. Complex processes and high standards in the manufacturing industry lead to a competitive market. Traditional manufacturing has become outdated and obsolete because of dynamic customer demands and will not be able to compete in future. The future manufacturing factories are required to meet the volatile demand, highly customized products, and shorter product life cycle in a timely and efficient manner. The evolution of technologies like big data, Internet of Things (IOT), automation, cloud computing, and virtualization has put the manufacturing sector in the spotlight and thus led to the emergence of Industry 4.0 in which all these technologies are being integrated. Therefore, smart factories are a new emerging concept in the manufacturing sector. It is an intelligent system wherein the various machines and components integrate and interact with each other. Green manufacturing is being adopted by the manufacturing industry for sustainable development. The green engineering focuses on new materials, processes, and products suitable for the environment and humans. The objective of green engineering is prevention of waste, management of materials, and the development of systems and processes for maximum utilization of energy and space. Through green engineering and applying smart factory concepts using the latest technologies, the manufacturing industry can considerably reduce waste, increase usage of nonhazardous materials, and employ cleaner alternatives in the industry. The industry has evolved over time, starting from the first industrial revolution to the current ongoing fourth industrial revolution. The various revolutions are discussed in detail.

8.1.1 First Industrial Revolution

The first Industrial Revolution started in 1760 and lasted until 1840 in Great Britain. The first revolution brought several major changes where daily life was influenced. It has impacted the standard of living of people, especially in the western part of the world. The transitions include the change from manual production methods to machines and machine tools and automated factories using steam power. This industrial change produced a lot of employment, output value, and capital investment, especially in the textile Industry. The invention of the steam engine in 1782 by James Watt was a major breakthrough. The other popular invention was the spinning jenny developed in 1764. The machine allowed workers to spin more wool, thus enhancing productivity. The first industrial revolution also witnessed the invention of sewing machines, thereby decreasing the sewing work in the textile industry, previously done manually. The sewing machine, invented by Thomas Saint in 1790, transformed the clothing industry, thus improving the production considerably at that time. Ashton (1948) highlighted the role of iron and textile industries in the industrial revolution. Britain conducted the first world fair in 1851 and showcased revolvers, sewing machines, steam hammers, reaping machines, and telegraphs to the world. The Industrial revolution resulted in better transportation through canals, roads, and rails. The banking and financial system also improved to run the business smoothly.

8.1.2 Second Industrial Revolution

The second industrial revolution happened around the year 1870. It was characterized by intense mechanization combined with electrification. The technological development in the industries of gas lighting, chemicals, transport, paper, and glass making played a central role during the second industrial revolution. The industry got stimulated after the invention of electricity, thus ensuring mass production and emergence of new markets. Sir Henry Bessemer invented the Bessemer process for the production of steel during the second industrial revolution. The furnace was able to convert molten iron into steel in this process. Henry Ford introduced the assembly line for the manufacture of cars in 1913. Agarwal and Agarwal (2017) concluded that a large number of banks increased in Britain during the second industrial revolution. The banking and finance sector during that time saw the emergence of clearing banks, declining of bills, and the cheques system.

8.1.3 Third Industrial Revolution

The third industrial revolution initiated around 1970 and the development of computers and microprocessors was done. It was the major change that developed the internet and set the base for the information age.

Markillie (2012) and Rivkin (2011) highlighted the major change during the third revolution with the help of new materials and processes like use of robots, 3D printing, and online manufacturing services. The third industrial revolution has impacted every aspect of society. This revolution has impacted various sectors like defense, advanced manufacturing, education, finance, health, and communication, etc. These sectors developed and continued to rise because of new innovations and discoveries of various services and products. The third industrial revolution also led to the development of numerically controlled machines. There was a huge shift of traditional manufacturing to the flexible and automated manufacturing system by the use of electronics and IT communication. This revolution began after 1970 and has also seen the emergence of fields like new materials, cleaner technology, biotechnology, etc. It also witnessed the internet, mobile telecommunication, and high-speed railways and renewable energies. The third industrial revolution had a huge impact on the developing nations and represented the importance of the global market. The high productivity was recorded during this period because of the use of information technology and advanced manufacturing techniques.

8.1.4 Fourth Industrial Revolution

The current industrial revolution involves transformation of traditional manufacturing and industrial products to automate the process using smart technology. Integration of the Internet of Things (IOT) and large-scale machine-to-machine communications improve self-monitoring and automation in Industry 4.0. This concept was first introduced by Klaus Schwab in 2015 by the executive chairman of the World Economic forum. Industry 4.0 integrates hardware, software, and cyber physical systems and emphasizes advances in communication and interconnectivity of devices. Internet of Things, artificial intelligence, robotics, nanotechnology, biotechnology, Industrial Internet of Things, quantum computing, and fully autonomous vehicles and 3D printing have emerged in this industrial revolution.

This current revolution is disrupting every industry throughout the world. Businesses are experiencing exponential growth during the fourth industrial revolution. It is impacting the complete business chain, like production, management, and governance of the enterprise. The new upcoming technologies like IOT, robotics, artificial intelligence, 3D printing, and fusion of these technologies are going to impact and transform life in a big way.

All machines, appliances, and devices, including systems and processes, are able to communicate with one another through digital technologies like artificial intelligence and Internet of Things. Machine-to-machine communication brings efficiency and extra security in the production system. The communication is without human intervention, making use of sensors and meters via a communication network to software, thereby converting the raw data into useful information. Various machines, appliances, industrial tools, and processes are connected with the internet in today's world.

8.2   INDUSTRY 4.0

The world is changing very rapidly and dynamically, and traditional manufacturing is unable to adjust with the increased production capacity and different varieties of products. The traditional manufacturing processes have the limitations of less monitoring and lack of automation to produce customized products at low-cost efficiently and profitably. The traditional manufacturing systems have fewer applications and poor integration between the various product life cycles and are unable to cope with the challenges generated by the changing and ever-growing technologies. The traditional manufacturing processes are non-profitable and getting obsolete with the blooming in the area of electronic technology. Automation and computerization is the need of the hour. The enterprises nowadays are using agile and lean management to optimize and improve the various production processes. The business enterprises are using various concepts to maximize the utilization of resources and assets. With the growing expectation of customers, especially during turbulent times in the era of globalization, there is an urgent need to reassess the manufacturing and production capabilities.

To curb the limitations of existing manufacturing systems, many researchers proposed some advanced-manufacturing schemes like flexible and agile manufacturing. Xu et al. (2016) highlighted that the manufacturing is experiencing the transformation from traditional practices to intelligent manufacturing, thus impacting the manufacturing throughout the world. The term Industry 4.0 was used at the beginning of this decade.

8.2.1   Definition of Industry 4.0

Industry 4.0 was denoted as the fourth industrial revolution by many researchers and business enterprises. Some of the definitions are summarized below:

Sanders et al. (2016) defined that the principles of cyber physical systems, the internet, and future technologies are applied with enhanced human-machines interaction paradigms.

Koch et al. (2014) defined Industry 4.0 as controling the complete value chain of the product lifecycle. It elevates the organization for a customized requirement.

Pfohl et al. (2015) defined Industry 4.0 wherein all the innovations implemented in a value chain check the change in digitization, network collaboration, automation, modularization, transparency, mobility, and socialization.

8.2.2 KEY PILLARS OF INDUSTRY 4.0

Hermann et al. (2016) considered cloud computing, big data, CPS, IoT, and smart factories as main constituents of Industry 4.0. The main components are discussed in detail.

8.2.2.1 Big Data

With the emergence of novel technologies and the internet, there is a continuous production and growth of digital data. The term "big data" can be referred to the heterogeneous mass of digital data produced by companies and individuals. It requires sophisticated computer storage and analytical tools. Big data provides the user the right information at the right time from a large amount of data. The United Kingdom Government office for Science (2014) stated that around 20–100 billion devices will be connected by 2020, leading to more data collection, thereby highlighting the need for applying big data analytics. Big data emphasizes the following three issues (Boyd and Crawford, 2012):

1. **Analysis:** Exploring large data sets to identify data patterns
2. **Technology:** Using computational tools to analyze large data sets
3. **Mythology:** The large data generated leading to better knowledge and intelligence

Big data is turning out to be an important application in the IT Industry. Big data application is generally in the fields of finance, medical field, telecommunication, and commerce industries where a large amount of data is generated (Song and Lee, 2015). Integration of big data with the manufacturing industry is required since the manufacturing industry is complex and diverse. It is essential to improve flexibility and quality of manufacturing process, reduction of cost, and improvement of equipment effectiveness. Therefore, integration of big data analytics and the manufacturing sector is required for the smart manufacturing process and achieving manufacturing excellence. Big data is defined as large data sets, and those can't be captured, stored, managed, and analyzed by typical software (Yin and Kaynak, 2015). Big data is getting attention because of huge data generation by machine and devices, and it is expected that more than 1000 exabyte data is generated annually and is growing by 20 times in the next ten years. The recently emerging areas like cloud, cyber physical systems, and Internet of Things has further necessitated the need for big data. Kagermann et al. (2013) emphasized that big data can play a role in Industry 4.0. Big data can achieve cost-effective and fault-free running of the

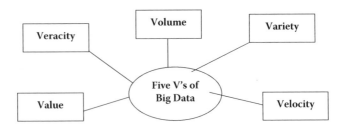

FIGURE 8.1   Five V's of big data.

process in industrial applications. Mckinsey suggested that manufacturers could reduce 50% cost in product development and assembly. It also claimed that working capital gets reduced by 7% by using big data. The optimum and efficient use of big data can enhance productivity and better managed supply chain, thus resulting in increased competitiveness in the various industrial sectors. The key description of the 5 V's of big data are discussed below and displayed in Figure 8.1.

Volume: Katal (2013) highlighted the issue of huge data generation currently existing in petabytes and the increasing use of mobile and social media. It would increase to zettabytes in the next few years. International data corporation (IDC) has estimated that 79.4 zettabyte (ZB) data will be generated by 2025 because of 41.6 billion connected IOT devices.

Velocity: Velocity is the rate at which the data flow and data is captured.

Variety: Data generated and collected is not from a single source. It is in the unstructured form obtained from web, sensors, texts, emails, etc.

Veracity: This is the ambiguity within the data because of existing noise and abnormalities in the data.

Value: The unstructured data available need to be structured so as to add value to the system.

Figure 8.1 indicates 5 V's of big data, i.e. high velocity, high value, high volume, high variety, and high veracity. The better analysis of large volumes of data can help in making advances in various disciplines and thereby improve productivity and profitability of the enterprise.

8.2.2.2   Cloud Manufacturing

Improvement in the domain of information and communication technology (ICT) has impacted the various sectors and is also playing a considerable role in modern manufacturing. Cloud manufacturing applies the basic concept of cloud computing and manufacturing domain.

Cloud manufacturing (CM) is an industrial version of cloud computing, as shown in Figure 8.2. In Industry 4.0, the enterprises need increased data sharing across the various channels. The main advantage of cloud manufacturing is that the data is stored in an internet service provider, and it can be easily accessed and retrieved remotely (Yu et al., 2017). Cloud is applied in Industry 4.0 to achieve the reaction time in milliseconds or even faster, thereby increasing the efficiency and productivity of the system (Rüßmann et al., 2015). Cloud facilities can be employed, thereby enabling the offerings in various ways, which include private,

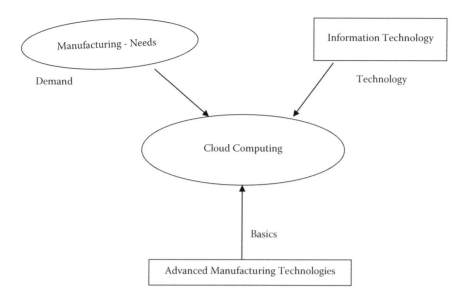

FIGURE 8.2   Cloud manufacturing.

public, and hybrid and community cloud (Luo et al., 2012). Each model of cloud provides consumers with different services, e.g. product offering of end consumer can be found on public cloud and organization employees can find the details on private cloud.

8.2.2.3   Internet of Things (IOT)

Internet of things (IoT) is considered as a major founding technology in the manufacturing sector in Europe, especially in Germany in Industry 4.0. Trappey et al. (2017) explained that actuators and sensors are implanted in each device in the manufacturing system and also connected with the internet. Sensors and actuators interact with the manufacturing system. Sensors collect the data and useful information is later processed. The stored data is processed and then sent to a remote server. Suitable feedback is shared to the system on a results basis. Sethi and Sarangi (2017) explained that servers, actuators, sensors, and the communication network are the main information of the IoT. Shariat zadeh et al. (2016) defined it as a future wherein every physical object like people and system (things) are connected in the shop floor by the internet to build the services necessary in the manufacturing. Kevin Ashton first introduced the term IoT in 1999 at the auto-ID center at the Massachusetts Institute of Technology (MIT). He expressed the use of radio frequency identification (RFID) to tag and track products in the Procter & Gamble supply chain.

a. Definition of Internet of Things (IoT)

Many researchers discussed the Internet of Things (IoT) in the literature.

Atzori et al. (2010) explained that IoT consists of two words: internet and things. The first part is the internet, or the network component, and the second one is the component.

Dorsemaine et al. (2015) explained it as "Collection of Infrastructures inter-connecting objects and access of the generated data to the management."

Gubbi et al. (2013) defined Internet of Things (IoT) as sharing information of platforms collected from sensors and actuators through a common network.

Xu et al. (2014) described the Internet of Things (IoT) network as having more essential layers:

Sensing layers for integrating various forms of "things" like RFID tags, sensors, and actuators. It is the physical layer that senses and gathers information about the environment. Networking layers to connect the smart things, net-working devices, and servers, and support the information transfer either from wireless or through wired network. A service layer is added to integrate service and application through middleware technology. Interface layer is used to display information to the user.

IoT has potential applications in practically all areas of our daily life. The ap-plication covers areas such as transportation, building, lifestyle, agriculture, emergency, healthcare, factory, environment, etc. IoT is very useful in the detection of leakage of gasses and chemicals in industry and mines. The IoT sensors can monitor the toxic gas and oxygen level inside the chemical plant. It also finds good applications in maintenance and repair, especially in predicting the equipment malfunction well ahead of actual failure of the machine component.

b. Sensor

Sensors in the Internet of Things (IoT) are the main component. It changes a nonelectrical signal input into an electrical signal, which is then sent to the electrical circuit. Institute of Electrical and Electronic Engineers (IEEE) defined sensors as: "An electronic device producing electrical, optical or digital data derived from a physical condition or event. The data produced obtained from sensors is then electronically transformed by another device into information i.e. output which is useful for decision making."

c. Characteristics of Sensors

Some of the few important factors that define the suitability of the sensors for a particular specification are:

1. Accuracy
2. Repeatability
3. Range
4. Noise
5. Resolution
6. Selectivity

d. Types of Sensors

TABLE 8.1

Different Types of Sensors Used for Different Applications

Sl. No.	Type of Sensors	Description	Sensor Example
1.	Position	The position sensor measures the position of the object.	Proximity sensor, Potentiometer.
2.	Occupancy and Motion	It detects the presence of living being in a particular area.	Electric Eye, RADAR (Radio detection and Ranging)
3.	Velocity and Acceleration	It measures the movement of object.	Accelerometer, Gyroscope
4.	Force	It detects the physical force.	Bourdon gauge, Barometer,
5.	Flow	The flow sensors detect the flow of liquid.	Piezometer, mass flow sensors, Anemometer.
6.	Acoustics	These sensors measure the sound level.	Microphone, hydrophone
7.	Humidity	It detects the humidity level in the air.	Humistor, soil moisture sensor, Hygrometer.
8.	Light	It detects the presence of light.	Photo detector, Flame detector, Infrared sensors
9.	Radiation	It detects the radiation in the environment.	Scintillator, Neutron detector.
10.	Temperature	It measures the amount of hotness or coldness present in the system.	Temperature gauge, Thermometer, Calorimeter.
11.	Chemical	It evaluate the concentration of chemicals present in the system.	Breathdyer, Smoke detector
12.	Biosensors	These detect the biological entities such as cells, organisms, nucleic acids, proteins and tissues.	Electro cardio graph, Biosensors, Oximeter, Blood glucose

8.2.2.4 Cyber Physical System (CPS)

The cyber physical system originated in the United States in 2006 at the National Science Foundation (NSF). Wolf (2007) described it as a digital system that collaborates various components that can communicate, control, and engage with the physical system.

Wikipedia (2020) described a cyber physical system (CPS) as a system in which a computer-based algorithm is being monitored or controlled in a computer.

Shi et al. (2011) described it as a system that imbeds control capabilities and communication into physical devices to monitor and control the various physical activities. Cyber capabilities through actuators, sensors, and embedded systems integrate with physical systems in the cyber physical system.

Lee (2006) explained it as "Integration of Computation with the physical processes."

Gunes et al. (2014) described CPS as a large complex multidisciplinary physical system that integrates physical and cyber systems.

Poovendran (2010) highlighted that CPS is the integration of physical and cyber elements and that they achieve a lot of improvement on scalability, interactivity, intelligence, and re-configurability. It resulted in remarkable advances in the field of manufacturing, healthcare, aerospace, defense, etc.

This is a need for a cyber-physical system because of the increased connectivity between various physical objects and, as a result, some secure and reliable system is essential to protect the critical industrial system and manufacturing systems from various cyber attacks. The strong communicating link of the physical object and services considerably improves the quality that is required for various operational and administrative activities of the manufacturing system. CPS is the main production unit of the Smart Factory having two components, i.e. physical system and the other component is at cyber level (Ergunova et al., 2017).

Gurjanov et al. (2019) discussed the model of a cyber-physical system consisting of following components:

1. Controller
2. Actor model
3. Work chamber model of technological process
4. Sensor model

Cyber physical systems are interconnected systems. These systems communicate with each other, and the complete environment or system will perform the operation intelligently. This is also called a smart system consisting of a warehousing system, machines, and production facilities that are integrated digitally. The smart factory conception of industry 4.0 is that in which the physical process is monitored by CPS and a similar virtual copy of the physical system is created and thereby decentralized decisions are taken. Lee and Seshia (2011) concluded that CPS is the integration of computer networks and physical systems. CPS is considered as major support to the various smart-manufacturing strategies proposed by various countries. CPS is an emerging area having great potential for the future manufacturing systems.

8.3   SMART FACTORY

The emerging technologies like IoT and CPS, along with their implementation in the manufacturing system, have added new capabilities and assisted in managing the complex manufacturing system so as to cope up with the rapid changes in production volume. These new technologies help in assisting people and machines to complete and execute the various tasks. The German government initiated the technology concept in 2010 with an aim to develop and encourage innovation in the manufacturing systems (Zuehlke, 2010). Smart factory is considered to be the main concept of Industry 4.0 (Mabkhot et al., 2018; Wagire et al., 2019; Shi et al., 2020). Smart factory is synonymously termed as digital factory, digital manufacturing, innerconnected factory, integrated industry, advanced manufacturing, and smart production

(Nieuwenhuize, 2016). Radziwon et al. (2014) explained it as a flexible and adaptive production process to solve issues arising out of manufacturing facilities because of increased complexity and dynamic changes in business conditions. Automation of the manufacturing system will lead to optimization, thereby reducing labor and waste of resources. Park (2016) explained that the smart factory is an integrated manufacturing system that collects all the data of the manufacturing system in a real time thus implementing an optimized and ever dynamic production system. Lu (2018) defined smart factories as intelligent production systems in which fault detection and the various issues of troubleshooting collected through data analysis are supervised through assistance of cloud and self-diagnosed manufacturing systems. In the smart factory all the physical entities are connected with each other thereby exchanging information and assessing the situation. The integration of the physical and cyber world is shown in Figure 8.3.

8.3.1 DESIGN PRINCIPLE OF A SMART FACTORY

Hermann et al. (2016) designed the principle of the smart factory in Industry 4.0. This will surely help architects to create smart factories or upgrade existing ones into a smart factory.

The various principles proposed are:

i. Modularity

It is the ability or the ease of the components to separate or combine quickly. Weyer et al. (2015) emphasized that the smart factory must have high modularity thereby allowing fast integration. High modularity results in better responsiveness to the

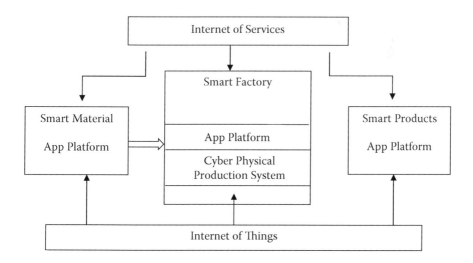

FIGURE 8.3 The concept of smart factory.

changing customer requirement and thereby overcoming the malfunction of the system.

ii. Interoperability

Interoperability refers to the ability wherein technical information of various components is being shared within the system. It also shares the related information between the manufacturing organization and the end customer.

iii. Decentralization

It is the ability of the system and its elements to make decisions on their own. Cyber-physical systems are adjusted to make decisions independently as per the design of the model. It was ensured that while changing decisions, the organizational goal is not affected. In this type of system, the employee makes various decisions without wasting time and implements a new strategy that considers the business environment. Tantik and Anderl (2017) highlighted that these interconnections assist in adapting and enabling low-cost and customized products.

iv. Virtualization

It is the ability of the system in simulating an artificial factory environment with cyber-physical, which is virtual to the actual physical environment. Lee (2015) explained that virtual systems analyze the physical system and the actual data in real time. The virtual system implements the design, digital prototype that is very similar to the real physical system (Zawadzki and Żywicki, 2016). Virtualization helps in checking and modifying the design, and the virtual product needs to be tested before making a physical system. Virtualization is also helpful in assisting and training the workforce, assisting in diagnosing, and predicting and guiding maintenance workers to fix any malfunction.

v. Service Orientation

Service orientation will transform the manufacturing system to sell the combination of products and services. Fischer et al. (2012) highlighted that in the smart factory system, the organization should focus on selling the services rather than concentrate on profit from product sale. Product and services need to be integrated and sold together.

vi. Real-Time Capability

It is the responsiveness of the system during changes immediately. The system should respond to customer requirements and can be analyzed in real time. In this type of system, the current situation in the enterprises is permanently and continuously monitored, and any deviation or machine failure is taken accordingly without wasting any productive time.

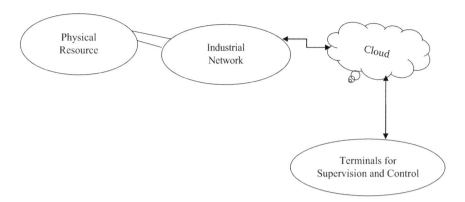

FIGURE 8.4   Basic structure of smart factory for industry 4.0.

8.3.2   SMART FACTORY ARCHITECTURE

The smart factory concept is going to bring changes in business models and consumer behavior because of flexible manufacturing, dynamic reconfigurations, and production optimization. It is basically an engineering system concept that focuses on three main parameters, i.e. interconnections, collaboration, and execution. The architecture of a smart factory is important as it is different from the traditional manufacturing industry. Chen et al. (2017) described that architect of smart factory consist of four layers including (Figure 8.4):

1. Network layer
2. Physical resource layer
3. Terminal layer
4. Data application layer

i. Network Layer

The industrial network layer is an important kind of infrastructure that not only connects but also communicates with the physical resource layer and the cloud layer. Industrial wireless networks (IWN) must have high reliability and accuracy with low latency in smart factories. Li et al. (2017) highlighted that latency and reliability are the core requirements of industrial network communication. The network layer has the advantage of improved resource utilization and efficiency. The data-transmission performance also gets improved as it is implemented in different network tasks. Chen et al. (2018) highlighted that with the use of artificial technology and cognitive computing, the learning and cognitive ability of the system improved considerably, as the number of data nodes and volume data increased in the smart factory.

ii. Physical Resource Layer

Wang et al. (2016) explained the physical resource layer consisting of various components such as smart machines, smart products, smart conveyors, etc. These physical devices collaborate and communicate smartly with each other through the network to achieve system goals. Chen et al. (2017) suggested that smart factories can work intelligently and independently with flexible scheduling. The manufacturing unit has equipments like robot, mechanical arm, and machining center, etc., to improve dynamic scheduling of the system. The modular manufacturing unit must coordinate and cooperate with each other to complete the common organizational goal. The production line of the smart factory must be flexible and have the ability to reconfigure the process path. These manufacturing enterprises must be able to manufacture the products with different varieties in small batches. The production line of the smart factory must have a capability of manufacturing different products with variable schedule and scalability as an important basis for the flexible manufacturing in the smart factory. Wireless sensor networks (WSN) are generally used in the smart factory for gathering and monitoring the data. Some of the commonly used WSN are ZigBee, Radio Frequency Identification (RFID), and Bluetooth. These sensors collect data in manufacturing areas since they are having low prices and consume less energy (Choudhury et al., 2015).

iii. Terminal Layer

Chen et al. (2017) discussed the use of the terminal layer in devices like computers, electrical boards, and smart phones. They are distributed in workshops, offices, and other regions. The purpose of the terminal layer is to analyze the result of cloud processing and remotely monitor the operation and maintenance of the system. Wang et al. (2016) elaborated the basic structure of the smart factory concept in the Industry 4.0 context.

iv. Data Application Layer

The smart factory generates a lot of data, and thus, application of big data is developing fast, along with the cloud platform in the smart factory. Big data in smart factories have data generated in real time from sensors, machine logs, and manufacturing. (He et al., 2018; Sharp et al., 2018) highlighted that the machine-learning application in smart manufacturing is rapidly increasing with the advancement of big data, especially in the field of extraction of knowledge, decision-support system, and lifecycle management of product and equipment.

Basically, the terminal layer connects the people to the smart factory. Chen et. al. (2017) highlighted the three layers, i.e. network layer, physical resource layer, data application layer, and the terminal layer are the components of smart factory architecture. Therefore, it is essential that all these components must be connected with each other, exchange relevant information, and integrate the physical system with the cyber system.

8.3.3   System Integration in Industry 4.0

The main purpose of Industry 4.0 is to utilize upcoming latest technologies to implement the Internet of Things (IOT) and other services to the business processes.

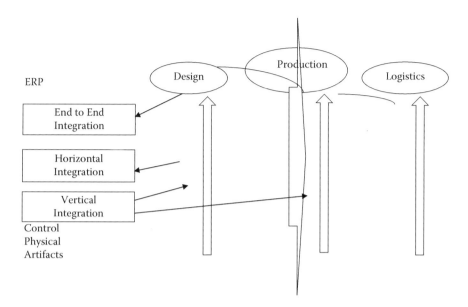

FIGURE 8.5 Depicts the three integration of smart factory.

These processes are connected with each other and work in a flexible and efficient manner to produce products having better quality and low cost. Wang et al. (2016) described that there are three kinds of integration of Industry 4.0, which are essential for making a smart factory.

1. Vertical Integration
2. Horizontal Integration
3. End-to-End Integration

The three types of Integration were illustrated by Wang et al. (2016). (Figure 8.5).

i. Vertical Integration

Smart factories have many physical systems, such as sensors, actuators, and various control systems. Vertical integration is essential to integrate various physical systems across different levels so as to have a flexible manufacturing system. The main objective of integration is to collect the massive information and thereby make the production process transparent and flexible.

ii. Horizontal Integration

Wang et al. (2016) discussed that one corporation should act together with other related corporations. By horizontal integration, the information related to material and finance can be exchanged among corporations, thereby producing an efficient ecosystem.

iii. End-to-End Integration

These are a number of activities that are involved, like design and development of product, planning of production, services, and maintenance. The integration will help in the development of a consistent product model that can be revised at every level. By integration, it is very easy to assess the impact of product design on production systems and services using software so that customized products can be manufactured at a low cost with high quality.

The emerging technologies like cloud computing, Internet of things, artificial intelligence, and big data are the backbone of Industry 4.0. With the integration of these technologies with automation, businesses can bring a considerable improvement to the industry. Through the use of artificial intelligence and microprocessors, the products and machines become smart, and these are connected with each other as well as with the internet. The IoT and services is the main foundation of all three kinds of integration. In vertical integration, the physical artifacts can be reconfigured in a dynamic way, generate a lot of data, and then added on to the cloud. Software tools are used for integration.

These emerging technologies are employed and integrated in industry across all the three verticals. The integration can help in producing high-quality products.

8.4   CHALLENGES FOR IMPLEMENTING SMART FACTORY

Various authors highlighted many issues for implementation of the smart factory concept in Industry 4.0. Some of the common challenges faced in implementing smart factory are:

8.4.1   Employee Resistance

Resistance to change is the biggest challenge in implementing smart factory. Most of the staff and other employees are not certain about the advantages and benefits of smart factory. They have a fear of losing their job. Sjödin et al. (2018) highlighted the issue of factory staff lacking common vision for understanding and implementing smart factories.

8.4.2   High Cost

The smart factory involves usage of the latest technologies. Park (2016) emphasized the horizontal and vertical integration of the manufacturing system, thus producing customized production. The smart factory must require customized machine tools and workstations, highly skilled workforce, real-time monitoring, and control. All these equipment and elements require huge investments, thus increasing the cost of implementing the smart factory.

8.4.3   Technical Issues

Many researches highlighted the different technical issues that need to be addressed to implement the concept of smart factory. Wang et al. (2017) discussed that communication and high-speed bandwidth networks are the main requirements in implementing smart factories. Chen et al. (2017) highlighted the issue of generation of large volumes of unstructured and real-time data in the smart factory. This data cannot be utilized directly for better performance. So, (Song et al., 2017; Wang et al., 2017) emphasized the need of huge storage space, better collection, and quick analysis of data along with the high bandwidth as the major requirement of producing high-quality products and the efficient working in the smart factory. There is a lack of funds and resources in the traditional organization, especially in small and medium-size enterprises.

8.4.4 Security Issues

Wan et al. stressed that smart factories are at higher risk compared to traditional enterprises. This is because of limitations of proper architecture in smart factories and massive amounts of critical data and equipment are vulnerable to cyber attacks. RÜßMANN et al. (2015) addressed that the critical industrial systems and manufacturing lines need to be protected from cyber threats.

8.4.5 Integration Issues

Smart factory requires the integration of various technologies. Sjödin et al. (2018) highlighted that the traditional work processes need to be digitized. All the verticals of the business enterprise from product design and development, manufacturing, marketing, sales and supply chain need to be integrated.

8.5 REQUIREMENT OF SMART FACTORY

Mabkhot et al. (2018) highlighted 26 factors that are the major requirement of smart factories. These factors mainly included tools, equipments, infrastructural facilities, modern architecture, and skilled workforce. Besides these, the need for online data analysis and monitoring, cloud computing, cloud connection were also emphasized.

8.5.1 Relevance of Smart Factory with respect to Green Engineering and Sustainability

Smart factory, or intelligent factory, is the future of manufacturing in Industry 4.0. In manufacturing, green engineering resulted in the reduction of waste in the overall supply chain, like selection of material, manufacturing process, and logistics. In green engineering, there is a better usage of energy, raw material, and other resources. The smart factory concept focuses on 3R, i.e. reuse, recycle, and reduce. The green manufacturing in the digital factory concept of Industry 4.0 uses appropriate technologies and materials and focuses on modern production processes to reduce pollution and minimization of waste. The concept of smart factory is being adopted in various industries. Few case studies are discussed in detail:

8.5.1.1   Case Study 1: Petrochemical Industry

Li (2016) discussed the smart factory perspective in the petrochemical industry. The petrochemical Industry is facing new challenges because of rules regarding the environmental protection and conserving energy, cost reduction, and efficient management of the supply chain of various raw materials.

A smart factory concept in the petrochemical industry is characterized by greenness, high efficiency, high safety, better management of supply chain, and in-depth integration of industrialization to achieve the high operational excellence of the factory. Smart factory concept in petrochemical industry is based on the following points:

1. Integrating management and production verticals for the refining and chemical industries
2. Integrating the supply chain of the petrochemical industry
3. Integrating the design and operation for the petrochemical industry

TABLE 8.2

Strategic Framework of Petrochemical Industry (Li, 2016)

Sl. No.	Framework	Content
1.	Objective	Operational excellence achieved
2.	Three Main Integration	• Integrating the management and production. • Integration of supply chain • Integrating the design and operation
3.	Key Capabilities	• Forecast and early warning • Optimization • Location awareness • Scientific decisions
4.	Characteristics	• Digitization • Automation • Visualization • Modeling • Integration
5.	Business Domain	• Supply chain management • Management of production and control • Energy management • Assest management • Decision making • Health safety and environment management

The intelligent and smart petrochemical industry features integration of human and computer, and thus, improving the decision-making ability of the petrochemical industry. The various trends in the field are:

1. High dimensional and heterogeneous manufacturing and massive information.
2. 3-D display and simulation of design, production, and operations based on virtual reality.
3. Human-machine collaborative decision-making assistive technology in smart manufacturing.
4. Utilization of human computer integrating techniques supporting visual analysis in production and manufacturing environments.

8.5.1.2 Case Study 2: Fabrication of Glass Reinforced Polymeric Centrifuged Pipes

Aversa et al. (2016) presented the case study of fabrication of glass-reinforced polymeric (GRP) centrifuged pipes using the experimental verification of intelligent processes. The centrifuged pipe manufacturing must maintain quality control and smooth wall thickness. Any variation from the said parameters leads to increased deflection, distortion, and buckling of the pipe and chance of joint leakage. These pipes are generally used in water and sewage transport. Mathematical models were used to analyze the thermo-kinetic and chemorheological features of the polymerization of unsaturated polymers. The study used an electronically controlled automatic centrifugal casting system. The process controlled the various parameters like material quantities, mould rotation speed, and temperature to achieve the desired result.

8.5.1.3 Case Study 3: Management of Workshop of Aircraft Jet Engine Production

Shpilevoy et al. (2013) presented the design principle of smart factory for resource management of aircraft jet engine production in real time. Digital factory agent is employed to enhance the efficiency and productivity of the factory by automating scheduling, forecasting, supply chain, and control of the manufacturing in machine assembly. Resources of the workshop adopted the dynamic scheduling on the basis of multi-agent technology, architecture, and user-interface employed. The traditional resources optimizations have a large number of limitations like interdependency of orders and resources, issues in adaptive planning in real time, lack of adjustment of decisions. All these issues reduce the efficiency and productivity of the existing process and system. However, the adaptive method developed for synchronous scheduling is built on the multi-agent technology that includes the equipment, product specification, and the technological processes as well as skill set of the workers. Productivity of the aircraft engine workshop increased by 10%–15%. The work on task allocation, coordination, scheduling, coordination, and monitoring was reduced by 3%–4%. The efficiency of resources increased to more than 15%. The adopted technology in the workshop reduced response time by 2%–3%.

8.6 CONCLUSION

The chapter discussed in detail the history and inventions occurred in various industrial revolutions. The Industry 4.0 perspective is also discussed, where disruption is happening in every industry and the new technologies, like IoT, artificial intelligence, robotics, and 3D printing, are impacting the complete business chain and governance of the enterprise. The chapter highlighted the concept of smart factory with the perspective of green engineering, which is a main concept of Industry 4.0 where all the machines and equipments are interconnected and exchanging information. The smart factory is considered an intelligent system because it integrates the services and manufacturing. The chapter elaborated in details the challenges and requirements of the smart factory by discussing the various case studies in different sectors.

REFERENCES

Agarwal, H, and Agarwal, R 2017. First industrial revolution and second industrial revolution: Technological differences and the differences in banking and financing of the firms. *Saudi Journal of Humanities and Social Sciences*, 2(11): 1062–1066.

Ashton, TS 1948. The Industrial Revolution (1760–1830). London and New York: Oxford University Press.

Atzori, L, Iera, A, and Morabito, G 2010. The internet of things: A survey. *Computer Networks*, 54(15): 2787–2805.

Aversa, R, Petrescu, RV, Petrescu, FI, and Apicella, A 2016. Smart-factory: Optimization and process control of composite centrifuged pipes. *American Journal of Applied Sciences*, 13(11): 1330–1341.

Boyd, D, and Crawford, K 2012. Critical questions for big data in information. *Communication and Society*, 15(5): 662–679.

Chen, M, Herrera, F, and Hwang, K 2018. Cognitive computing: Architecture, technologies and intelligent applications. *IEEE Access*, 6, 19774–19783.

Chen, B, Wan, J, Shu, L, Li, P, Mukherjee, M, and Yin, B 2017. Smart factory of industry 4.0: Key technologies, application case, and challenges. *IEEE Access*, 6: 6505–6519.

Choudhury, S, Kuchhal, P, and Singh, R 2015. ZigBee and bluetooth network based sensory data acquisition system. *Procedia Computer Science*, 48: 367–372.

Dorsemaine, B, Gaulier, JP, Wary, JP, Kheir, N, and Urien, P 2015. Internet of things: A definition & taxonomy. *Ninth International Conference on Next Generation Mobile Applications, Services and Technologies*, Cambridge, UK.

Ergunova, OT, Lizunkov, VG, Malushko, EY, Marchuk, VI, and Ignatenko, AY 2017. Forming system of strategic innovation management at high-tech engineering enterprises. *IOP Conference Series: Materials Science and Engineering*, 177(1): 012046. IOP Publishing.

Fischer, T, Gebauer, H, and Fleisch, E 2012. *Service business development: Strategies for value creation in manufacturing firms*. Cambridge, U. K.: Cambridge University Press.

Gubbi, J, Buyya, R, Marusic, S, and Palaniswami, M 2013. Internet of things (IoT): A vision, architectural elements, and future directions. *Future Generation Computer Systems*, 29(7): 1645–1660.

Gunes, V, Peter, S, Givargis, T, and Vahid, F (2014). A survey on concepts, applications, and challenges in cyber-physical systems. *KSII Transactions on Internet & Information Systems*, 8(12): 4242–4268.

Gurjanov, AV, Zakoldaev, DA, Shukalov, AV, and Zharinov, IO 2019. Formation principles of digital twins of Cyber-Physical Systems in the smart factories of Industry 4.0. *IOP Conference Series: Materials Science and Engineering*, 483(1): 012070.

Hans-Christian, Pfohl, Burak, Yahsi, and Ta 2015. The Impact of Industry 4.0 on the supply chain. *Proceedings of the Hamburg Inter Innovations and Strategies for Logistics*, 20: 30–58.

He, X, Wang, K, Huang, H, Miyazaki, T, Wang, Y, and Guo, S 2018. Green resource allocation based on deep reinforcement learning in content-centric IoT. *IEEE Transactions on Emerging Topics in Computing*, 8(3): 781–796.

Hermann, M, Pentek, T, and Otto, B 2016. Design principles for industrie 4.0 scenarios. In *2016 49th Hawaii International Conference on System Sciences (HICSS)* (pp. 3928–3937). IEEE.

Kagermann, H, Wahlster, W, and Helbig, J 2013. *Recommendations for implementing the strategic initiative Industrie 4.0: Final report of the Industrie 4.0 Working Group.* Berlin: Forschungsunion.

Katal, A 2013. Big data: Issues, challenges, tools and good practices. *Challenges, Tools and Good Practices*, 1: 404.

Koch, V, Kuge, S, Geissbauer, R, and Schrauf, S 2014. *Industry 4.0: Opportunities and challenges of the industrial internet. Tech. Rep. TR 2014-2*, New York: PWC Strategy GmbH, United States.

Lee, A 2006. Cyberphysical systems-are computing foundations adequate, in: *Position Paper for NSF Workshop on Cyber Physical Systems: Research Motivation, Techniques and Roadmap, Austin, Texas, USA*, 2: 19.

Lee, J 2015. Smart factory systems. *Informatik-Spektrum*, 38(3), 230–235.

Lee, EA, and Seshia, SA 2011. *Introduction to embedded systems. A cyber –Physical systems approach* (pp. 502). California: Berkeley University of California.

Li, D 2016. Perspective for smart factory in petrochemical industry. *Computers & Chemical Engineering*, 91: 136–148.

Li, X, Li, D, Wan, J, Vasilakos, AV, Lai, CF, and Wang, S 2017. A review of industrial wireless networks in the context of industry 4.0. *Wireless Networks, 23*(1): 23–41.

Lu, Y 2018. Blockchain and the related issues: A review of current research topics. *Journal of Management Analytics*, 5(4): 231–255. 10.1080/23270012.2018.1516523

Luo, F, Dong, ZY, Chen, Y, Xu, Y, Meng, K, and Wong, KP 2012. Hybrid cloud computing platform: The next generation IT backbone for smart grid. In *2012 IEEE Power and Energy Society General Meeting* (pp. 1–7). IEEE.

Mabkhot, MM, Al-Ahmari, AM, Salah, B, and Alkhalefah, H 2018. Requirements of the smart factory system: A survey and perspective. *Machines*, 6(2): 23.

Markillie, P 2012. A third industrial revolution. Economist, Special Report.

Nieuwenhuize, G 2016. Smart manufacturing for Dutch SMEs: Why and how. *Rotterdam School of Management–Erasmus University.*

Park, S 2016. Development of innovative strategies for the Korean manufacturing industry by use of the Connected Smart Factory (CSF). *Procedia Computer Science*, 91: 744–750. 10.1016/j.procs.2016.07.067

Poovendran, R 2010. Cyber–physical systems: Close encounters between two parallel worlds [point of view]. *Proceedings of the IEEE*, 98(8): 1363–1366.

Radziwon, A, Bilberg, A, Bogers, M, and Madsen, ES 2014. The smart factory: exploring adaptive and flexible manufacturing solutions. *Procedia Engineering*, 69: 1184–1190.

Rivkin, J 2011. *The third industrial revolution*. New York: New York Times.

Rüßmann, M, Lorenz, M, Gerbert, P, Waldner, M, Justus, J, Engel, P, and Harnisch, M. 2017. Industry 4.0: The Future of Productivity and Growth in Manufacturing Industries. [online] www.bcgperspectives.com. Available at: https://www.bcgperspectives.com/content/articles/engineered_products_project_business_industry_40_future_productivity_growth_manufacturing_industries

Rüßmann, M., Lorenz, M., Gerbert, P., Waldner, M., Justus, J., Engel, P., & Harnisch, M. (2015). Industry 4.0: The future of productivity and growth in manufacturing industries. Boston consulting group, 9(1), 54–89.

Sanders, A, Elangeswaran, C, and Wulfsberg, JP, 2016. Industry 4.0 implies lean manufacturing: Research activities in industry 4.0 function as enablers for lean manufacturing. *Journal of Industrial Engineering and Management (JIEM)*, 9(3): 811–833.

Sethi, P, and Sarangi, SR 2017. Internet of things: Architectures, protocols, and applications. *Journal of Electrical and Computer Engineering.*

Shariat Zadeh, N, Lundholm, T, Lindberg, L, and Franzén Sivard, G 2016. Integration of digital factory with smart factory based on Internet of Things. In *26th CIRP Design Conference, 2016, KTH Royal Institute of Technology Stockholm*, Sweden.

Sharp, M, Ak, R, and Hedberg, T, Jr. 2018. A survey of the advancing use and development of machine learning in smart manufacturing. *Journal* of *Manufacturing Systems*, 49: 170–179. 10.1016/j.jmsy.2018.02.004

Shi, Z, Xie, Y, Xue, W, Chen, Y, Fu, L, and Xu, X 2020. Smart factory in Industry 4.0. *Systems Research and Behavioral Science*, 37(4): 607–617.

Shi, J, Wan, J, Yan, H, and Suo, H 2011. A survey of cyberphysical systems. *International Conference on Wireless Communications and Signal Processing (WCSP)* (p. 16).

Shpilevoy, V, Shishov, A, Skobelev, P, Kolbova, E, Kazanskaia, D, Shepilov, Y, and Tsarev, A 2013. Multi-agent system "Smart Factory" for real-time workshop management in aircraft jet engines production. *IFAC Proceedings*, 46(7): 204–209.

Sjödin, DR, Parida, V, Leksell, M, and Petrovic, A 2018. Smart factory implementation and process innovation: A preliminary maturity model for leveraging digitalization in manufacturing moving to smart factories presents specific challenges that can be addressed through a structured approach focused on people, processes, and technologies. *Research-Technology Management*, 61(5): 22–31.

Song, YM, and Lee, CS 2015. A study on the big data analysis system for searching of the flooded road areas. *Journal of Korea Multimedia Society*, 18(8): 925–934.

Song, Z, Sun, Y, Wan, J, and Liang, P 2017. Data quality management for service-oriented manufacturing cyber-physical systems. *Computers and Electrical Engineering*, 64: 34–44.

Tantik, E, and Anderl, R 2017. Integrated data model and structure for the asset administration shell in industrie 4.0. *Procedia Cirp*, 60: 86–91.

Trappey, AJ, Trappey, CV, Govindarajan, UH, Chuang, AC, and Sun, JJ 2017. A review of essential standards and patent landscapes for the Internet of Things: A key enabler for Industry 4.0. *Advanced Engineering Informatics*, 33: 208–229.

UK Government Office for Science. 2014. The internet of things: Making the most of the second digital revolution.

Wagire, AA, Rathore, APS, and Jain, R 2019. Analysis and synthesis of Industry 4.0 research landscape. *Journal of Manufacturing Technology Management*, 31: 31–51.

Wahlster, W 2014. *Cyber-Physical AI Systems for Resource-Efficient Living*, The way ahead new directions at FBK ICT IRST encounter outstanding visions from across the eld trento. 49: 681.

Wang, S, Ouyang, J, Li, D, and Liu, C 2017. An integrated industrial ethernet solution for the implementation of smart factory. *IEEE Access*, 5: 25455–25462.

Wang, S, Wan, J, Li, D, and Zhang, C 2016. Implementing smart factory of industrie 4.0: An outlook. *International Journal of Distributed Sensor Networks*, 12(1): 3159805.

Weyer, S, Schmitt, M, Ohmer, M, and Gorecky, D 2015. Towards industry 4.0-Standardization as the crucial challenge for highly modular, multi-vendor production systems. *IFAC-Papersonline*, 48(3): 579–584.

Wikipedia. 2020. Cyber Physical-System. https://en.wikipedia.org/wiki/Cyber-physical_system

Wolf, W 2007. Embedded computing – The good news and the bad news. *Computer-IEEE Computer Magazine*, 40(11): 104–105.

Xu, L, Cai, L, Zhao, S, and Ge, B 2016. Editorial: Inaugural issue. *Journal of Industrial Integration and Management*, 1(1): 1601001. 10.1142/S2424862216010016.

Xu, LD, He, W, and Li, S 2014. Internet of things in industries: A survey. *IEEE Transactions on Industrial Informatics*, 10(4): 2233–2243.

Yin, S, and Kaynak, O 2015. Big data for modern industry: Challenges and trends [point of view]. *Proceedings of the IEEE*, 103(2): 143–146.

Yu, W, Liang, F, He, X, Hatcher, WG, Lu, C, Lin, J, and Yang, X 2017. A survey on the edge computing for the Internet of Things. *IEEE Access*, 6: 6900–6919.

Zawadzki, P, and Żywicki, K 2016. Smart product design and production control for effective mass customization in the Industry 4.0 concept. *Management and Production Engineering Review*, 7.

Zuehlke, D 2010. Smart factory—Towards a factory-of-things. *Annual Reviews in Control*, 34(1): 129–138. 10.1016/j.arcontrol.2010.02.008

9 Electric Vehicle Research: Need, Opportunities, and Challenges

Indu K. and Aswatha Kumar M.

Department of Electronics and Communication Engineering,
CHRIST (Deemed to be University), Bangalore, Karnataka,
India

CONTENTS

9.1 INTRODUCTION TO ELECTRIC VEHICLE

As highlighted in the book series title "Green Engineering and Technology," the relevance of electric vehicle research has a compelling demand for future generations. Even though there is general awareness and evidence that electric vehicle technology can deliver significant economic, environmental, and health benefits, often there are

DOI: 10.1201/9781003127819-9

debates and discussions on whether electric vehicles are really to be considered as green enough. Mostly these are consequences of misinformation or incorrect observations of extending arguments related to thermal-power generations.

This chapter recommends listing the relevance of electric vehicle research by concentrating more on sustainable development aspect. In the automobile field, this sustainability can be accomplished vitally through environment friendly vehicles or, more specifically, through electric vehicles. The subject of electric vehicles is more relevant today than any time in history. This chapter on electric vehicles will focus on aiding students around the globe to apprehend the current trends, opportunities, and challenges in electric vehicle research.

9.1.1 OVERVIEW ON ELECTRIC VEHICLE

Electric vehicles (EVs) are slowly gaining significance in the automobile industry for various reasons when compared to an internal combustion engine (ICE) vehicle. Benefits of EVs include their lower maintenance needs, fewer parts, reduced carbon footprint, and zero-emission features, to name a few. Research helps in identifying the unknown facts and information related to the field of interest in a larger perspective. Essentially, EV research spans across a variety of areas due to its multidisciplinary nature. This is due to the work and involvement of prime engineering and science domains. The research focuses may vary from working on motor and battery characteristics, battery materials, charging infrastructure, energy-management systems, controller design, EV design & development, and vehicle control. But, before indulging into these areas, it's always beneficial to begin with a comparison, as stated in Table 9.1, between EV & ICE for catering to the doubts in any beginner researcher.

Fully electric EV powers up the electric motor through the rechargeable battery. The number of moving parts in a battery-powered EV is less than 20 when compared to 2000+ components (Drive Electric, 2020) in a conventional ICE. Main components in a battery-powered EV are depicted in Figure 9.1. Functionalities of each of these components (EV Reporter, 2019) are detailed in the following section of this article.

TABLE 9.1

Comparison between EV & ICE

Parameters	Electric Vehicles	Internal Combustion Engines
Input	Plug in port – electricity	Fuel – petrol/diesel
Storage	Batteries	Fuel tank
Conversion	Electric motor	Internal combustion engine
Transfer	Gear box with fixed ratio	Gear box with varying speeds
Refill	Charging (at home or at charging stations)	Refuel the tank at petrol/diesel pumps

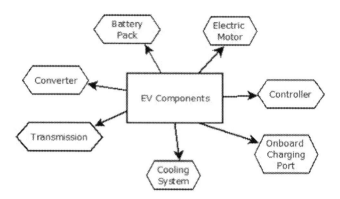

FIGURE 9.1 Major components in an electric vehicle.

1. **Battery Pack:** Batteries are the vital power source for an EV. Most commonly used batteries are lithium-ion batteries. The unit for measuring the battery capacity is Kilo-Watt hour (kWh). The smallest unit of lithium-ion battery (power source) can be defined as a "cell." Modules are few cells when arranged or connected together. A normal battery pack will consist of several modules connected in series or parallel used as the main energy source to drive the electric motor.

2. **Electric Motor:** The motor converts the electrical energy to mechanical energy, and then this energy is transmitted to the wheels via shaft for the vehicle to start. Motors that are capable of delivering the required torque to the wheels are known as traction motors. When compared to an ICE, EVs have lesser noise and other vibrations due to the use of motors. Different types of motors used in an EV include DC motor, 3-phase induction motor (IM), brushless motors, permanent magnet synchronous motors (PMSM) and switched reluctance motors (SRM).

3. **Controller:** This works to monitor and control the motor and battery by regulating the energy flow. These controllers are even capable of controlling the speed and acceleration of the EV according to the situation. They can be used in monitoring and controlling the dynamicity of EV, torque control, braking control, optimisation, and energy-management purposes as well. Also known as control unit.

4. **Onboard Charging Port:** Converts alternating current (AC) to direct current (DC) (Hyundai Motor Group Newsroom, 2020) while the vehicle is in charging mode. However, these are not used in fast-charging mode since DC is directly supplied to batteries in the fast-charging mode. Different fast-charger ports ar available within the EV for fast charging.

5. **Converter:** For the smooth functioning of different electronic sub-systems in EV, high-voltage electricity from the battery needs to be converted to a lower voltage (acceptable by the various sub-systems). This functionality is normally performed by a converter.

6. **Transmission:** This helps to transfer the generated motor power to the wheels for driving the EV. Different types of transmission systems normally seen in a production vehicle include the manual, automatic, and continuously varying, as well as infinitely varying transmission systems.
7. **Cooling System:** This system helps to constantly maintain the temperature of the inbuilt components of the EV. These systems help to maintain the battery's operating temperature within the limits. While discharging, the battery generates heat; if this heat is not reduced, temperature within battery increases beyond its threshold. If the battery temperature varies uncontrollably, the battery will stop working; hence, cooling systems are required for preventing such unexpected situations.

In general, EVs can be of different types (Ergon Energy, 2020), namely hybrid electric vehicle (HEV), plug-in HEV (PHEV), and battery electric vehicle (BEV) and fuel cell electric vehicle (FCEV). HEVs use the electric motor to start the vehicle and then switch to engine mode for increasing speed; they incorporate the load variations as per the speed profile. Example for HEV is Toyota Camry Hybrid. In case of PHEV, the vehicle can be powered by both petrol and electricity since it has an ICE alongside the electric powertrain setup within the vehicle, which works by switching modes. An example is Mitsubishi Outlander PHEV. FCEV uses fuel cell technology (Community Environmental Council, 2020) to run the EV. Examples are Hyundai Tucson FCEV, Toyota Mirai, and Honda Clarity Fuel Cell.

Further to the types & different components in an EV, let's understand the main specifications to look through while purchasing an BEV (from an Indian vehicle market point of view). Table 9.2 gives a brief comparison between EVs currently popular and available in the Indian market. State of charge (SOC) determines the amount of charge left in the battery when used after being charged to its full capacity. It's normally specified in terms of percentage. Regular charging (RC) refers to the slow-charging method, which can be carried out at home or the office with the normal electric socket available. This method may take 6–12 h approximately, depending on the battery capacity. The fast-charging (FC) method is practiced at the charging stations or outlets where charging occurs at a faster rate. With higher charger rating, this type of charging normally takes anywhere from 0.5 to 1 h approximately.

9.1.2 Organisation of the Chapter

This chapter explores and covers the fundamental facts and figures every researcher working in this field needs to be aware of. The introduction section is a brief overview on EV, with a simple Indian EV market study detailed for creating inquisitiveness in the reader. In the upcoming section of the chapter, need and relevance of EV is highlighted in detail. Moving further, the sustainability aspect in EV, with special emphasis on the sustainable development goals, are looked into. In Section 9.4, modelling aspects of EV design are elaborated from an embedded-system perspective. Few concepts relating embedded systems and EV are discussed in detail. Main modelling blocks of EV design, followed by testing approaches and

TABLE 9.2
EV Models in Indian Market

Vehicle Brand Model	Motor (Type)	Battery Capacity (kWh)	Driving Range on Full Charge (km)	Power (kW)/ Torque (nm)	Charging Time (hours) Regular Charging – RC Fast Charging – FC
Mahindra e-Verito	3 Phase IM	21.2	181	31/91	RC: 0%–100% SOC in 11.5 h FC: 0%–80% SOC in 1.5 h
Tata Tigor EV	3 Phase IM	21.5	213	30/105	RC: 0%–100% SOC in 11.5 h FC: 0%–80% SOC in 2 h
Tata Nexon EV	PMSM	30.2	312	129/245	RC:10%–90% SOC in 8.5 h FC: 0%–80% SOC in 1 h
Hyundai Kona EV	PMSM	39.2	452	100/394.9	RC: 0%–100% SOC in 6.16 h FC: 0%–80% SOC in 0.95 h (57 min)
MG ZS EV	PMSM	44.5	340	104.9/353	RC: 0%–100% SOC in 6.5 h FC: 0%–80% SOC in 0.66 h (40 min)

the concept of hardware in loop, are also discussed. In the last section, safety aspects related to EV design are explained.

9.2 NEED AND RELEVANCE ASPECTS

9.2.1 NEED FOR ELECTRIC VEHICLES

The main source of power in an EV is the electric motor, driven through the battery. In an EV, unlike ICE, dependency (Energy Sage, 2016) on the fossil fuels is lesser. Continuous visits to refuel the tank can be reduced; ever-increasing crude oil prices urge the public to look for effective alternatives. EVs are a greener option since burning of fossil fuels is reduced, lowering the tailpipe emissions and leading to cleaner air to breathe. Advantages have been illustrated in the Figure 9.2.

Due to fewer components when compared to an ICE, noise and vibrations within the EV are reduced significantly. EVs are easy to maintain due to the reduced wear and tear of parts. Due to independency of non-renewables, tailpipe emissions are almost zero. Both fast-charging and slow-charging plugs are available within the EV. They can be charged even from homes, thereby increasing flexibility in refuelling. It is interesting to note that, additionally in EVs, the energy-conversion efficiency is 77% (U.S. Department of Energy, 2020a) when compared to an ICE of 12%–30%. This indicates that as 77% of the energy generated for driving using an EV is transferred to the wheels through the grids, charging infrastructure and powertrain drive systems. In case of ICE, the conversion efficiency is quite low due to various losses that occur while transferring the fuel. More specifically, it means, between the extraction of crude oil, its related processes, reaching fuel stations through pipelines, fuel tank of the vehicle, and then finally to drive the wheels, several losses occur, thereby reducing the efficiency.

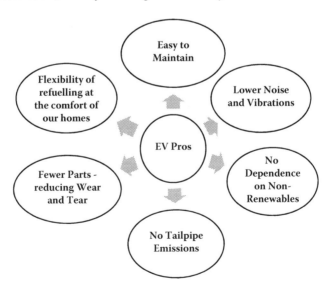

FIGURE 9.2   EV advantages.

In brief, the need for EV can be justified due to the following aspects:

1. High cost of fossil fuels reflects directly on all transportation needs, purchase price of all commodities, and thereby overall cost of living.
2. Overall, two-thirds of petroleum produced in the world is used for transportation. Lack of diversity in the fuel choice for transportation is a threat due to vulnerability on price variations and supply limitations.
3. Ecological damage due to emissions impacts climate change, smog, public health, and all living beings on the planet.
4. Utilisation of renewable energy like solar, wind, and other sources are becoming much more feasible now compared to the past due to technological innovations and developments in recent years.
5. Use of renewable energy sources are highly beneficial for ecosystem of the planet compared to exceedingly high dependency on fossil fuels for transportation around the globe.
6. Better monitoring, instrumentation, navigation, guidance, control, communication, and entertainment are feasible today at lower energy demands due to improved products and embedded systems for almost all of these subsystems in recent years, continuing for the better in the future as well.

Perhaps the only disadvantage against EV so far are the limitations in overall range due to constraints on battery and lack of charging facilities. The inadequate charging facilities and the higher cost of batteries lead to costlier EVs. However, these limitations have gradually been minimized due to technological breakthroughs in battery systems and renewable energy-related products in the global market.

9.2.2 Relevance of EV Research

The transportation sector globally contributes heavily to the emissions. Among the different modes of transport, road vehicles contribute tremendously due to the burning of fossil fuels. When it comes to vehicle emission, main classification as per (U.S. Department of Energy, 2020b) is as shown in Figure 9.3. Emissions are considered direct or due to the lifecycle involving the vehicle production process, processing stage, distribution and utilisation, recycling of the parts, or disposal, in some cases. Due to fewer components in an EV, the lifecycle-based emissions are reduced considerably. Furthermore, in an EV, direct harmful emissions can be reduced significantly since there are no tailpipe emissions in them. Tailpipe emissions from vehicles are one of the main reasons for the increase in carbon dioxide emissions globally.

Slow-moving traffic and congested roads in cities add to the cause. Annual CO_2 emissions in India contribute to around 7% (Ritchie & Roser, 2020) when compared to the world average. Figure 9.4 depicts the annual CO_2 emissions for the years 1998–2018, as retrieved from the world CO_2 emission database.

From the figure, it can be perceived that CO_2 emissions measured in metric tons have been increasing in the Indian subcontinent. As seen from the data, in 1998, nearly 900 metric tons of CO_2 emissions were observed. Gradually, these emissions

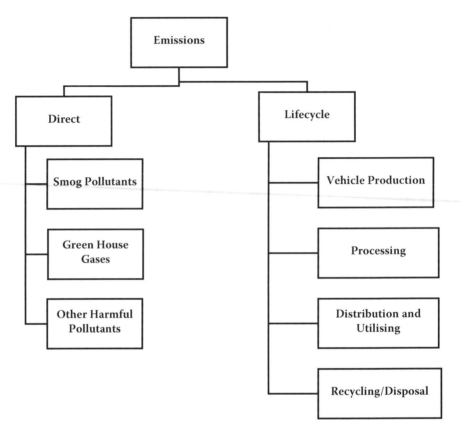

FIGURE 9.3   Types of emissions.

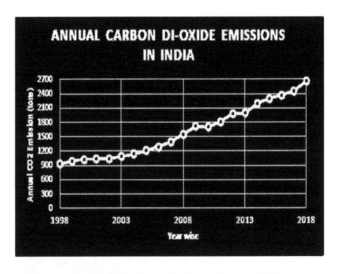

FIGURE 9.4   Annual emissions in India 1 (Ritchie & Roser, 2020).

have increased, recently touching 2700 metric tons in terms of emissions. Burning of fossils fuels (in terms of vehicles) is alone considered in this data collection. Even though electric mobility is being accepted slowly in many developed countries, developing countries may take even longer for a wider acceptance. Many policies and incentives are being announced in developing nations like India to promote EV. The National Electric Mobility Mission Plan 2020 is the latest initiative by the government of India for adopting and promoting the EV culture.

9.3   SUSTAINABLE DEVELOPMENT

9.3.1   Sustainable Development Concepts

Sustainable development aims toward identifying and solving human developmental issues without affecting the natural ecosystem in the region or area. In 2015, all the United Nations member states had agreed upon 17 integrated and independent sustainable development goals (United Nations, 2015) for ensuring peace and prosperity by 2030. These goals are represented in the below Figure 9.5.

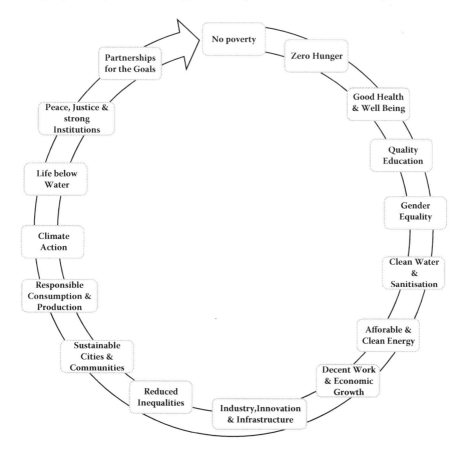

FIGURE 9.5   Sustainable development goals (United Nations, 2015).

EV research, being interdisciplinary in nature, enables researchers to work in line with the direction considering these goals, as well to incorporate sustainability.

9.3.2 ELECTRIC VEHICLE AND SUSTAINABLE DEVELOPMENT

EV research and sustainable-development needs go hand in hand for achieving better social-economic conditions on par with the sustainable-development goals. Let us now look into what EV researchers have contributed toward sustainability, and their views and observations in the following section of this article. Authors in (Faria et. al., 2012) have done well-to-wheel analysis to compare economic and environmental aspects between ICE, EV, HEV and PHEV. Contrary to this approach, a Monte Carlo technique is used to evaluate the consumer behaviour based on probabilistic distribution (Tran et. al., 2013) to support the energy policy. The work elaborates on the alternative fuel market considering the European Union, United States of America, China, and Japan. However, in another research work, use-phase efficiency of an EV (Casals et. al., 2016) is calculated for different driving scenarios using the Monte Carlo method. Well-to-wheel analysis is normally done in two phases, namely well-to-tank (depicts the analysis from the extraction of fuel until filling the tank) and then tank-to-wheel (filling the tank to the wheels to initiate motion when driven) phases.

Slightly different from the approach discussed until now, work in (Schoch et. al., 2018) deals with interpretations on how extended battery life aids sustainability. Authors have considered the basic quadratic programming model to arrive at an optimal charging strategy. In (Yi and Bauer, 2019), weather-based optimisation technique is introduced to study how does weather conditions impact the driving. The study tries assuming the wind speed and other weather conditions, incorporating dimensional optimisation models for testing the energy-saving aspect. Shared-mobility models (Santos and Georgina, 2018) for enhancing the sustainability are discussed. The GREET approach is named after greenhouse gases and regulated emissions; energy use in transportation (Almeida et. al., 2019) is an assessment model that evaluates lifecycle emissions, air pollutants of BEV with chemical reactions background, and types in varied vehicle segments. In a recent research article, sustainability aspects are detailed using fuzzy TOPSIS approach (Samaie et. al., 2020) from BEV adoption point of view. Latest technological changes in the field of EV battery and motor and charging infrastructure is summarised in article.

9.4 MODELLING, DESIGN AND DEVELOPMENT

This section of the chapter focuses on discussing the concepts of embedded systems. Embedded systems' importance in EVs, automobile applications, different modelling blocks in an EV design phase, researchers' perspectives, testing approaches, and finally, performance evaluation are based on hardware in loop (HIL) concept. Embedded system can be defined as a system capable of performing specific tasks for a particular application. Their applications may vary from consumer electronics, industrial equipment, or machinery to autonomous vehicles. These are done by incorporating microcontroller chips with control and computational capabilities to perform a particular task or a set of tasks. Now, let us look into ways in which embedded-system concepts are associated with vehicles, specifically EVs.

9.4.1 EMBEDDED SYSTEM CONCEPTS IN EV

Any embedded system's capability is evaluated based on, but not restricted to, reliability, response time of a particular application following a cost effective and optimised approach. Power consumption, flexibility of the system, precision, and physical size (Talukdar et al., 2019) of the system also matters in certain situations. In case of vehicles, these are termed as control units (CU)/electronic control units (ECU)/vehicle control unit (VCU). Modern vehicles consist of several ECUs performing specific tasks in the subsystems of the vehicle. They are capable of monitoring and interpreting different situations based on the complexity level at which these are used. They interact with each other through communication protocols as well. These ECUs are capable of handling electrical, mechanical, or even discrete parameters or variables. ECUs in an EV can be broadly classified as shown in the Figure 9.6 below.

It is estimated in a survey (Research And Markets.Com, 2018) that, ECU market is currently [for a period starting from 2019] witnessing a steady and faster growth rate due to increased awareness about its uses and benefits. One of the main reasons for this market drift is due to increased use of automotive electronics in vehicles. Some original equipment manufacturers (OEMs) incorporate the centralised ECU architecture, contrary to the decentralised or multiple-ECU based design in others. Let us further look into the details of these major ECUs:

Motor ECU: Also known as motor controller, this controls the speed and acceleration of the EV. Motor ECU helps in balancing the energy flow battery and motor. Supports in reverse motion of the vehicle by reversing the motor rotation. Additionally, supports the regenerative braking feature in an EV. Earlier controllers had a resistor-based design with increased energy loss at low speeds. However, pulse width modulation techniques are used in currently available motor controllers/motor ECU.

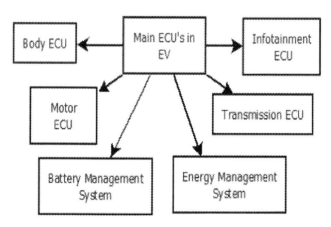

FIGURE 9.6 Major ECU's in EV.

Battery Management System (BMS): This electronic-embedded system consists of controllers, sensors, and actuators to monitor and control the functions of the battery. Individual cell-monitoring and cell-balancing algorithms are inbuilt for continuously analysing the battery behaviour.

Energy Management System: This control system has sensors, controller, and actuator systems capable of monitoring the energy consumption of the EV. Sometimes, this module is combined with the BMS of the vehicle. For improving the range-related features in an EV, energy-generated, transmitted, and regenerated data are very crucial. These systems are installed to obtain the continuous data related to a particular drive cycle analysis.

Transmission ECU: Responses from the sensors, motor controller, and battery controller are monitored and analysed. The speeds are adjusted based on the input from motor unit. The current is drawn from the battery; power and other vital parameters are adjusted for ensuring safe and sound drivability using an EV.

Body ECU: This typically functions as an interlink between different electronic control units within the EV. Communication is enabled through the different networking and communication protocols. It interacts by transmitting and receiving real-time signals from the various ECUs.

Infotainment ECU: All the multimedia and information-related features are incorporated in this ECU for better in-vehicle experience. This ECU enables the vehicle-2-infrastructure communication and connectivity features of modern vehicles. It includes display panels, audio/video inputs and output interfaces, variety of panels & other luxury enhancing features.

In the next subsection, critical modelling blocks that are significant in the EV design and development process are elaborated in detail.

9.4.2　Modelling Blocks of EV Design

During the modelling process, EV components discussed in Section 9.1, can be divided into subsystems. These subsystems can be individually designed and developed as simple embedded systems. EV researchers can choose their field of interest among these subsystems to work further on. This makes EV domain multi-disciplinary in nature. Subsystems in an EV can be classified as depicted by Figure 9.7 for designing purposes:

During the process of EV design, components can be perceived as a part of these individual subsystems, modelled separately, and then integrated together. Several researchers have worked in these areas and are still working toward evolving greener solutions. A few research areas have been discussed in Table 9.3 but are not limited to these alone. After the EV design stage, testing and integration of the subsystems are essential. Different testing approaches are elaborated in the upcoming sections.

9.4.3　Approaches for Testing and Integration

The main attribute of a testing procedure is its adaptiveness. Adaptability of testing procedures means to fit in, for varying complexities that occur time to time

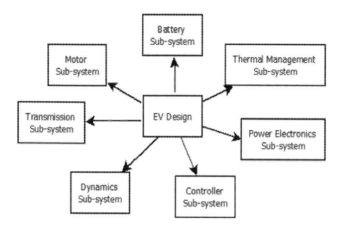

FIGURE 9.7   Subsystems in EV design.

throughout the vehicle development and commercialisation stages. Main complexity drivers of 21st century are shown in Figure 9.8.

As mentioned in the figure, main drivers can vary from big data applications that involve huge volumes of data to automotive electronics applications that comprise mostly of hardware testing.

Different ways of in-loop testing approaches include software-in-loop (SIL), hardware-in-loop (HIL), and vehicle-in-loop (VIL). In the early stage of design and development, SIL, or even known as the model-in-loop approach, can be used to simulate the design via software tools. These are proven to be cost effective and robust since models can be validated by following a step-by-step approach from the initial stages.

In HIL approach, the hardware-software co-design method is used, wherein designs can be simulated through software tools and implemented on the hardware models for better validation. Similar to HIL, when the prototype vehicle is used to validate a model, this testing will be called VIL testing. The effect of HIL testing can be further improved by incorporating different optimisation (Kariem et al., 2020) techniques.

Learning-based testing (Prosvirnova et al., 2013) is mostly used for safety-related applications and can be considered as a part of MIL; it functions on the basis of machine-learning concepts. In the automotive field, the significance of closed-loop controllers is highlighted in the work (Justyna, 2009) due to its improved rejection of disturbances, guaranteed performance, stability and reduced sensitivity over systems, or controllers without proper feedback. Driving simulators (Siddartha et al., 2015) are another method used to test systems. They play a vital role in safety critical applications, where the response time of the system is an essential parameter. Different test and validation methods can be divided as shown in Figure 9.9.

Simulated testing is popularly known as model in loop or software in loop. Prototype testing is also known as a hardware-in-loop approach. Module-wise testing involves application testing done in terms of separate functional modules, finally integrated into a system. Once these modules are integrated into a single system, testing process for verifying the overall behaviour as a system will be called as system testing.

TABLE 9.3
Research Activities

Sub-system	Research Activities
Motor	Drive system design and selection-based studies
	Motion and slip control of motors
	Torque and speed control
	Efficiency analysis and fault diagnosis
	Different motor combinations-based performance analysis
Battery	Battery characterisation studies
	Lithium-ion cell studies
	SoC estimation techniques
	Battery management systems
	Charging infrastructure related studies
	Slow and fast-charging techniques
	Battery storage and swapping techniques
	Diagnostics and prognostics of battery
Dynamics	Vehicle handling
	Stability and optimisation
	Tyres and wheel-based studies
	Suspension
	Yaw moment control
	Fault diagnosis
	Braking and anti-skid control techniques
Power Electronics	Converter design and analysis
	Inverter design and analysis
	Grid-based studies
Transmission	Automatic & manual transmission
	Performance analysis
	Variable transmission-based studies
	Energy consumption and cost effectiveness-based studies
Thermal Management	Temperature monitoring and analysis
	Battery health monitoring
	Cooling system-based studies
	Safety studies related to battery
Controller	Performance analysis based on terrain, gradients, transmission and drive train variants
	Optimisation for battery charge estimations, regenerative braking control, fuel cell-based studies, driver assistance techniques, vehicle stability improvement
	Motor controller designing
	Control system design and development

Testcase generations by manual cross-checking fall under the manual testing methods. Contrary to this method, in an automated approach, the whole testing procedure is automated. However, in the black-box approach, internal program and interface-related information will not be known when compared to the white box testing method.

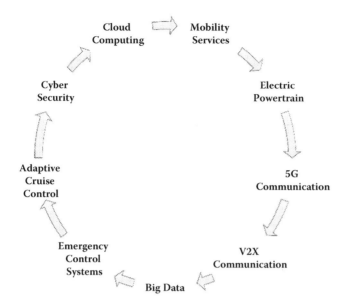

FIGURE 9.8 Complexity drivers in 21st century (Lauber et al., 2018).

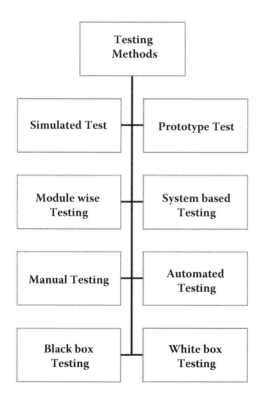

FIGURE 9.9 Different testing methods (Tatar and Mauss, 2014).

Furthermore, test specification and test methodology for embedded systems in automobiles (TEMEA), as discussed in (Grossmann et. al., 2009), tests continuous data streams and real-time behaviour with focus on quality improvement.

9.4.4 PERFORMANCE EVALUATION CONCEPTS (HILs)

When an VCU/ECU is connected to the system under test via the communication network, real-time responses will be monitored and tweaked as per the application and design requirements. Before the HIL evaluation phase, few steps have to be followed, namely:

- Prepare the requirement specification as per the application
- Start modelling on the simulation software tool
- Test the modelling and simulation stage results to know if these are in-line with the requirements specifications
- Configure the simulated model for real time simulation
- Identify a target system, ECU, its allied software's, connectivity with any other systems, input/output interfaces.

Basic workflow of a hardware-in-loop system can be described as shown in Figure 9.10.

Preliminary configuration setup will be initiated within HIL software application; embedded-code generation is the next step, followed by compilation for identifying the errors. Rectify the errors, if any, by crosschecking with the functional requirements until the code is compiled with zero errors. Download the compiled code on the system under test. Execute the same on the system under test. Furthermore, compare results from the simulated model and HIL configured model to estimate the percentage change in simulation time between the two. Evaluate the response time of the application, and finally, keep iterating the process until promising results are achieved for a particular application. Scenarios and manoeuvres can also be tested in the HIL setup in case of automotive applications.

9.5 SAFETY AND ADVANCES FOR FUTURE

9.5.1 SAFETY FEATURES

In any vehicle, safety measures and considerations are vital while considering passengers inside the vehicle or the pedestrians outside the vehicle. In an EV, the roll-over prevention can be improved and better ride quality can be achieved, due to its lower centre of gravity than an ICE (Office of Energy Efficiency and Renewable Energy, 2019). Detailed review on the vehicle control and safety standards for an electric vehicle has been reported in (Indu and Aswatha, 2020). Authors have elaborated on International Organisation for Standardisation (ISO) – ISO 26262, ISO 21488 & Underwriters Laboratories (UL) Standards. Most commonly seen safety features in any modern EV are shown in Figure 9.11.

FIGURE 9.10 Basic HIL workflow.

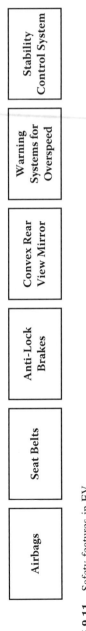

FIGURE 9.11 Safety features in EV.

Safety enhancement for real-time application is another recent area research focus that can be applied. Contrary to the backward-safety enhancement approach discussed in (Xie et al., 2018), authors have worked on forward-safety enhancement approach (Xie et al., 2020) for safety critical applications. In the backward approach, backward recovery from exit tasks to entry tasks will be checked, and ECU will be assigned tasks accordingly. However, in the forward approach, the recovery process happens from the beginning to the end, as per the task allocations in the ECU. Reliability and response time of the system are considered as safety-critical attributes in this study. Additionally, authors have tried to detail about pros and cons of these approaches when used separately or in combination. In the combined repeated backward and forward approach, the ECU, which is free at a particular instant, will be assigned a particular task. Additionally, a stable stopping approach consisting of both backward, forward, and repeated backward and forward techniques, are also elaborated in the same study.

9.5.2 Trends and Advances for Future

Research trends may not be limited to but includes the following in brief:

1. Inter- and intra-vehicle connectivity to enable vehicle-to-infrastructure communications
2. Lightweight vehicle architectures
3. Balancing the active and passive safety features with improved crashworthiness
4. Reducing the number of accidents with sophisticated safety features
5. Battery management systems with improved monitoring capabilities
6. Accurate and reliable state estimations for battery health monitoring
7. Efficient motor-drive configuration with improved vehicle-drive parameters
8. Costeffective in terms of battery pack, internal and external features
9. Scalable computer architectures for ECUs
10. Energy saving solutions with range extensions features
11. Ensuring the automotive software reliability feature
12. Improving automotive software security and privacy features

9.6 CONCLUSION

This chapter, as the title suggests, is an attempt to perceive the need for electric vehicles and discuss the fundamentals involved in electric vehicle research. With EV research, which is interdisciplinary in nature, concept grasping from a root level is required for bringing out innovative ideas in the field. This chapter tries to incorporate the basics of EV research by considering different subsystems, followed by interpretations on electronic control unit design. Relevance of electric vehicle research for a cleaner and greener environment is highlighted in this chapter as well. Different areas of EV research, sustainability aspects, are also elaborated. Furthermore, testing and performance-based evaluation concepts are detailed. Safety aspects in EV research are explained, and the chapter concludes with the research trends for the future.

ACKNOWLEDGEMENTS

Authors would like to convey their sincere gratitude to CHRIST (Deemed to be University) for providing the required support for carrying out the research on electric mobility.

REFERENCES

Almeida, A, Sousa, N, and Coutinho-Rodrigues, J 2019. Quest for Sustainability: Life-Cycle Emissions Assessment of Electric Vehicles Considering Newer Li-Ion Batteries. *Sustainability (Switzerland)*, 11(8), pp. 1–19. doi: 10.3390/su11082366.

Casals Lluc, C, Martinez-Laserna, E, García, BA, and Nieto, N 2016. Sustainability Analysis of the Electric Vehicle Use in Europe for CO2 Emissions Reduction. *Journal of Cleaner Production*, 127, pp. 425–437. Elsevier Ltd. doi: 10.1016/j.jclepro.2016.03.120

Community Environmental Council. Fuel Cell Electric Vehicles. 2020. Online from https://www.cecsb.org/fcev/ retrieved on 02 October 2020.

Drive Electric Incorporated. What is an EV? | Drive Electric. 2020. Online from https://driveelectric.org.nz/individuals/what-is-an-ev/ retrieved on 01 October 2020.

Energy Sage. 2016. Advantages of Electric Vehicles | Energy Sage. Energysage.Com. Online from https://www.energysage.com/electric-vehicles/advantages-of-evs/ retrieved on 01 October 2020.

Ergon Energy. 2020. Types of Electric Cars. Online from https://www.ergon.com.au/network/smarter-energy/electric-vehicles/types-of-electric-vehicles retrieved on 01 October 2020.

EV Reporter. 2019. EV Powertrain Components – Basics. Online from https://evreporter.com/ev-powertrain-components/ retrieved on 22 September 2020.

Faria, R, Moura, P, Delgado, J, and De Almeida, AT 2012. A Sustainability Assessment of Electric Vehicles as a Personal Mobility System. *Energy Conversion and Management*, 61, pp. 19–30. Elsevier Ltd. doi: 10.1016/j.enconman.2012.02.023.

Grossmann, J, Serbanescu, D, and Schieferdecker, I 2009. Test Specification and Test Methodology for Embedded Systems in Automobiles. *Test Specification and Test Methodology for Embedded Systems in Automobiles*, no. May 2014. pp. 1–16.

Hyundai Motor Group Newsroom. 2020. EV A to Z Encyclopedia - 1: Understanding EV Components. Online from https://news.hyundaimotorgroup.com/Article/Understanding-EV-Components#:%7E:text=Instead%2C%20EVs%20carry%20several%20components,that%20drives%20the%20EV%20forward retrieved on 01 October 2020.

Indu, K and Aswatha K, M 2020. Electric Vehicle Control and Driving Safety Systems –A Review. *IETE Journal of Research*. Taylor and Francis: doi: 10.1080/03772063.2020.1830862.

Justyna, KZ 2009. Model-Based Testing of Real-Time Embedded Systems. In*PhD Dissertation Technical University* Berlin.

Kariem, H, Touti, E and Fetouh, T 2020. The Efficiency of PSO-Based MPPT Technique of an Electric Vehicle within the City. *Journal of Measurement & Contro*, 53(3–4), pp. 461–473.

Lauber, A, Guissouma, H, and Sax, E 2018. Virtual Test Method for Complex and Variant-Rich Automotive Systems. *IEEE International Conference on Vehicular Electronics and Safety*, ICVES 2018. IEEE, 1–7. doi:10.1109/ICVES.2018.8519599.

Office of Energy Efficiency and Renewable Energy. 2019. Electric Car Safety, Maintenance, and Battery Life. Energy.Gov. Online from https://www.energy.gov/eere/electricvehicles/electric-car-safety-maintenance-and-battery-life#:%7E:text=EVs%20must%20undergo%20the%20same,the%20high%2Dvoltage%20system%20to retrieved on 02 October 2020.

Prosvirnova, T, Batteux, M, and Rauzy, A 2013. Model-Based Safety Assessment. *Safety, Reliability and Risk Analysis*, 1, pp. 1129–1136. doi: 10.1201/b15938-170.

Research and Markets. 2018. Global Electric Vehicle ECU Market (2019 to 2024). Online from https://www.businesswire.com/news/home/20200618005644/en/Global-Electric-Vehicle-ECU-Market-2019-to-2024---Advent-of-Autonomous-Mobility-Services-Presents-Lucrative-Opportunities---ResearchAndMarkets.com. retrieved on 01 October 2020.

Ritchie, H, and Roser, M 2020. India: CO2 Country Profile. Our World in Data. Online from https://ourworldindata.org/co2/country/india?country=%7EIND retrieved on 01 October 2020.

Samaie, F, Meyar-Naimi, H, Javadi, S, and Feshki-Farahani, H 2020. Comparison of Sustainability Models in Development of Electric Vehicles in Tehran Using Fuzzy TOPSIS Method. *Sustainable Cities and Society*, 53(November 2019), p. 101912. Elsevier. doi: 10.1016/j.scs.2019.101912.

Santos, G 2018. Sustainability and Shared Mobility Models. *Sustainability* (Switzerland), 10(9), pp. 1–13. doi: 10.3390/su10093194.

Schoch, J, Gaerttner, J, Schuller, A, and Setzer, T 2018. Enhancing Electric Vehicle Sustainability through Battery Life Optimal Charging. T*ransportation Research Part B: Methodological*, 112(2018), pp. 1–18. Elsevier Ltd. doi: 10.1016/j.trb.2018.03.016.

Siddartha, S, Birrell, G, Dhadyalla, and Jennings, P 2015. Identifying a Gap in Existing Validation Methodologies for Intelligent Automotive Systems: Introducing the 3xD Simulator. *IEEE Intelligent Vehicles Symposium, Proceedings*, 2015 August (Iv), pp. 648–653. doi: 10.1109/IVS.2015.7225758.

Talukdar, J, Mehta, B, and Gajjar, S 2019. Computational Intelligence in Embedded System Design: A Review. *Smart Innovation, Systems and Technologies*. Satapathy, Suresh Chandra , and Joshi, Amit . Vol. 106, pp. 473–484, Springer, Singapore. doi: 10.1007/978-981-13-1742-2_47.

Tatar, M, and Mauss, J 2014. Systematic Test and Validation of Complex Embedded Systems. *Embedded Real Time Software and Systems*, pp. 5–7.

Tran, M, Banister, D, Bishop, JDK, and McCulloch, MD 2013. Simulating Early Adoption of Alternative Fuel Vehicles for Sustainability. *Technological Forecasting and Social Change*, 80(5), pp. 865–875. Elsevier Inc. doi: 10.1016/j.techfore.2012.09.009.

U.S. Department of Energy. 2020a. All-Electric Vehicles. US Department of Energy and US Environment Protection Agency. Online from https://www.fueleconomy.gov/feg/evtech.shtml retrieved on 01 October 2020.

U.S. Department of Energy – Office of Energy Efficiency & Renewable Energy. 2020b. Reducing Pollution with Electric Vehicles. Online from https://www.energy.gov/eere/electricvehicles/reducing-pollution-electric-vehicles retrieved on 01 October 2015.

United Nations. 2015. Sustainable Development Goals. UNDP. Online from https://www.undp.org/content/undp/en/home/sustainable-development-goals.html retrieved on 01 October 2020.

Xie, G, Zeng, G, and Li, R 2020. Safety Enhancement for Real-Time Parallel Applications in Distributed Automotive Embedded Systems: A Stable Stopping Approach. *IEEE Transactions on Parallel and Distributed Systems*, 31(9), pp. 2067–2080. IEEE. doi: 10.1109/TPDS.2020.2984719.

Xie, G, Zeng, G, Liu, Y, Zhou, J, Li, R, and Li, K 2018. Fast functional Safety Verification for Distributed Automotive Applications during Early Design Phase. *IEEE Transactions Industrial Electronics*, 65(5), pp. 4378–4391.

Yi, Z, and Bauer, PH 2019. Energy Aware Driving: Optimal Electric Vehicle Speed Profiles for Sustainability in Transportation. *IEEE Transactions on Intelligent Transportation Systems*, 20(3), pp. 1137–1148. doi: 10.1109/TITS.2018.2839102.

10 Sustainable Developments through Energy-Efficient Buildings in Smart Cities: A Biomimicry Approach

Divya Sharma and Hima C. S.
School of Architecture, KLE Technological University, Hubli, India

CONTENTS

10.1 INTRODUCTION

The research examines the prototype of responsive building skin using biomimicry principles to reduce the energy footprint. Overall global energy consumption and carbon emission have grown by 36% and 40%, respectively. Thus, to tackle the climate crisis and conserve Mother Earth, the energy load of the edifice must be reduced. Energy efficiency can be enhanced by the uplifting design of the building façade with the application of sustainable materials. The exterior skin pattern of

DOI: 10.1201/9781003127819-10

built space becomes dominant in the case of a highly glazed built form in which the glass on the exterior of the building protects from the quantum of heat transfer, with the result being that the consumption of energy will be very high.

In the literature study, various buildings are discussed with the biomimicry façade wherein each building façade aims at different energy-saving techniques. Every design is adaptable, to a certain extent, as buildings can be adapted physically in some way. Hence, this type of architecture is concerned with buildings that are intentionally designed to adapt to their environment. As per the research study, surface components and modules, spatial features, and technical systems are identified as a core aspect to make a building adaptable. To increase energy efficiency in buildings, adaptive skins are analyzed with smart materials; to design the adaptive façade, biomimicry is the epitome (Bar-Cohen, 2005). In this research paper, we try to explore the morphological and adaptive properties of the Oxalis Oregana plant, which tracks the magnitude of solar rays through photoreceptors. As sun rays hit the leaf surface, they tend to wrap downward and resume their original shape in shade. This property enables leaves to interrupt the surplus sunlight and orient it consistently, thus intensifying the prevailing light. The main concern of the authors is to imitate this physical behavior of the leaf to develop a façade integrant wherein a shading mechanism is the second skin of a building, which can change its position regarding solar angles. The criteria regarding the proposed biomimetic adaptive façade are as mentioned below.

First, a dynamic adaptive facade is designed to increase energy efficiency in highly glazed buildings. Second, an advanced building skin truncates procuring solar energy, thus increasing daylighting through less glare. To cut down the solar heat earnings of buildings, the façade must block the view on the whole; the perceptible serenity of the occupant is impacted. In this proposed façade, we try to control daylight, irrespective of blocking the exterior notion. The proposed shading device of the second skin folds into a bi-axial direction, basically along the x and y axes, enhancing the block of solar rays from dawn to dark.

To conclude, we apply the proposed adaptive façade on a substantial built space and perform the simulation process as required. The prototype of the recommended built envelope is considered as an actual two story library with a highly glazed facade at Hubballi, Karnataka. Energy-analysis software, including Revit 2019, Sketchup plugin, and Ecotect software, is used to perform the simulation. The heating and cooling analysis indicate that 40% of the building's energy load can be minimized by modifying the building envelope. As per lighting analysis data, 60% of the entresol is flooded with light (as opposed to 65% before the modification); it still has an incandescence degree within the standard range of 110–450 lux. after proposing an adaptive second skin. The research aims at the proposal of a bio-inspired built envelope that will minimize the energy load and carbon emission with structural glazing. The rest of the paper is organized as follows: Section 10.2 discusses biomimicry as an energy-management tool and smart-building envelope. Section 10.3 presents smart city buildings using a biomimetic façade and conveys an understanding of adaptive. Section 10.4 proposes the design strategy of the proposed bio-inspired built envelop and prototype building. Section 10.5 examines numerical results performed by simulation. Eventually, the conclusion is addressed in Section 10.6.

10.2 BIOMIMICRY AS AN ENERGY-MANAGEMENT TOOL

Building skin can be derived in different ways. Rankouhi derives it as the junction where the building interplays with the external surroundings. It comprises various overlays that harmonize with extrinsic parameters like sound, the intensity of heat, and moisture (Rankouhi, 2012). The habitual aspect is the proximity to sustain advisable inner circumstances that acknowledge the tasks they tend to. The building envelope is the intermediate junction between the inward and outward conditions of a built space, where most energy exchange occurs.

10.2.1 BIOMIMICRY AND BUILDING SKIN

To understand the resemblance between built envelops and bio-inspired architecture, it is important to examine the similarities of both. Building skin is a lean layer that shelters the bones (building), synchronizes the organs (services), and construes its inner array. The persistent character encompassed by the natural skin of the built environment is to prolong inner conditions that are flexible to its task. Building skin behaves as a filter and allows what can escalate in, like light, moisture, sound, air, and heat, and what is exposed out to reduce the energy utilization of the building (Faber et al., 2018). By adopting natural processes, employing an intelligent and eco-friendly built envelop known as bio-inspired façade, and applying design to it, one of the latest innovations in building science can be built to curtail energy consumption in cities through building skin.

10.2.2 SMART BUILDING ENVELOPE

Certain parameters like air, shade, and climate that correlate with thermal and visual consolation are to be considered while designing a biomimetic smart façade. Along with this, a sustainable, green-built envelope should consider strategies such as solar radiations, wind pace, humidity, and materials used. Advanced material is derived, like material that is sustainable, energy efficient, and responsive to climatic conditions in an effective way that self-operates to furnish shade, light, and air. Smart material like Thermo bimetals allows a built envelope to self-ventilate. Smart materials wrap and reshape themselves to align with the position of the sun to furnish shade, irrespective of static power. A smart building tends to ventilate, or aerate, on its own, and it manages heat control, eventually minimising carbon emissions and increasing overall energy efficiency. Adaptive bio-responsive additional skin is a facade layered between 0.2 and 4.5 m to the environment barrier (Yowell, 2011). Inner layout modulates heat, light, wind, noise, and environmental strain. A variety of technologies are being used as second skins, such as screens, vertical rotators, blinds, metal mesh, and green facades, which help in achieving energy-efficiency smart city criteria.

TABLE 10.1

Buildings with Different Adaptive Façade

Sl. No.	Name of the Building	Façade Type	Concept	Inspiration	Material	Energy-Saving Technique
1	Articulated cloud, Pittsburgh	Active façade	Building Envelop self-adjust to building environmental changes	Cloud	Aluminium frame, wind-activated acrylic panels/plastic leaves	Cools the building by diffusing sunlight and reflects heat.
2	The Algae House, Hamburg	Bio-inspired façade	Energy generated by algae biomass by its façade	Algae	The second outer shell of microalgae	Production of biomass
3	Mexico City Hospital	Intelligent façade	A network of antimicrobial tiles cleans the air entering the building.	Air filters	Pro solve tiles of titanium dioxide	Eliminate pollutants, eats smog to improve air quality
4	InDeWaG, Bayreuth	Smart façade	Circulating water between layers of glass which in turn collects solar radiation & supplies captivated heat for heating, preheating, domestic hot water.		Fluid flow glazing façade	Zero energy building minimizes HVAC & PV installations.
5	Al-Bahar towers, Abu Dhabi	Responsive façade	The solar screen responds dynamically to the angle of the sun and improves control over solar radiation, glare.	Mashrabiya	Triangular fiberglass	Reduces solar gain, reduces energy need, and controls daylighting.
6	Green pix wall, Xicui entertainment center, Beijing.	Interactive façade	Interactive skin stores energy by day and uses to illuminate the screen after dark	Flickering light on the ocean's undulating surface	Photovoltaic cells integrated into the glass curtain wall	Reduces heat gain and transforms excess solar radiation into energy.
7	Kiefer technic showroom, BadGleichenberg	Kinetic façade	Panels expand and contract which controls the number of solar rays penetrating the building.	Window arrangement & axes	Perforated folding aluminium panels	Minimizes heat. Keeps the temperature cool.

10.3 ADAPTIVE ARCHITECTURE

Adaptive architecture is an integrated field of architecture that deals with buildings designed to adapt to their existing environments and occupants, and buildings that are based on inner statistics. Regardless of the functioning of the façade, which may be a kinetic, green, or dynamic façade, façades create nature-oriented chemistry rather than a computer-sustained adaptation. Eventually, to figure out how successful a building is responding to adaptive architecture, we need to analyze the elements adapted, a method for adaptation, and the effects of that adaptation. To program a building for adaptive architecture, there is a certain paradigm that may be cultural, societal, organizational, or communication (Romano et al., 2018). De facto few salient principles evolved in design make space or façade as adaptive architecture. Environment, objects, and inhabitants are determined as major elements to make architecture adaptive. Several other elements can be adapted by a building; they are selected based on the need of the inhabitants (Benyus, 1998). Researchers have figure out various façades based on their material usage and function. Following is the related study analyzed (Table 10.1), which has applied different adaptive biomimetic façade.

The advanced building skin, eco-efficient, and sustainable envelope regulates energy flow and also becomes a system fit to process energy, electricity, and the heat, finally circulating these in a built space. Air, shade, and the climate are considered as guidelines for an adaptive smart façade. Additionally, an eco-friendly built envelop must mimic sunlight, the pace of wind, and humidity.

10.4 DESIGN METHODOLOGY

In this section, we discuss bridging between biology and architecture. Then we discuss the methodology for the proposed adaptive façade inspired by the Oxalis oregana leaf (Figure 10.1). Further, we examine the kinematics and control

FIGURE 10.1 Oxalis oregana leaf has the physical behavior of tracking sunlight.

mechanism of the façade. Nature remodels accordingly, swapping with in its habitat; similarly, a built form can be feasible by imitating the process nature follows. Three levels are identified in biomimicry: organism, behavior, and ecosystem level (Zari, 2007). The approach to the proposed biomimicry façade is derived from the Oregana leaf at the organism and behaviour level. The substantial module has been developed from the Y shape perspective of veins in the leaf at an organism level (Figure 10.2). The working of the proposed component arises through the behaviour property of the leaf to track sunlight with intensity through photoreceptors in only 6 minutes and change its angle according to the intensity of sunlight. The plant identifies light and aligns itself vertically to avoid photoinhibition of photosynthesis or horizontally when light is scarce. These properties of a leaf are mimicked to design shading devices that can change their angle according to sun angles.

10.4.1 WORKING OF THE PROPOSED ADAPTIVE FACADE

The fundamental element of the proposed built envelop is a rhombus-shaped shading device (Figure 10.3). Modules move on the x-axis and y-axis, supported by an aluminium frame structure that is connected to the façade by hinges and guide rails that work on an intelligent controller that controls the mechanism. Sensors can scrutinize the discrepancy of a superficial stimulant that conducts instruction to the actuator, until finally, a change in one of its properties is observed. Shape memory polymers can be used as sensors or actuators for solar-shading materials. The proposed module has a shape-morphing shading system with chronicle fluctuation that executes two-dimensional transpose in shape (Figure 10.4 to Figure 10.7). It is slender, which permits adaptation on an external envelope by splitting proportions or protruding layers. The concept of sustainable energy initiation is adapted potentially in the adaptive second skin, integrating a binary advantage: shading and electricity generation. As per researchers, a building integrated with PV on external shading generates electricity, thus minimizing cooling loads by 10% (Al Dakheel and Tabet Aoul, 2017).

10.5 CASE STUDY

A library building is singled out for this case study (Figure 10.8). The library building is a two-story building located in Hubballi, Karnataka (Figure 10.9). Hubballi experiences a tropical wet and dry climate with hot summers. The prototype is slightly oriented to the north-south axis by 52°. The reading arena is positioned at each corner of the building exposed to sunlight, with a seminar hall on the first floor and stacking in the northeast corner. The floor height is 4 m and the clear height is 3.8 m. The parameters below are considered to choose the library building as a prototype.

1. Building facade has structural glazing of more than 60%.
2. Single panel glass is used.
3. The façade looks monotonous.
4. Built envelop experiences harsh solar rays.

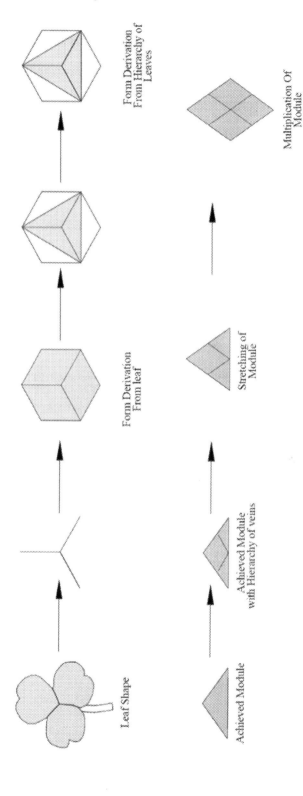

FIGURE 10.2 Derivation of module from basic Y-shape of leaf.

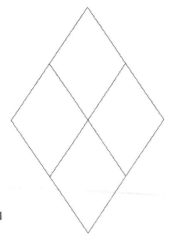

FIGURE 10.3 Oxalis oregana leaf has the mechanical property of trapping sunlight.

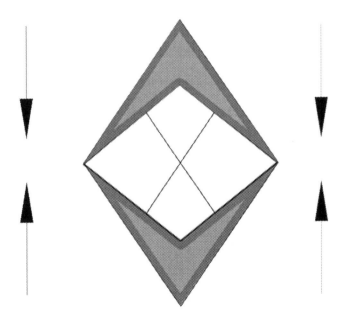

FIGURE 10.4 Rhombus-shaped primary unit of the proposed façade, which wraps bilaterally according to external conditions.

10.5.1 NUMERICAL RESULTS

Here we discuss in detail the comparative analysis of the interpretation of the prototype before and after modifying the proposed façade. The application of an adaptive biomimetic façade (Figure 10.10 and Figure 10.11) is evaluated in detail. Daylight analysis, solar insolation analysis, and shading analysis are performed for the case study, with the existing structure and after retrofitting the façade. Parameters like solar angles, building orientation, solar irradiance levels, and

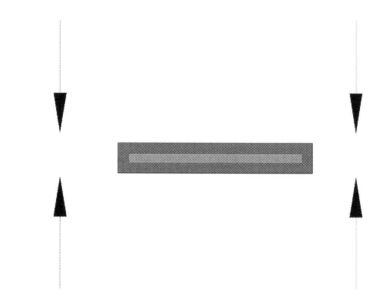

FIGURE 10.5   Horizontal wrapping of the proposed module facade forming a rectangular shape.

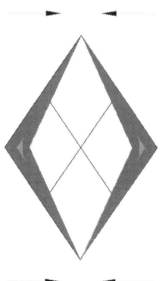

FIGURE 10.6   Working mechanism in vertical direction retrofitting the façade, which has a bilateral movement made up of thermo bimetal.

thermal performance of materials are considered. Energy-analysis software, such as Radiance Synthetic Imaging system software, Ecotect 2012, Revit 2019, Sketchup, and Insight 360, is habituated. The existing façade material is analyzed thoroughly and then re-created digitally with a second skin of adaptive material (Figure 10.12 and Figure 10.13).

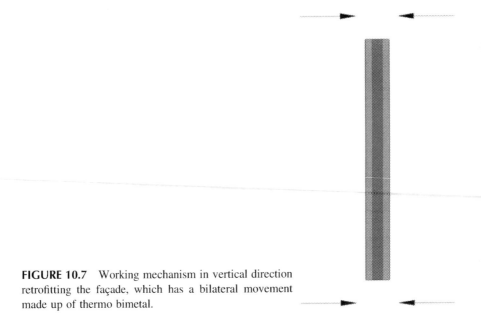

FIGURE 10.7   Working mechanism in vertical direction retrofitting the façade, which has a bilateral movement made up of thermo bimetal.

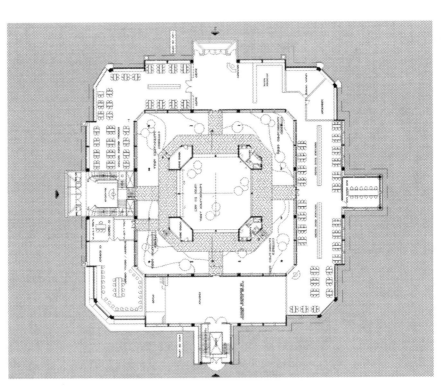

FIGURE 10.8   Plan of existing library building selected as prototype retrofitting façade.

FIGURE 10.9   View of an existing building that experience much glare in existing conditions.

10.5.2   ANALYSIS OF DAYLIGHT

As per standard guidelines for interior lighting, suggested lighting levels for a library building vary between 300–400 lux (Mohammed et al., 2019). For research purposes, daylight analysis is conducted on the ground and first floor with the help of radiance-synthetic software. Daylight analysis specifies that the existing building experiences more than 50% of glare conditions (Figures 10.14a–c) in the regularly occupied area, making the user thermally uncomfortable. Daylight analysis after retrofitting a responsive façade results in 40% of space being illuminated through daylight extent between 110–450 lux (Figures 10.15a–d). A detailed view of the library was generated in Ecotect software and then imported to radiance software to perform daylight simulation. Radiance synthesis-imaging software traces a detailed graph for daylight in a given space. Radiance is supported by a simulation technique that figures out the degree of light and renders a drawing with a detailed illustration of the 3-dimensional superficial configuration, materials, and reference of light in a built space. The illumination graph was created to document the quantity of illumination accessible in the given area. The glass visible light transmission and shading devices were simulated to arrive at the lux levels in the interiors. Daylight has been modeled for 21 March at noon, as it is considered a crucial period of the year.

10.5.3   INSOLATION ANALYSIS

The insolation simulation has been performed using Ecotect software. A detailed view of a prototype was digitalized in Autodesk Ecotect Analysis and imported to

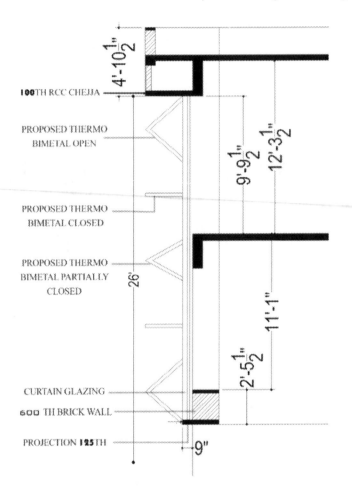

FIGURE 10.10   Section of the proposed building that shows horizontal, vertical folding of module.

solar access to examine the radiant impact. Insolation has been attained for 31 March for a solar day for four-side orientations. The output demonstrates average radiation and comparative analysis for insolation simulation. Insolation analysis performed on the east side of the existing facade shows that the horizontal surface receives maximum radiation, which exceeds more than 3000 W/sqm. Whereas the vertical surface on the east side ranges from 1110 to 1920 W/sqm, which contributes to internal heat gain (Figure 10.16). Due to the biomimetic façade design, radiation on the east surface gets reduced. Thus, a reduction in internal heat gain results. Radiation ranges from 840 to 1310 W/sqm. On simulation upon the north surface, the horizontal surface received more than 3000 W/sqm. The vertical surface on the north side ranges from 1110 to 1650 W/sqm, which contributes to internal heat gain. Due to biomimic façade, radiation on the north surface gets

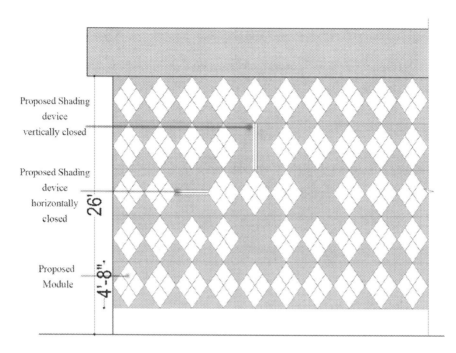

FIGURE 10.11 Elevation of proposed prototype where bi-axial movement is observed on a responsive façade.

FIGURE 10.12 View of a proposed prototype with few modules open and closed regulating internal temperature.

FIGURE 10.13   View of a proposed prototype with responsive façade mechanism allowing required daylight into interiors.

reduced. Thus, reduction in internal heat gain. Radiation ranges from 570 to 840 W/sqm. By simulating the existing west façade, it was observed that the maximum amount of radiation falls on the west and south side which ranges from 1650 to more than 3000 W/sqm. Thus, the south side plays a crucial role. By applying simulation on adaptive skin on the west side, radiation on the west surface gets reduced. Thus, reduction in internal heat gain. Radiation ranges from 570 to 1640 W/sqm. Considering the simulation on the existing south façade, it is observed that it is the second-worst affected façade due to radiation on the south façade, which ranges from 1380 to 2400 W/sqm. As a result of adaptive façade simulation on the south façade, it is observed that radiation on the south surface gets reduced, reducing internal heat gain. Radiation ranges from 1110 to 1380 W/sqm.

10.5.4   SOLAR SHADE ANALYSIS

Solar shadow range helps to understand the relationship between site and building.

Two parameters could be achieved by performing solar-shadow analysis. First, the building shadows the surrounding amenities present onsite, which reduces the urban heat-island effect. Second, the self-shading of the building helps to achieve a reduction in façade radiation (Figure 10.17). As a result of shading analysis, positions are identified to intensify solar gain by reviewing the impact of shading and periodic modification in solar radiation. We can benefit from passive solar design strategies to maximize efficient energy, relaxation, and economic features of built space. Solar-shadow analysis concluded that building may be self-sustainable if designed, oriented, and placed properly, and if certain measures are taken prior, we can achieve a building energy efficiency that is a step closer to a smart city.

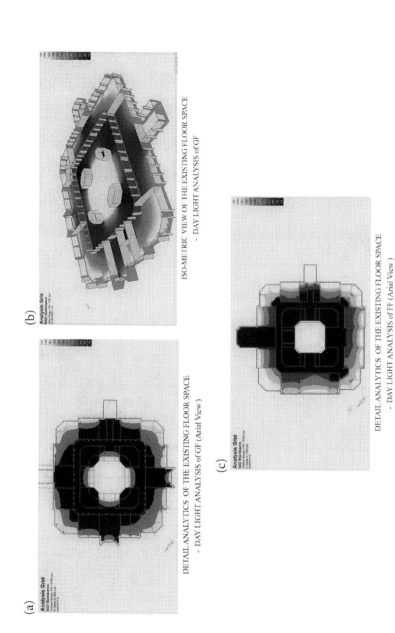

FIGURE 10.14 (a) An aerial view showing daylight analysis for existing ground floor with illumination graph. (b) Isometric view showing daylight analysis for existing first floor with illumination graph. (c) Aerial view showing daylight analysis for existing ground floor with illumination graph.

FIGURE 10.15 (a) An aerial view showing daylight analysis for proposed ground floor with illumination graph. (b) Isometric view showing daylight analysis for proposed ground floor with illumination graph. (c) Isometric view showing daylight analysis for proposed first floor with illumination graph. (d) Aerial view showing daylight analysis for proposed first floor with illumination graph.

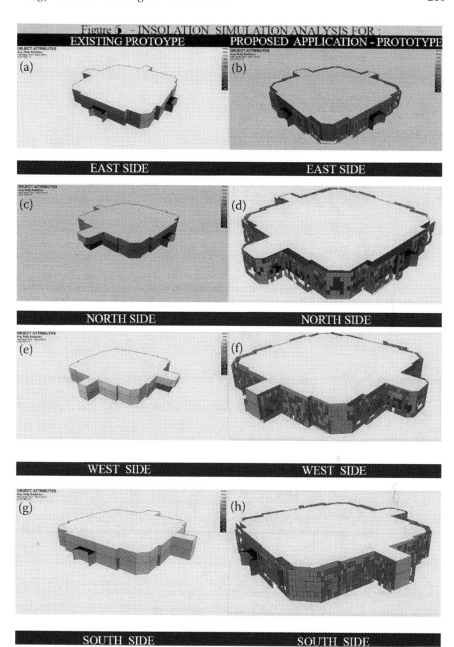

FIGURE 10.16 Insolation analysis on existing and proposed façade on all four sides. (a and b) East side. (c and d) North side. (e and f) West side. (g and h) South side.

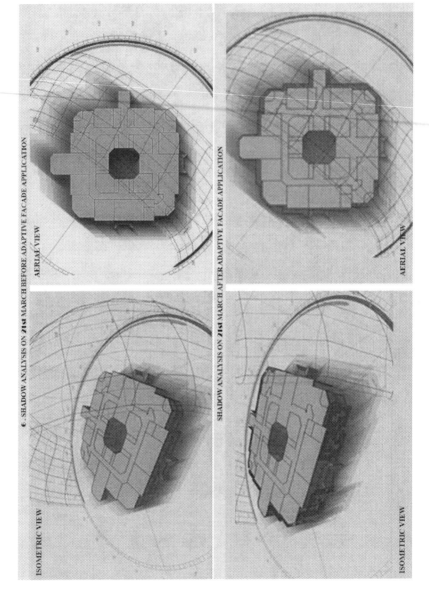

FIGURE 10.17 Solar shadow analysis on 21 March before and after retrofitting.

10.6   CONCLUSION

The research scrutinizes retrofitting of a responsive bio-inspired external envelope on structural glazing-built space. Authors attempt to install a responsive adaptive façade, which truncates solar heat gain such that energy consumption of built space shows less depletion in the perceptible comfort of occupants. Stimulation performed on buildings specifies that a building's substantial energy load can be reduced up to 40% by modifying the façade. By daylight simulation, we achieved a variation of 40% in energy consumption. Due to the adaptive façade, a drastic change in glare reduction was observed, which reduces the heat load of the building, ultimately reducing in AC tonnage and achieving indoor thermal comfort, hence achieving a sustainable smart building. The insolation analysis gave a 42% reduction in radiation on the east side, 49% reduction in radiation on the north side, 45.3% reduction in radiation on the west side, and 31.7% reduction in radiation on the south side. This reduction in the building after retrofitting the façade can help us to edge off energy consumption and design adaptive smart buildings. Finally, shadow analysis helps to identify the location to maximize solar gain by adapting passive biomimic principles. All these parameters of analysis help understand the reduction of energy consumption in buildings and reduce the carbon footprint, providing a sustainable methodology for buildings with a glazed structure in a smart city. In addition to these various applications, façade can be provided for the consumption of energy. As building an integrated photovoltaic grid is harmonized with a bio-inspired responsive adaptive-shading system, the built envelop is competent to liberate its electricity in addition to solar shading. Different adaptive facades, like light control, energy generative, kinetic façade, algae façade, and so on, can be adapted for future design to achieve an eco-friendly green building with less waste achieving steps toward a smart city.

ACKNOWLEDGEMENTS

The authors acknowledge AKASH S R student of eighth semester School of Architecture for his sincere efforts in contributing an excellent retrofitted view of the proposed building. All photographs & drawings are from the School Collection.

REFERENCES

Al Dakheel, J, & Tabet Aoul, K (2017). Building Applications, Opportunities and Challenges of Active Shading Systems: A State-of-the-Art Review. *Energies*, 10(10), 1672. https://doi.org/10.3390/en10101672

Bar-Cohen, Y 2005 (Ed). *Biomimetics: Mimicking and being Inspired by Biology*. CRC Press, pp. 505.

Benyus, JM 1998. *Biomimicry: Innovation Inspired by Nature*. Perennial (Harper Collins).

Faber, JA, Arrieta, AF, & Studart, AR (2018). Bioinspired spring origami *Science Robotics*, 359(6382), 1386–1391.

Mohamed, NA, Bakr, AF, and Hasan, AE 2019. Energy Efficient Buildings in Smart Cities: Biomimicry Approach. *Real Corp*, (5–11). http://www.biomimeticarchitecture.com

Rankouhi, A 2012. Naturally Inspired Design. *An Investigation into the Application of Biomimicry in Architectural Design*.

Romano, R., Aelenei, L., Aelenei, D., & Mazzucchelli, E. S. (2018). What is an adaptive façade? Analysis of Recent Terms and definitions from an international perspective *Journal of Facade Design and Engineering*, 6(3), 65–76. https://doi.org/10.7480/jfde.2 018.3.2478

Yowell, J 2011. Biomimetic Building Skin (Online) Available at http://tulsagrad.ou.edu/ studio/biomimetic/jy-FINAL-thesis.pdf

Zari, MP 2007. Biomimetic Approaches to Architectural Design for Increased Sustainability. Sustainable Building Conference. Auckland.

11 Biomass Derived Activated Biochar for Wastewater Treatment

Ashish Pawar and N. L. Panwar

Department of Renewable Energy Engineering, College of Technology and Agriculture Engineering, Agriculture University, Udaipur, Rajasthan, India

CONTENTS

11.1 OVERVIEW

Green energy is becoming a basic need for all nations and societies due to the rapid growth in human population. Given the increasing demand for energy and the wide use of fossil fuels in the region, many countries are looking for a secure energy supply and are increasingly concerned about finding alternative sources. Many countries, including India, have identified biomass as an abundant, easily available energy resource that could play a significant role in reducing their

DOI: 10.1201/9781003127819-11

energy deficit. India is an agriculture-presiding country that produces a total biomass about 600 Mt annually, among which 150–200 Mt is considered as additional biomass waste (UNDP 2011). This surplus residue could be used as a feedstock for animals and as an energy source for domestic as well industrial thermal application. About 85 Mt of unutilized biomass, especially agro residues, includes straw and stubbles burnt in open fields by some peasants (Pawar and Panwar, 2020). Therefore, appropriate waste recycling facilitates the production of energy-rich and efficient green fuel, whether solid, liquid, or gas, and has become an important issue throughout the world. There are different biomass-conversion technologies, and out of them, pyrolysis is one of the economically feasible thermochemical conversion routes; it produces mainly three end products, biochar, bio-oil, and syngases, respectively. Biochar is carbonaceous material mostly produced from a carbonization process in absence of air as per its end application (Panwar et al., 2019). Further, the produced biochar is further modified by adopting two activation methods (physical and chemical) for the preparation of activated biochar. The feedstock material, production conditions, and activation mode significantly effect the physical and chemical characteristics of activated biochar. The activated biochar possesses a larger surface area, surface functionality, porous structure, and the availability of oxygenated functional groups, which pose improvement in adsorption performance for different contaminants in wastewater. The adsorption process mainly depends on the type of activated biochar and the properties of targeted pollutants. The heavy metals, organic, and inorganic pollutants, and phenols, etc., are present in the water and enter the human body via the food chain, which may expose humans to many dangerous diseases. In addition, the organic contaminates (mostly phenol) present in the water can change the test and odour of drinking water. Activated biochar have huge prospects for the adsorption of contaminates from wastewater due to its strong adsorbing affinity and capacity. Initially, coal or grain were mostly utilized as precursors for making granular-activated carbon; however, they were not economically sustainable, requiring more money to the manufacturers for procurement. Activated biochar prepared from the synthesis of biochar shows remarkable properties, mainly high internal surface area and porosity. Due to its microporous structure, it acts as an adsorbent for the elimination of different contaminants. In addition, activated biochar becomes a reliable renewable material because of its successful phase application, such as deodorization, color removal, separation, purification, storage, and catalysis. Activated biochar derived from biomass as a precursor acts as a pollutant removal and has many applications in industrial sectors, such as for treating wastewater, chemical processing, controlling the air pollution, petroleum refining, and adsorption of volatile organic compounds. The literature reports that different kinds of agro wastes, such as groundnut shell, rice husk, corn cob, straw, wheat, and cotton residues, have been utilized appropriately as a precursor from many years to obtain a low-cost activated biochar, which acts as an adsorbent. Therefore, conversion of waste into energy-rich activated biochar, which is a sustainable and cost-effective adsorbent material, plays an important role in different environmental applications.

The present book chapter focuses on the synthesis of activated biochar from different feed stocks, its operating condition, physical and chemical characteristics of activated biochar, and its applications in wastewater treatment. An attempt has been made to divide the current chapter into two parts: the first parts shows the preparation of activated biochar using different precursors and the influence of different experimental processes, and the second part comprises the classification, properties, and applications in wastewater treatments.

11.2 BIOMASS RESOURCES

Biomass is simply an organic material that is available on Earth's surface. The plant absorbs the solar energy and stores it in chemical form through the photosynthesis process. The plant species or animal waste is considered as organic biomass material, which is renewable in nature (Fahmy et al., 2020). The biomass sources are mainly classified into two categories: the first is natural, and the other is derived materials. Generally, raw materials like woody waste, forest waste, agriculture residues, and by-products, MSW, waste generated from agricultural-processing industries, animal waste, and aquatic waste, etc., are considered as main resources of biomass (Pawar and Panwar 2022).

- Agriculture wastes: wheat straw, rice straw, rice husk, groundnut shells, cobs, sugarcane waste, herbaceous crops, etc.
- Forest residues: woody biomass, sawdust, logging residues, shrubs, trees, bark etc.
- Food-processing industrial waste: dairy waste, industrial waste, oil residue, brewery waste, fruit waste, marine waste, bakery waste etc.
- Animal waste: animal dung, poultry waste, animal carcasses, animal excreta etc.
- Municipal solid waste: kitchen waste, vegetable waste, garden waste, waste from office, homes etc.
- Aquatic waste: Algae waste, marine waste etc.

Figure 11.1 shows the different biomass resources available for energy generation.

11.3 BIOMASS POTENTIAL FOR ENERGY GENERATION IN INDIA

The total available land in India is about 328 Mha (million hactares). Among this total, nearly 39 Mha is covered by marginal land, which is 12% of the whole area. Marginal land is simply defined as the agricultural land initially used for sowing and cultivation purposes, but presently, is not in use due to reduced productivity. Generally, marginal land is not recognized for agricultural practices because of the high risk of lower production, which simultaneously affects the agricultural economy. Therefore, reforming marginal land for good agricultural production requires large amounts of chemical fertilizer dosages, raising the economics and affecting the environment. However, many researchers reveal that marginal land can produce 230–720 MT of energy crops annually. These energy crops have a

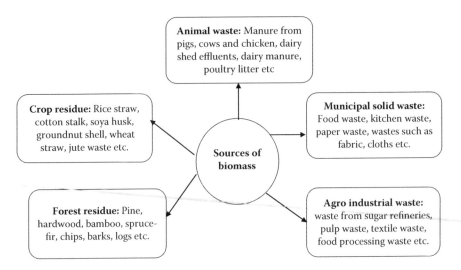

FIGURE 11.1 Resources of biomass.

potential between 1–13 EJ annually, which covers approximately 11% to 160% of the energy demand of the transport sector. The energy crops have a total bio-fuel potential between 3–103 GGE/y, which covers about 5% to 160% of the gasoline demand in the transport sector. The total availability of the agriculture and forest residues in India is about 439 and 19 MT, respectively, which has the potential to generate 1.8–7.2 EJ of energy annually, which fulfills the 21%–84% energy demand in the transport sector. In terms of biofuel, energy crops have the potential to produce 6–37 GGE annually, which covers 10%–56% of the gasoline demand.

11.4 BIOMASS CONVERSION ROUTES

Various biomass conversion routes are available, including: thermochemical, biochemical, direct combustion, and chemical conversion, as shown in Figure.11.2. However, the most commonly used biomass routes for sustainable biofuel production are thermochemical conversion and biochemical conversion. The thermal-conversion process consists of biomass gasification via pyrolysis, while the biochemical-conversion process includes anaerobic decomposition of organic waste for biogas production. The trans-esterification method is also a chemical conversion, but of vegetable oil into biodiesel, with the direct combustion of forest waste or woody biomass used to generates heat for thermal applications.

11.4.1 THERMOCHEMICAL CONVERSION

Multiple thermo-chemical conversion routes are available, including combustion, liquefaction, pyrolysis, gasification, etc. However, the pyrolysis process has gained significant attention due to its direct biomass conversion into solid, liquid, and

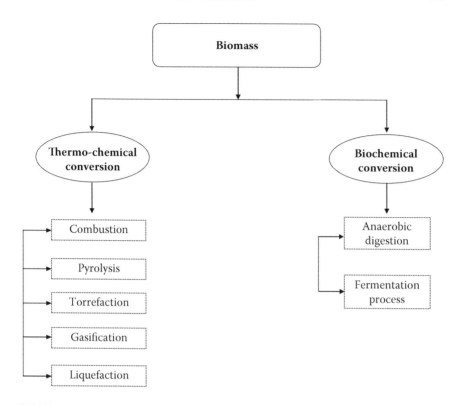

FIGURE 11.2 Schematic representation of classification of biomass conversion processes.

gaseous fuel via thermal decomposition in anaerobic conditions. These can occur at low or high temperatures.

Biomass Combustion: This is the direct burning of biomass in aerobic conditions, readily converting chemical energy into thermal, electrical, or mechanical energy. Such combustion processes face several disadvantages, such as requiring pre-treatment and causing environmental pollution.

Gasification: This is the thermochemical conversion of biomass at a very high temperature (800°C–1000°C) in a controlled air supply, resulting in partial oxidation of organic waste into a producer gas, which is combustible in nature, and which has a calorific value ranging from 900–1100 kcal/m^3. The producer gas is mainly composed of some combustible gas components, such as CH_4, H_2, and CO, as well as other inert gases like N_2, and CO_2.

Liquefaction: The thermochemical conversion of biomass takes place under high pressure, at a minimum temperature, in the presence of hydrogen, and with the use of a catalyst.

Pyrolysis: Pyrolysis is one of the more economically feasible thermochemical conversion processes in which waste material is indirectly heated in the range of 400°C to 700°C in absence of air, yielding primary end products like bio-oil, biochar, and syngases. Pyrolysis processes are broadly categorised into three classes: slow, fast, and flash pyrolysis.

11.4.2 BIOCHEMICAL CONVERSION

Biological or biochemical conversion includes both anaerobic digestion and the fermentation process. In the case of the anaerobic digestion process, organic waste material can be transformed into gaseous fuel by breaking the larger molecules into smaller molecules in the presence of bacteria, offering economically viable applications worldwide. While, in the case of the fermentation process, sugar feedstock is converted into ethanol in the presence of yeast. However, produced diluted alcohol can't be used directly without distillation; it also suffers a high cost with lower performance of the plant. This process takes longer to complete as compared to the thermo-chemical conversion process, although it needs considerable external energy.

11.5 WHAT IS ACTIVATED BIOCHAR

In India, 686 Mt gross residues are available from agriculture crops, in which 23.5 Mt signifies the surplus residues reported by Hiloidhari et al. (2014). The surplus residues can be converted into biochar through the thermochemical conversion process. In a carbonization process, biomass is thermally decomposed at higher temperature in the absence of air/oxygen for a specific duration, resulting in the production of biochar. The produced biochar can be further activated using physical and chemical activation methods to obtain an activated biochar. Biomass is thermally decomposed at a higher temperature in the absence of air for a specific duration, resulting in the production of biochar. Activated biochar have an ability to minimize the environmental problems like accretion of agricultural residues or waste, air pollution, as well as water pollution. Activated biochar is also known as activated carbon or charcoal (Hiremath et al., 2012) and sporadically known as a solid sponge (Vaghela et al., 2022). A lignocellulosic biomass is mainly composed of cellulose, hemicelluloses, and lignin. However, lignin content in biomass feedstock is found a very useful component for the preparation of activated biochar (Carrott and Carrott, 2007).

Crop residues or biomass obtained from agriculture waste have become a promising type of precursor because of their minimal cost and abundant availability for making the activated biochar. Any agriculture waste with a greater percentage of carbon and lesser availability of inorganic components may be preferred for making activated biochar (Tsai et al., 1997). In addition, rather than fossil resources like coal and other petroleum by-products, activated biochar obtained from lignocellulosic biomass or crop residues helps reduce the greenhouse effect. Therefore, distribution of carbon and adsorption process occurs simultaneously, referred to as a carbon neutral cycle (Nor et al., 2013). (Figure 11.3).

11.6 PREPARATION OF ACTIVATED BIOCHAR

The first stage in the preparation of activated biochar is the indirect heating of biomass at higher temperatures in anaerobic condition. This is simply a carbonization or pyrolysis process (Suhas et al., 2007). During the biomass carbonization, due to higher process temperatures, all moisture and volatile components present in

FIGURE 11.3 Activated biochar.

biomass are released. The produced biochar at higher pyrolysis temperatures can be further activated using physical and chemical activation. In most cases, chemical activation is more widely preferred over a physical activation. The chemical activation increases the yield of activated biochar, showing a larger surface area without any need of higher temperatures and auxiliary energy (Gratuito et al., 2008). Therefore, at the moment, the production of activated biochar is popular because of its environmental friendly characteristics (Hernández-montoya et al., 2012; Zhang et al., 2013).

11.6.1 CARBONIZATION

A good quality biochar is essentially required to produce activated biochar. For the purpose of quality biochar, biomass is heated or thermo chemically decomposed at a temperature ranging from 400°C–850°C in inert atmosphere for a specific residence time to enrich the percentage of carbon. The type of biomass feedstock, particle size, pyrolysis temperature, and heating rate, etc., significantly affect the physicochemical composition of biochar. At moderate temperatures, the lignocellulosic composition of biomass is thermally degraded, and kinetic reactions take place, such as fragmentation, depolymerisation, and crosslinkings, resulting in a yield of biochar followed by bio-oil and syngases. The solid carbon-rich endproduct and a brown liquid endproduct are mainly referred to as "biochar" and "bio-oil," whereas the condensable and noncondensable composition of gases are known as syngases, respectively. The syngas is mainly composed of some combustible gas components, i.e. the mixture of hydrogen, methane, and carbon monoxide. Biochar produced at higher carbonization temperatures (600°C–700°C) showed a positive relationship relative to its aromatic nature and a well-improved carbon layer on its surface. The higher carbonization temperature showed a negative relationship with hydrogen and oxygenated functional groups. Therefore, it possesses low ion-exchange capacity, which might be due to the dehydration and deoxygenation reactions taking place during biomass carbonization at a high temperature. On the other hand, the produced biochar at moderate carbonization temperatures from 300°C to 400°C possess large availability of aliphatic, C=O, C functional groups, and celluloses on biochar surface. Therefore, from above discussion, it was concluded that the produced biochar up to 700°C have some limited properties, mainly the initial porosity and surface area, and it restricts its direct application in waste-water treatment. During the carbonization process, the pores inside the biochar are filled with some tars and other volatile components, so it showed minimum surface area; to overcome this barrier, there is a need of the secondary process for the preparation of well-developed porous-structured biochar. In addition, it is necessary to produce biochar at higher carbonization temperatures above 700°C because the biochar at a mentioned temperature possesses higher surface area, electrical conductivity, porosity, pH, microscopic structure, and the availability of micro and macro nutrients. Therefore, it could be used for further activation to prepare a good quality activated biochar.

11.7 BIOCHAR ACTIVATION

The biochar activation using the physical and chemical modification methods could increase the surface area, porosity, pore volume, and diameter of pore (Al-Swaidan and Ahmad, 2011). These treatments are responsible for or influence the physical characteristics of biochar, mainly increasing the size and shape. In case of physical activation, biomass is pyrolyzed or carbonized (up to 800°C), followed by modification using steam or CO_2 (Ioannidou and Zabaniotou, 2007). Therefore, physical activation mainly involves carbonization. And it means that physical activation has two main steps: a carbonization and a modification step. On the other hand, in chemical activation (also called as wet oxidation), the raw biomass is impregnated by using some chemicals (reported later), followed by pyrolysis in inert atmospheric conditions (Giraldo and Moreno-Piraján, 2012).

11.7.1 PHYSICAL ACTIVATION

The activated biochar was prepared from precursors using mild oxidizing gases. Carbonization of raw feedstock and then modification of biochar has been done using steam, and CO_2, among others. The use of carbon dioxide as an activation agent is mostly preferred due to its significant properties and the fact that it may provide control in the activation process. Compared to carbon dioxide, steam is effective in forming greater surface area in the resulting biochar. In addition, the conversion process, activated using steam, was found two to three times faster than using carbon dioxide (Nowicki et al., 2010). Many researchers have studied and reported that various raw materials like corn cob, hulls, and stover, oak, peanut hull, and rice hull, are also suitable for creating activated biochar through the physical activation process (Ahmedna et al., 2000). Carbonization temperature usually ranges from 400°C to 850°C, and up to 1000°C, while modification temperature varies from 700°C to 900°C (Ioannidou and Zabaniotou, 2007). The end product of carbonization is charcoal with high carbon content but less surface area, so it is inactive material. McDougall (1991) has studied that due to biomass pyrolysis, all noncarbon elements, as well as volatile matter (including tar, moisture content and volatile matter), were eliminated and resulted in charcoal having a maximum percentage of fixed carbon. So, it becomes more suitable for activation purposes. In the first stage of physical activation, the initial porosity appeared on charcoal (Daud et al., 2000), in addition to further activation by the presence of oxidizing gases on charcoal that will help to prepare activated carbon with high porosity (Baseri et al., 2012). The category and degree of modification in physical treatment take priority in determining the physio-chemical properties of activated biochar. Zhang et al. (2004) obtained activated biochar from different agricultural residues including corn hulls, corn stover, and oak wood waste by using the physical activation process with carbon dioxide as an activating agent and observed that resultant product had larger surface area (400–1000 m^2/g) and micropore volume (0.38–0.66) at a higher activation temperature (700°C and 800°C). Authors also reported that activated biochar prepared from oak wood waste required more time for activation, and activated biochar prepared from oak wood waste shows high adsorption capacity as compared to other materials. Apparently, the biomass with lower ash content and high fixed carbon content should be preferred in the gasification process for the production of activated biochar Haykiri-Acma et al. (2006).

Ahmedna et al. (2000) used a physical activation process for granular-activated biochar from different agro waste, such as rice hulls, rice straw, and sugarcane bagasse, etc. The granular-activated biochar prepared from sugarcane bagasse found more desirable in sugar decolorizing than rice straw and hulls. Hernandez et al. (2007) used cotton gin trash as a agro waste for making activated biochar at different pyrolysis temperatures 600°C, 700°C, and 800°C for 30, 45, and 60 min, and then activated using steam for a temperature range from 250°C–600°C for 60 min. During the experiment, the authors observed that as pyrolysis temperature and time were increased, then the yield of char significantly reduced. The iodine number increased value from 200–427, when pyrolysis temperature was 700°C for 45 min.

11.7.2 Chemical Activation

Impregnation of a catalyst or chemical agent into a biomass, followed by pyrolysis, produces an end product such as an activated biochar. Compared to physical activation, chemical activation provides two salient advantages: one is that it requires lower temperature, and another is that this process produces a global yield of activated biochar (Chen et al., 2011). According to Molina-Sabio and Rodríguez-Reinoso (2004), the temperature required for wet or chemical activation varies from 400°C to 600°C. Many studies report that chemical activation needed a lower temperatures in comparison with physical activation, varying from 300°C–700°C (Giraldo and Moreno-Piraján, 2012), 400°C–700°C (Girgis et al., 2002), or 400°C to 800°C (Alhamed, 2006), and 500°C to 800°C (Hu and Srinivasan, 2001). The variation in temperature range during activation is due to impregnation of inorganic additives and their action for degradation or dehydration of cellulosic materials present in the precursor. Activating or chemical agents involved in chemical activation act as oxidants or dehydrating agents, assisting with pyrolytic decomposition. The maximum carbon percentage is recorded in activated biochar obtained from chemical activation and may be due to the used chemical oxidants inhibit the formation of tar and ash. The following chemical or oxidizing agents act as a catalyst and take part in activation: zinc chloride (Z_nCl_2), phosphoric acid (H_3PO_4) (Cruz, et al., 2012) sulphuric acid (H_2SO_4), potassium sulphide (K_2S), potassium thiocyanate (KSCN) (Demiral et al., 2008), nitric acid (HNO_3), hydrogen peroxide (H_2O_2), potassium permanganate (KM_nO_4) (Al-Qodah and Shawabkah, 2009), sodium hydroxide (N_aOH), potassium hydroxide (KOH) (Zhengrong and Xiaomin, 2013), and potassium carbonate (K_2CO_3) (Adinata, et al., 2007). Zinc chloride is the most preferred chemical for the degradation of lignocellulosic material. However, zinc chloride does have some environmental drawbacks (Teng et al., 1998). In addition, phosphoric acid is being used for activation because it does not have any environmental effect compared to zinc chloride, but it does have an inefficient chemical recovery and minimizes the problems regarding corrosion. Many reviewers report that any precursors used in chemical activation should be of cellulosic or lignocellulosic origin, such as coconut, peach stone, coal, hard wood, or peach stone. These materials are used for making modified or activated biochar by impregnating the chemical agents into precursors. Among all the oxidizing chemical agents, zinc chloride and phosphoric acid are the most preferred for the modification of lignocellulosic biomass, while potassium hydroxide is mostly used on char as a virgin material to make activated biochar. According to Srinivasakannan and Bakar (2004), activated biochar prepared by impregnation of zinc chloride into a precursor produces a high surface area compared to that produced by phosphoric acid.

The final, and one of the most important stages in activated biochar preparation using chemical treatment, is washing. It helps to determine the porosity of the resulting sample. Generally, some acids or alkali are used to wash activated biochar, followed by washing it with water. Basically, the process of washing removes chemical contaminants from the surface of biochar. Kalderis et al. (2008) have made activated biochar in a single-stage chemical-activation process by using raw material bagase and rice husk. The authors examined the surface area of activated biochar by using three different chemicals ($ZnCl_2$, N_aOH, H_3PO_4) at three different temperatures (600°C, 700°C, 800°C) and reported that activated biochar prepared from rice husk showed greater surface area up to 750 m^2/g with an equal concentration of zinc chloride and rice husk (1:1). It was also observed that zinc chloride was more effective oxidizing agent than N_aOH or H_3PO_4. Sharath et al. (2017) used rice husk for the preparation of activated biochar by using sodium hydroxide (N_aOH) as a chemical. The modification was carried out at three different temperatures 650°C, 700°C and 800°C and noted a higher yield of about 48.2% at 650°C and corresponding yields at 47.65% and 45.95%, respectively. Mahamad et al. (2015) produced biochar activated from pineapple waste (including leaves, crown, and stem) for the application of dye removal from wastewater. The activated biochar was prepared by using impregnation of $ZnCl_2$ into biomass, and activation was conducted at 500°C for duration of 1 h. The resulting product showed a greater surface area (914.6 m^2/g) and indicated the highest dye removal ability due to its adsorption capacity (288.34 mg/g). The impregnation ratio and process temperature could effect the porosity, as well as the surface chemistry of the resulting sample investigated by Prahas et al. (2008). Further, the authors prepared activated biochar from jack fruit peel waste using H_3PO_4; chemical activation took place in the temperature range of 450°C–550°C and found that the resulting end product had a maximum surface area (907–1260 m^2/g) and pore volume in between 0.525 and 0.733 cm^2/g, respectively.

11.8 CHARACTERISTIC OF MODIFIED BIOCHAR

The physical and chemical characterization of activated biochar is significantly influenced by kind of biomass, pyrolysis temperature, residence time, heating rate, and activating agents, etc. The properties of modified biochar-like mineral contents, molar ratio, pH, surface area, porosity, surface charge, and elemental composition significantly varied according to activation condition (Panwar and Pawar, 2020). Many pyrolysis technologies, such as fluidized bed reactor, vacuum pyrolyzers, etc., play a significant role while determining the morphology and structural composition of activated biochar. In addition, the activated biochar obtained via physical activation of biomass showed larger surface area and microporous structure, while chemically (H$_3$PO$_4$) modified biochar showed surface area up to 600 m^2/g, respectively. The spouted bed reactor was found most promising for activated biochar preparation instead of fixed and fluidised bed reactor. Spouted bed reactor promotes uniform distribution of heat inside the pyrolysis chamber and mixing of raw material. Owing to this Niksiar and Nasernejad (2017) found that the physically modified biochar using pistachio shell activated biochar showed larger surface area 2596 m^2/g at a activation of 850°C. This might have happened due to significantly less heat loss and resistance between the biomass and heat inside the reactor.

Activated biochar prepared using steam activation exhibit the enriched textural properties of activated biochar. Steam modification build up the better properties of activated biochar, mainly larger surface area, pore volume, and higher porosity, etc. (Pallarés et al., 2018). Besides, this physical activation using carbon dioxide creates greater microporous structure as compared to steam activation. The activated biochar obtained through the steam-activation process contains less percentage of hydrogen to carbon, oxygen to carbon, and nitrogen to carbon molar ratios than chemically activated biochar. This might be due to steam modification containing less oxygen, carbon, and hydrogen (Rajapaksha et al., 2015).

In cases of acid activation, the presence of oxygenated functional groups on activated biochar with a larger surface area is greater. The phosphoric acid-based pine sawdust modified biochar showed larger surface area up to 950 m^2/g, respectively (Iriarte-Velasco et al., 2016). In addition, the cucumber waste biochar, which was treated using sulphuric and oxalic acid, also showed larger surface area and pore volume. In some cases, however, acid-activated biochar showed lower surface area. This could be due to breakage of the pore structure inside the biochar. In addition, Vaughn et al. (2017) found that oxalic and sulphuric acid-activated biochar possess minimum surface area, pH, electrical conductivity, surface area, and cation exchange capacity compared to pristine biochar. The acid-activated biochar has low pH and therefore has a potential application in alkaline soil. In some cases, acid-activated biochar possesses an average value of hydrogen to carbon, oxygen to carbon, and nitrogen to carbon molar ratio, which might be caused due to the produced acid-activated biochar containing a lower percentage of carbon. Recently, Zhang et al. (2019) made modified biochar from organic waste using acid agents HCL, H_2O_2, NaOH for the adsorption of nitrobenzene from the wastewater. Among the three different chemicals, the hydrochloric acid-modified biochar was observed to be more efficient for effective remediation of NB. This might be possible because of larger surface area, higher electron transfer rate, and the availability of acidic functional groups as compared to other activated biochar.

Alkali-activated biochar showed good physico-chemical properties for wastewater treatment. Potassium hydroxide biochar possesses a larger surface area, almost 50 times greater than raw biochar. The alkali-activated biochar can enhance the cation exchange capacity, porosity, thermal stability, surface area, and also improve surface functional groups, therefore resulting in modified biochar showing many environmental applications. Sodium hydroxide-treated modified biochar produced using *Hickey* biomass showed good mineral composition on the surface of biochar (Ding et al., 2016). (Table 11.1).

11.9   APPLICATION IN WATER AND WASTEWATER TREATMENT

Open water sources, such as ponds, lake, rivers, contains many toxic pollutants, heavy metals, and other organic and inorganic pollutants that causes water pollution. These harmful pollutants primarily originate with certain industries, such as mining, leather and battery industries, as they discharge pollutants into the water (Clemens and Ma, 2016). Heavy metals are recognized as the most toxic and dangerous pollutants, and the potential for bioaccumulation and carcinogenicity is extremely harmful for living organisms in the environment. Therefore, these toxic pollutants that threaten the environment also create health issues for humans (Zou et al., 2016). There are various traditional and modern wastewater treatment technologies, such as membrane removal, flotation, ion exchange, and chelation (Fu and Wang, 2011; Huang et al., 2016; Kaya, 2016). These techniques, however, are not economically feasible or effective, and they produced toxic secondary waste material like volatile organic and inorganic compounds and persistent pollutants, etc. Therefore, there is an urgent need to optimise environmentally friendly treatments that effectively remove heavy metals. The availability of functional groups like carboxyl, hydroxyl, and carbonyl shows a remarkable potential for the adsorption of contaminants for wastewater. Among the different adsorbent materials is modified biochar, which is prepared with a different activation technique that has attracted widespread attention (Li et al., 2010). The adsorption of heavy metals and pollutants from wastewater is significantly attributed to targeted pollutants and activated biochar. For example, the Mg-activated biochar showed enhanced adsorption capacity (ranging from 6.4% to 98.9%) to climinate Pb^{2+} from wastewater (Wang et al., 2015). Magnesium-modified biochar showed good adsorption performance due to the availability hydroxyl groups on the biochar surface. Similarly, MgO and citric acid activated biochar also reported Pb(II) removal ability of 121.8 mg/g and 159.9 mg/g, respectively (Cerino-Córdova et al., 2013). In addition, zinc nano-composites based magnetic biochar showed good sorption performance for a targeted Pb(II) heavy metal of 367.6 mg/g. This might be due to the availability of zinc nano particles on the activated biochar surface (Yan et al., 2015). The biochar-produced rice husk and MSW were further modified using ferrous oxide

TABLE 11.1

Activation Conditions and Its Affect on Characteristics of Activated or Modified Biochar

Precursor/Particle Size	Carbonization Condition	Activation Agent	Activation Condition	C (%)	H (%)	O (%)	N (%)	Molar H/C	Molar O/C	Molar (O+N/C)	Surface Area (m²·g⁻¹)	Pore volume (cm³·g⁻¹)	Pore diameter (nm)	References
Burcucumber plants (< 1.0 mm)	700°C/2h, 7°C/min	Steam	700°C, 5 mL/min, 45 min	50.55	1.66	44.88	2.54	0.39	0.67	0.71	7.10	0.038	8.393	(Rajapaksha et al., 2015)
Giant *miscanthus* (4 cm)	N$_2$, 500°C/1h, 10°C/min	Steam	800°C	82.1	2.67	11.0	0.31	0.0325		0.138	322	n.a.	n.a.	(Shim et al., 2015)
Barley malt bagasse (< 4 mm)	N$_2$, 800°C/1h, 10°C/min, 0.25 mL/min	CO$_2$	900°C/1h, 10°C/min, 0.15 L/min	68.32	1.46	28.38	1.74	0.21			80.5	0.0468		(Franciski et al., 2018)
Corn straw	500°C/1.5h	KOH	800°C/0.5h, 2 g: 500 mL (Biochar: KOH)	67.20	0.80	6.8	0.537	0.144	0.076	0.082	466.37	0.081	4.40	(Tan et al., 2016)
sesame straw	600°C/2h, 5°C/min	ZnCl$_2$	600°C	74.1	1.8	21.1	2.7	0.29			319.4	0.2270		(Park et al., 2015)
Sugarcane residue	n.a.	MgO	N$_2$, 550°C/1h	27.17	2.37		1.47				40.6	0.372	22.38	(Li et at., 2017)

and calcium oxide for the removal of Cr^{6+} and As^{5+}. Due to electrostatic interaction between the targeted pollutants and activated biochar, it reported a 95% removal capacity (Agrafioti et al., 2014). Similarly, the haematite-activated biochar also showed good adsorption capacity for the sorption of arsenic for potable water. The impregnation of Fe^{3+} on the biochar surface significantly affected the $(O+N)/C$ and oxygen-to-carbon ratio, which results in enriched adsorption capacity of activated biochar for targeted As^{5+} sorption. Thus, the impregnation of haematite, magnate, calcium, and manganese oxide on biochar surface assists in creating the magnetism, adsorption sites, and its widespread distribution.

Several organic contaminants create a serious result when they enter the animal body via water and the food chain. Mostly, the organic pollutant like phenol mainly affect the taste and odour of the drinking water. The activated biochar prepared from different activating agents possess good affinity and adsorption capacity for the adsorption of such types of contaminates. The alkali- and acid-modified biochar also showed higher furfural removal. Langmuir sorption capacity varies between 93.5 to 109 mg/g, respectively. The properties like basicity and hydrophilicity for alkali- and acid-derived biochar showed a good adsorption performance. However, the bamboo waste-modified biochar reported the highest furfural removal capacity of 253.20 mg/g, due to the hydrophobicity property of activated biochar (Li et al., 2014). Saw dust biochar activated using citric acid showed the higher adsorption ability of 158.6 mg/g for removal of methylene blue from wastewater, which could be achieved because of the presence of carboxyl groups on the modified biochar surface. The hydrogen peroxide, acidic, and potassium permanganate-activated biochar prepared at higher activation temperature o 550°C showed moderate phenol removal capacity of about 93.5 mg/g, respectively. Sometimes, alkaline groups like OH- ions and $-NH_2$ reacted with functional groups that present on the modified biochar surface, which results in improved removal of negatively charged organic contaminants from wastewater. The Cr (VI) adsorption was recorded in the range of 46.9 to 94.4 mg/g using sugarcane bagasse-derived activated biochar using zinc oxide. It was observed that alkali-activated biochar showed the highest adsorption capacity 435.7 mg/g for Cr (VI) from wastewater. The plentiful availability of amino groups was present on alkali-activated biochar surface, which became one of the favorable conditions for the removal of heavy metals.

11.10   CONCLUSION

The production of biochar from crop residues via the thermochemical conversion process and further preparation of activated biochar provide a new platform for sustainable solid-waste management. Different activation methods could be used to prepare a good quality activated biochar. Chemical activation substantially improves the surface physicochemical properties of biochar. The acid and alkali agents help modify the physical characteristics and improve the availability of functional groups on activated biochar surfaces. However, the alkali-modified biochar possesses higher surface morphology, aromaticity, and N/C molar ratio. Heavy metals are toxic and dangerous, and they have extremely harmful effects on living organisms and the environment due to their bioaccumulation and carcinogenicity. The activated biochar is considered a promising material for the adsorption of heavy metals and toxic contaminants from wastewater.

REFERENCES

Adinata, D, Daud, WMAW, & Aroua, MK (2007). Preparation and characterization of activated carbon from palm shell by chemical activation with K2CO3. *Bioresource Technology*, 98, 145–149.

Agrafioti, E, Kalderis, D, & Diamadopoulos, E (2014). Ca and Fe modified biochars as adsorbents of arsenic and chromium in aqueous solutions. *J Environ Manag*, 146, 444–450.

Ahmedna, M, Marshall, W, & E, Rao (2000). Production of granular activated carbons from select agricultural byproducts and evaluation of their physical, chemical and adsorption properties. *Bioresource Technol*, 71, 113–123.

Alhamed, YA (2006). Activated carbon from dates' stone by ZnCl2 activation. *JKAU Eng Sci*, 17, 75–100.

Al-Qodah, Z, & Shawabkah R (2009). Production and characterization of granular activated carbon from activated sludge. *Brazilian Journal of Chemical Engineering*, 26, 127–136.

Al-Swaidan, HM, & Ahmad, A (2011). Synthesis and characterization of activated carbon from Saudi Arabian dates tree's fronds wastes. In *3rd International Conference on Chemical, Biological and Environmental Engineering* (Vol. 20, pp. 25–31).

Baseri, JR, Palanisamy, PN, & Sivakumar, P (2012). Preparation and characterization of activated carbon from Thevetia peruviana for the removal of dyes from textile waste water. *Adv. Appl. Sci. Res*, 3, 377–383.

Carrott, PJM, & Carrott, MR (2007). Lignin–from natural adsorbent to activated carbon: A review. *Bioresource Technology*, 98(12), 2301–2312.

Cerino-Córdova., FJ, Díaz-Flores, PE, & García-Reyes, RB (2013). Biosorption of Cu (II) and Pb (II) from aqueous solutions by chemically modified spent coffee grains. *Int J Environ Sci Technol*, 10, 611–622.

Chen, Y, Z.hu, Y, Wang, Z, Li, Y, Wang, L, Ding, L, & Guo, Y (2011). Application studies of activated carbon derived from rice husks produced by chemical-thermal process—A review. *Advances in Colloid and Interface Science*, 163, 39–52.

Clemens, S, & Ma, F (2016). Toxic heavy metal and metalloid accumulation in crop plants and foods. *Annu. Rev. Plant Biol.*, 67, 489.

Cruz, G, Pirilä M, Huuhtanen., M, Carrión., L, Alvarenga, E, & Keiski, RL (2012). Production of activated carbon from cocoa (Theobroma cacao) pod husk. *J Civil Environment Engg*, 2, 2.

Daud, WMAW, Ali, WSW, & Sulaiman, MZ (2000). The effects of carbonization temperature on pore development in palm-shell-based activated carbon. *Carbon*, 38, 1925–1932.

Demiral, H, Demiral, I, Tümsek, F, & Karabacakoğlu, B (2008). Pore structure of activated carbon prepared from hazelnut bagasse by chemical activation. *Surface and Interface Analysis: An International Journal devoted to the Development and Application of Techniques for the Analysis of Surfaces, Interfaces and Thin Films*, 40, 616–619.

Ding, Z, H.u, X, & Wan, Y (2016). Removal of lead, copper, cadmium, zinc, and nickel from aqueous solutions by alkali-modified biochar: Batch and column tests. *Journal of Industrial and Engineering Chemistry*, 33, 239–245.

Fahmy, TY, Fahmy, Y, Mobarak, F, El-Sakhawy, M, & Abou-Zeid, RE (2020). Biomass pyrolysis: Past, present, and future. *Environment, Development and Sustainability*, 22, 17–32.

Franciski, MA, Peres, EC, & Godinho M (2018). Development of CO2 activated biochar from solid wastes of a beer industry and its application for methylene blue adsorption. *Waste Management*, 78, 630–638.

Fu, F, & Wang, Q (2011). Removal of heavy metal ions from wastewaters: A review. *J. Environ. Manag.* 92, 407–418.

Giraldo, L, & Moreno-Piraján JC (2012). Synthesis of activated carbon mesoporous from coffee waste and its application in adsorption zinc and mercury ions from aqueous solution. *Journal of Chemistry*, 9(2), 938–948.

Girgis, BS, Yunis SS, & Soliman, AM, (2002). Characteristics of activated carbon from peanut hulls in relation to conditions of preparation. *Materials Letters*, 57, 164–172.

Gratuito, MKB, Panyathanmaporn, T, Chumnanklang, RA, Sirinuntawittaya, NB, & Dutta, A (2008). Production of activated carbon from coconut shell: Optimization using response surface methodology. *Bioresource Technology*, 99, 4887–4895.

Haykiri-Acma, H, Yaman, S, & Kucukbayrak, S (2006). Gasification of biomass chars in steam – nitrogen mixture. *Energy Conversion and Management*, 47, 1004–1013.

Hernandez, JR, Aquino, FL, & Capareda, SC (2007). Activated carbon production from pyrolysis and steam activation of cotton gin trash. In *2007 ASAE Annual Meeting* (p. 1). American Society of Agricultural and Biological Engineers.

Hernández-montoya, V, García-servin, J, & Bueno-lópez, JI (2012) Thermal treatments and activation procedures used in the preparation of activated carbons – Characterization Techniques and Applications in the Wastewater Treatment, Dr. Virginia Hernández Montoya (Ed.), ISBN: 978, 953-51-0197-0. IntechOpen , pp. 19–36.

Hiloidhari, M, Das, D, & Baruah, DC (2014). Bioenergy potential from crop residue biomass in India. *Renewable and Sustainable Energy Reviews*, 32, 504–512.

Hiremath, MN, Shivayogimath, CB, & Shivalingappa, SN (2012). Preparation and characterization of granular activated carbon from corn cob by KOH activation. *International Journal of Research in Chemistry and Environment*, 2(3), 84–87.

Hu, Z, & Srinivasan, MP (2001). Mesoporous high-surface-area activated carbon. *Microporous and Mesoporous Materials*, 43, 267–275.

Huang, Z, Lu, L, Cai, Z, & Ren, Z (2016). Individual and competitive removal of heavy metals using capacitive deionization. *J. Hazard. Mater*, 302, 323–331.

Ioannidou, O, & Zabaniotou, A (2007). Agricultural residues as precursors for activated carbon production—A review. *Renewable and Sustainable Energy Reviews*, 11(9), 1966–2005.

Iriarte-Velasco, U, Sierra, I, & Zudaire, L (2016). Preparation of a porous biochar from the acid activation of pork bones. *Food and Bioproducts Processing*, 98, 341–353.

Kalderis, D, Bethanis, S, Paraskeva, P, & Diamadopoulos, E, (2008). Production of activated carbon from bagasse and rice husk by a single-stage chemical activation method at low retention times. *Bioresource Technology*, 99, 6809–6816.

Kaya, M (2016). Recovery of metals and nonmetals from electronic waste by physical and chemical recycling processes. *Waste Manag*, 57, 64–90.

Li, Y, Chen, B, & Zhu, L (2010). Single-solute and bi-solute sorption of phenanthrene and pyrene onto pine needle cuticular fractions. *Environ. Pollut*, 158, 2478–2484.

Li, H, Dong, X, & da Silva, EB (2017). Mechanisms of metal sorption by biochars: Biochar characteristics and modifications. *Chemosphere*, 178, 466–478.

Li, Y, Shao, J, & Wang, X (2014). Characterization of modified biochars derived from bamboo pyrolysis and their utilization for target component (furfural) adsorption. *Energy Fuel*, 28, 5119–5127.

Mahamad, MN, Zaini, MAA, & Zakaria, ZA (2015). Preparation and characterization of activated carbon from pineapple waste biomass for dye removal. *International Biodeterioration & Biodegradation*, 102, 274–280.

McDougall, GJ (1991). The physical nature and manufacture of activated carbon. *Journal of the Southern African Institute of Mining and Metallurgy*, 91(4), 109–120.

Molina-Sabio, M, & Rodrıguez-Reinoso, F (2004). Role of chemical activation in the development of carbon porosity. *Colloids and Surfaces A: Physicochemical and Engineering Aspects*, 241(1–3), 15–25.

Niksiar, A, & Nasernejad, B (2017). Activated carbon preparation from pistachio shell pyrolysis and gasification in a spouted bed reactor. *Biomass and Bioenergy*, 106, 43–50.

Nor, NM, Lau, LC, Lee, KT, & Mohamed, AR (2013). Synthesis of activated carbon from lignocellulosic biomass and its applications in air pollution control—A review. *Journal of Environmental Chemical Engineering*, 1(4), 658–666.

Nowicki, P, Pietrzak, R, & Wachowska, H (2010). Sorption properties of active carbons obtained from walnut shells by chemical and physical activation. *Catalysis Today*, 150, 107–114.

Pallarés, J, González-Cencerrado, A, & Arauzo, I (2018). Production and characterization of activated carbon from barley straw by physical activation with carbon dioxide and steam. *Biomass and Bioenergy*, 115, 64–73.

Panwar, NL, & Pawar, A (2020). Influence of activation conditions on the physicochemical properties of activated biochar: A review. *Biomass Conversion and Biorefinery*, 12, 925–947.

Panwar, NL, Pawar, A, & Salvi, BL (2019). Comprehensive review on production and utilization of biochar. *SN Applied Sciences*, 1(2), 168.

Park, JH, Ok, YS, Kim, SH, et al. (2015). Evaluation of phosphorus adsorption capacity of sesame straw biochar on aqueous solution: Influence of activation methods and pyrolysis temperatures. *Environmental Geochemistry and Health*, 37, 969–983.

Pawar, A, & Panwar, NL (2020). Experimental investigation on biochar from groundnut shell in a continuous production system. *Biomass Conversion and Biorefinery*, 12, 1093–1103.

Pawar, A., & Panwar, NL (2022). A comparative study on morphology, composition, kinetics, thermal behaviour and thermodynamic parameters of Prosopis Juliflora and its biochar derived from vacuum pyrolysis. Bioresource Technology Reports, 18, 101053.

Pawar, A, & Panwar, NL (2022). Analysis of biochar from carbonisation of wheat straw using continuous auger reactor. International Journal of Environment and Sustainable Development, 21(1–2), 218–225.

Prahas, D, Kartika, Y, Indraswati, N, & Ismadji, S (2008). Activated carbon from jackfruit peel waste by H3PO4 chemical activation: Pore structure and surface chemistry characterization. *Chemical Engineering Journal*, 140, 32–42.

Rajapaksha, AU, Vithanage, M, & Ahmad, M (2015). Enhanced sulfamethazine removal by steam-activated invasive plant-derived biochar. *Journal of Hazardous Materials*, 290, 43–50.

Sharath, D, Ezana, J, & Shamil, Z (2017). Production of activated carbon from solid waste rice peel (husk) using chemical activation. *Journal of Industrial Pollution Control*, 33, 1132–1139.

Shim, T, Yoo, J, & Ryu, C (2015). Effect of steam activation of biochar produced from a giant Miscanthus on copper sorption and toxicity. *Bioresource Technology*, 197, 85–90.

Srinivasakannan, C, & Bakar, MZA (2004). Production of activated carbon from rubber wood sawdust. *Biomass and Bioenergy*, 27, 89–96.

Suhas, PJM, Carrott, MML, & Carrott, R (2007). Lignin – from natural adsorbent to activated carbon: A review. *Bioresource Technology*, 98, 2301–2312.

Tan, G, Sun, W, & Xu, Y (2016). Sorption of mercury (II) and atrazine by biochar, modified biochars and biochar based activated carbon in aqueous solution. *Bioresource technology*, 211, 727–735.

Teng, H, Yeh, TS, & Hsu, LY (1998). Preparation of activated carbon from bituminous coal with phosphoric acid activation. *Carbon*, 36, 1387–1395.

Tsai, WT, Chang, CY, & Lee, SL (1997). Preparation and characterization of activated carbons from corn cob. *Carbon*, 35, 1198–1200.

Vaghela, DR, Pawar, A, & Sharma, D (2022). Effectiveness of Wheat Straw Biochar in Aqueous Zn Removal: Correlation with Biochar Characteristics and Optimization of Process Parameters. BioEnergy Research, 1–15.

Vaughn, SF, Kenar, JA, & Tisserat, B (2017). Chemical and physical properties of Paulownia elongata biochar modified with oxidants for horticultural applications. *Industrial crops and products*, 97, 260–267.

Wang, S, Gao, B, & Li, Y (2015). Manganese oxide-modified biochars: Preparation, characterization, and sorption of arsenate and lead. *Bioresource Technology*, 181, 13–17.

Yan, L, Kong, L, & Qu, Z (2015). Magnetic biochar decorated with ZnS nanocrytals for Pb (II) removal. *ACS Sustain Chem Eng*, 3, 125–132.

Zanzi, R, Bai, X, Capdevila, P, & Bjornbom, E (2001). Pyrolysis of biomass in presence of steam for preparation of activated carbon. *Liquid and Gaseous*.

Zhang, L, Candelaria, SL, Tian, J, Li, Y, Huang, Y, & Cao, G (2013). Copper nanocrystal modified activated carbon for supercapacitors with enhanced volumetric energy and power density. *J Power Sources*, 236, 215–223.

Zhang, D, Li, Y, & Sun, A (2019). Enhanced nitrobenzene reduction by modified biochar supported sulfidated nano zerovalent iron: Comparison of surface modification methods. *Science of the Total Environment*, 694, 133701.

Zhang, T, Walawender, WP, Fan, LT, Fan, M, Daugaard, D, & Brown, RC (2004). Preparation of activated carbon from forest and agricultural residues through CO2 activation. *Chemical Engineering Journal*, 105, 53–59.

Zhengrong, G, & Xiaomin, W (2013). Carbon materials from high ash bio-char: A nanostructure similar to activated grapheme. *Am Trans Eng Appl Sci*, 2, 15–34.

Zou, Y, Wang, X, Khan, A, Wang, P, Liu, Y, Alsaedi, A, & Wang, X (2016). Environmental remediation and application of nanoscale zero-valent iron and its composites for the removal of heavy metal ions: A review. *Environ. Sci. Technol*, 50, 7290–7730.

12 A Study on the Cause and Effects of Paddy Straw Burning by Farmers in Fields and Proposal of Using Rice Husk as a Novel Ingredient in Pottery Industry

Shikha Tuteja and Vinod Karar
CSIR-Central Scientific Instruments and Organisation, Chandigarh, India

Academy of Scientific and Innovative Research (AcSIR), Ghaziabad, India

Chandigarh University, Gharuan, Mohali, Punjab, India

Ravinder Tonk
Chandigarh University, Gharuan, Mohali, Punjab, India

CONTENTS

DOI: 10.1201/9781003127819-12

12.1 INTRODUCTION

In Haryana and Punjab, rice is grown under a well-irrigated system. It is one of the primary sources of earnings in these states. The majority of the population in these locations depends directly upon agriculture, which provides employment to the locals. The rice hull is considered as residual-manufactured goods during the rice-refining method. After this process, excess rice husk is dumped or burned, which causes pollution (Kumar et al., 2013). These rice husks can be used to benefit society and avoid the pollution or other respiratory diseases.

Agriculture in Haryana and Punjab states suffers from certain limitations, especially residue that is burned after the rice-refining process, causing several health ailments such as respiratory infections and air pollution, which converts into smog (smoke + fog) and leads to poor vision. According to the World Health Organization, the moderate and bad quality of air we breathe is accepted as the primary cause of an epidemic disease. In certain states, like Delhi and NCR, the problem of poor air quality becomes so severe that the suspended particles in air are 20 times more than what is safe for breathing.

The primary reason behind this poor air quality is the burning of the crop-residue left standing in fields after farmers have cut the crop of rice. This problem is most severe in the states of Punjab and Haryana in the north Indian region. This practice has been going on unabated for many years, with farmers stating that it is economically unviable for them to remove the residual rice stubble using machines (Muntohar, 2002; Rozainee et al., 2008). The government is yet to come up with any proper plans to support the farmers. As a result, agriculturalists are left with no choice nonetheless to burn the rice-stubble. But this practice has been causing severe respiratory infections in people living in Delhi and NCR.

TABLE 12.1

Organic and Inorganic Traits and the Range of Different Content Available in Rice Husk

Physical and Chemical Characteristics	General Range
Density	90–165 kg/m^3
Hardness (Mohr No.)	5.0–6.0
C – Content	35% approximately
Ash content	20.0%–30.0% approximately
H – Content	3.0%–6.0%
O – Content	30.0%–35.0%
N – Content	0.20%–0.30%
S – Content	0.04%–0.08%
Content of Moisture	7.0%–9.0%

12.1.1 ORGANIC AND INORGANIC PROPERTIES OF RICE HULL

It is obvious that rice husk is mainly composed of organic material, and the remaining part is made up inorganic materials and silica. Its inorganic properties include: its very high quantity of left ash when burned, a high porosity, its light weight, and the high surface area compared to its mass (Assureira, 2002; Giddel and Jivan, 2007). The high amount of silica content makes rice husk very suitable for few industrial uses (Table 12.1).

12.1.2 PROPOSED SOLUTIONS FOR EFFECTIVE/ALTERNATE USE OF RICE HUSK

Rice husk is used in various fields that are based on the type of one or more primary chemical or physical properties to be put in use for a purpose (Chungsangunsit et al., 2009). For example, rice husk is used directly in many power plants as a fuel for burning. It is very useful as a basic component while making a lot of silica/silicon compounds (Figure 12.1).

- It may be used as mixing agent in building materials.
- The gradual decomposition is gradual and slow.
- Resistance toward moisture diffusion is good.
- The overall decay rate of this material is very slow.
- It is a great insulator.
- It may act as a renewable source of energy. (Figure 12.2).

Rice hull can be further utilized in products (Ragadhita et al., 2019). Different by-products of rice husk are listed below.

- Rice Bran and Oil
- Rice Crunch

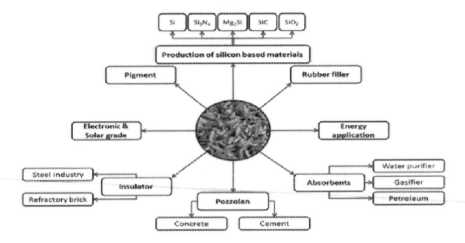

FIGURE 12.1 By products of rice husk.

Source: Rice husk – IRRI Rice Knowledge Bank.

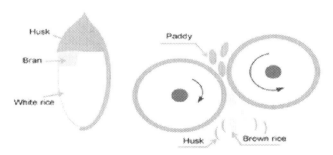

FIGURE 12.2 Configuration of the rice husk or paddy grain and its product after husking.

Source: Rice husk – IRRI Rice Knowledge Bank.

- Rice Husk
- Rice Starch
- Cinders from Husk
- Broken Rice
- Brewers Rice

12.1.3 As a Fuel for Burning in Power Plants for Producing Steam

Production of heat in any combustion process is a very basic process, but it is also dependent on the calorific value and ease of burning of the fuel. Rice husk, when in a dried/low-moisture state, is an excellent fuel for burning for the specific type of boilers where output required may be low (Nuzzolese et al., 2016). Also, it may be used in place of few conventional resources of fuel in poor households and rural/remote areas.

12.1.4 Preparing Active Adsorbent Carbon

It contains a great percentage of cellulose and other related components, which makes it ideally suitable as a raw material for preparing active adsorbent carbon. The desired products have a great micro-porosity type structure.

12.1.5 May Be Used as Fertilizer

One issue with rice husk is that it decomposes very slowly. However, there are a few processes through which decomposition can be accelerated and rice husk can be then used as an effective fertilizer (Matori et al., 2009). It may also find use in gardening when used as a substrate.

12.1.6 As the Ingredient of Pet-Foods

Although rice husk is not suitable for direct eating by domesticated animals like cows, buffalos, etc., after processing, it may become suitable for eating when mixed with other ingredients and serve as a wonderful fiber and a cheap food source also (Emmanuel et al., 2006).

12.1.7 Substrate

It can act as the surface or material on or from which an organism lives, grows, or obtains its nourishment for developing silica/Si compounds.

12.1.8 Construction Material

Another interesting application may be to use rice husk as a basic component for making construction material, such as special bricks. Potential problems are its high inflammability and strength (Mehta, 1978).

12.1.9 Few Other Applications

It may be used as a polishing agent, as cardboard material in the stationery industry, as insulating material, filler material, etc.

12.1.10 The Agricultural Drying Process

A great amount of energy is wasted in a few drying operations related to the agricultural industry, like drying of onions and paddy crop (Sarangi et al., 2009). The rice husk otherwise wasted may be utilized for these purposes in a controlled environment. It may reduce the time taken to a great extent.

To understand the reasons, we evaluate the adsorption isotherm of mesopore-free submicron silica debris. Mesopore-free submicron silica debris was organized from acid-base extraction of rice husk ash underneath a surfactant-free situation. The adsorption isotherm evaluation was accomplished within the borosilicate batch reactor gadget below, consistent pH condition in addition to room temperature and strain. Then, the results were compared to both Langmuir and Freund lich models. As a version of adsorbate, curcumin is utilized. To help the adsorption evaluation, numerous characterizations have been performed, including electron microscope, x-ray diffraction, and Fourier rework infrared (Ragadhita et al., 2019). The experimental results showed that the present mesopore-unfastened submicron silica particles have been efficiently adsorbing curcumin molecules. The adsorption isotherm confirmed that the equilibrium adsorption statistics of the present silica particles were fit to the Freund lich isotherm model, confirming the adsorption takes place on heterogeneous surfaces with multilayer adsorption. This informs molecule-molecule interaction at the adsorption layers. The low adsorption found is due to the lifestyles of mesopore-free shape on the silica adsorbent. This examination gives statistics for the importance of a variable influencing the adsorption capacity.

Graphene have become correctly synthesized through activating rice husk ash (RHA) using potassium hydroxide (KOH) at 800° Centigrade with 1:2 impregnation ratios. Raman spectroscopy evaluation showed the presence of graphitic structure. The tested approach makes use of RHA as carbon supply and used as cost to prevent oxidation in the route of synthesis method at the aggregate of KOH and RHA towards air at excessive temperature. The novelty of this synthesis methodology used environmentally-friendly biomass, a beneficial resource as a starting material, does not use catalysts, and shows that graphene can be synthesized at a really low synthesis temperature.

12.2 METHODOLOGY AND EXPERIMENTATION

In the current research work, rice husk has been used to make pottery products using a conventional pottery process in a village. The methodology adopted is discussed in detail below. The following physical properties of rice husk were measured to check its change in appearance and other important characteristics.

- The dry shrinkage rate
- The value of moisture level
- Level of porosity
- Change in coloration
- Time to dry
- Rate of absorption

The husk of rice is used as raw material, and lemon juice and carboxymethylcellulose is used as additive. All the samples were prepared by first sieving and washing the rice, and the husk was dried for 4–5 h in heating oven at 45° and converted to powdered state. Then husk paste is prepared using 3.5 g of the carboxymethylcellulose (CMC) mixed with 100 g water for preparing 3.5% glue; the glue is mixed uniformly with the rice husk powder in the ratio of 96.5%:3.5%, with a little bit of lime juice, and then the mixture is processed under autoclave for 20 min at 120°. After that, the mixture is sterilized with UV rays for 15 min at UV lamp. Then the sterilized paste is rolled in the pleated sheet. The rolled sheet is given shape by moulding in different types of moulders and dried at 80° in oven; then secondary shaping is given to eliminate cracks,

(a)   (b)   (c)

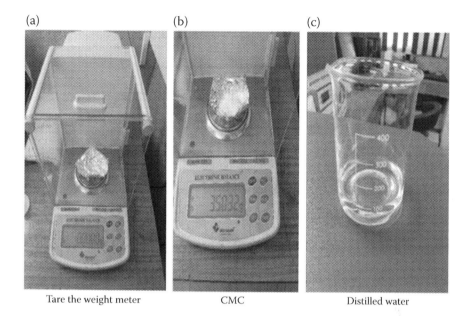

Tare the weight meter   CMC   Distilled water

FIGURE 12.3   Measurement of CMC and distilled water.

which is done by applying the husk paste on the cracks of the primary product and drying it at 80° in the oven (Figure 12.3).

Then, after the dried secondary product is rubbed with rough rubbing paper to flatten its layers, water-resistive paste is applied on its surface, completing the product formation (Figure 12.4).

12.2.1   PRODUCTS PLASTICITY (ASTM-D4318-10)

This test was done by rolling a coil of paste of husk and then wrapping it about the finger. The surface of husk paste was observed to have the most cracks.

(a)   (b)   (c)

The Powder sample   The mixture of Edible glue   Paste Sterilization

FIGURE 12.4   Mixing of rice husk with edible glue and sterilization process. (a) The powder sample. (b) the mixture of edible glue. (c) Paste sterilization.

12.2.2 SHRINKAGE RATE OF DRYING

First of all, the empty plate was weighed and then the tray with the 100 g husk paste was weighed. Then the plate with husk paste was heated and kept in a heating oven for 24 h with temperature maintained around 80°C. After that, a re-weighing of the dried sample was done and moisture content was calculated by

$$\text{Dry Shrinkage Rate}(\%) = (X - Y) * 100/X$$

Where: X = original weight of husk (g)
   Y = heated-oven dried husk weight (g)
   Dry Shrinkage level (%) = (6.80–6.00) * 100/6.8
   Dry Shrinkage Rate (%) = 11.7 (Figure 12.5).

12.2.3 MOISTURE CONTENT

For carrying out this test, the following procedure is followed:

(a)

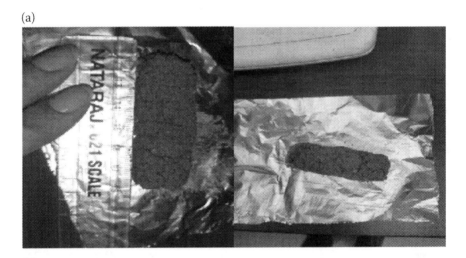

(b)

FIGURE 12.5 (a) The sample before putting under the microwave (b) The sample after putting under 105°C.

- Dried empty dish and lid in the oven at one hundred 5°C for 3 h. Transferred to the desiccator to cool off.
- Weighed the empty dish and the cap.
- Measure about 3.86 g of the sample of the dish. Spread the sample.
- Put the dish with the sample in the oven. Desiccated for 3 h at 105°C.
- After drying, transfer the dish with partially covered lid to the desiccator to cool. Reweigh the dish and its dried sample.

Calculation:

$$\text{Content } (\%) = (A1 - A2) * 100/A1$$

Where: A1 = weight (g) before drying
A2 = Weight (g) after drying
Content (%) = (3.86 – 1.91) * 100/3.86 = 50.51% (Figure 12.6)

12.2.4 COLOR

Color of pots was visually observed and compared with color chart.
At Wet Stage = Dark Brown
At Dry Stage = Light Brown (Figure 12.7).

12.2.5 DRYING TIME

First, the time of wetting was calculated and then the time of dryness was calculated by placing the pots at 80°C until it dries.

(a)   (b)

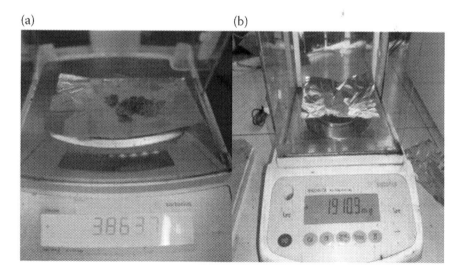

FIGURE 12.6 (a) Sample before Putting under the oven (b) Sample after oven from 80°C for 24 h.

(a)   (b)

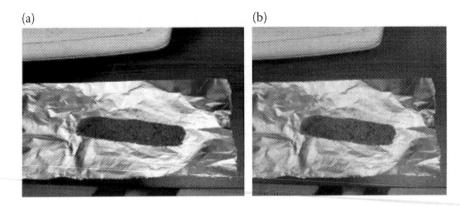

FIGURE 12.7   (a) Before drying the color of the sample (b) After drying the color of the sample.

Calculation:

$$\text{Drying Time} = X - Y$$

Where: X = Time Period of the pot after drying (hours.)
    Y = Time Period of the pot before drying (hours.)
    So, Drying Time = 12:50 – 3:50 at 105°C.

12.2.6 ABSORPTION RATE

To calculate the absorption rate, dried empty pot was weighed and then re-weighed the pot after it was boil at 40°C–45°C.
    Then calculate the absorption rate by using the given formula:

$$A.R = (X1 - Y1)/X1 \times 100$$

Where, A.R = Absorption rate (%)
    X1 = saturated weight (g), Y1 = dry weight (g)
    A.R = (153.43 – 151.48)/153.43 ∗ 100
    A.R = 1.27% (Figure 12.8).

12.2.7 POROSITY RATE

For the calculation of porosity rate, the sample of maturated husk was weighed and then it was put into the boiled water for 5 min and left it in water until it cools. It was re-weighed after it dried.
    Then measure the porosity rate by using the given formula:

$$\varphi = (X1 - Y1)/X1 \times 100$$

Where, φ = Porosity rate (%), X1 = weight of the piece of husk paste after boiling (g)

(a)   (b)

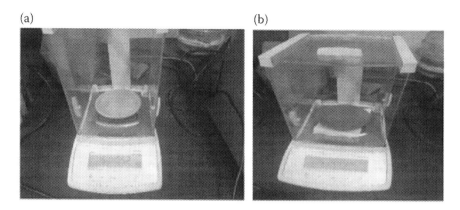

FIGURE 12.8   (a) Drying pot before dipping into boiled at 42°C (b) Pot after dipped into boiled for 1 h at 42°C.

Y1 = weight of piece of the husk paste before boiling (g)

$\varphi = (10.1900 - 4.2605)/10.1900 \times 100$

$\varphi = 97.7189$ (Figure 12.9).

12.3 RESULTS

The chemical composition of rice-husk paste changes after the magnetic stirrer solicited process is carried out at the temperature of 105°C. After this, it becomes smooth paste, and then it is molded into kitchen wares, pots, and utensils in our daily life. (Table 12.2) (Figure 12.10).

(a)   (b)

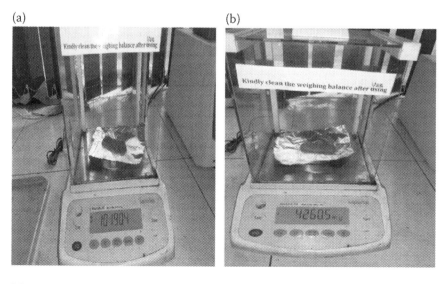

FIGURE 12.9   (a) Weight measurement of husk paste after boiling (b) Weight measurement of husk paste before boiling.

TABLE 12.2

Observed Values after Physical Testing of Rice Husk and Rice Husk Paste

Sl. No.	Test Undertaken	Observation
1	Dry Shrinkage Rate	11.7%
2	Moisture Content	50.51%
3	Color	Dark Brown to Light Brown
4	Drying Time	3 h at 105°C
5	Absorption Time	1.27%
6	Porosity Time	97.7%

FIGURE 12.10   The design of different pots using rice husk.

12.3.1 Cost and Benefit Analysis

The total cost of the collection of rice husk and grinding of the husk in powered form at Baroli and Batoli villages estimated to be:

- Sieving
- Washing the rice husk
- Dirt was removed from the husk
- Husk dried for 4–5 h in heating oven at 45°C
- The labour cost

Collection of rice husk and paste formation

Rice husk is collected and ground to make powder from its husk for paste formation. Approximate cost analysis is given below (Table 12.3):

Rice husk collection = 5 kg

Rice husk measurement in a beaker = 96.58 g.

TABLE 12.3

Approximate Cost of Preparing Samples

Sl. No.	Topic	Rate in Rs.
1	Rice husk (5 kg)	25
2	Grinding cost = 5 kg × 10 Rs. per kg	50
3	Carboxymethylcellulose	300
4	Lemon juice (few drops)	NA
5	Labour cost	600 per day
6	Total cost	975

Edible glue = 3.5%

Distilled water = 100 mL.

Testing and Return on Investment Cost

Production of the crop in one ton: 220 kg approximately

Calorific value = 3000 kcal/kg

Price of the crop per quintal: Rs. 1815 per quintal in Punjab

One-time investment cost of the Phoenix stainless steel ultrasonic cleaner system is Rs.19,500.00 (used for physical properties testing purpose)

SESW 28 L capacity laboratory hot-air oven with aluminium chamber – SESW – Rs. 3835.00 (used for sterilization purpose)

The total proposed cost of established set up: Rs. 24,310

Cost of average one utensil= Rs. 40

Total no. of utensils made by potter = 608 approximately

So, the cost will be recovered of the setup arranged for this pottery maker can regain after making 608 utensils. (Tables 12.4 and 12.5).

TABLE 12.4

Development Cost of Project

Sl. No.	Total Cost of the Project Development (series 1)	Cost at which Each Product Is Sold to the Market (series 2)	Total Product Designed (series 3)
1	24,310	40	608
2	24,310	35	694
3	24,310	50	486

TABLE 12.5
Overall Cost Consumption for Five Years for This Product

Sl. No.	Initial Cost for Project Establishment	Second Year Maintenance Cost	Third Year Maintenance Cost	Fourth Year Maintenance Cost	Fifth Year Maintenance Cost	Total Cost
1	24,310	2400	2400	2400	2400	33,910

12.4 CONCLUSIONS

It is clearly stated that rice husk can be used in numerous products. Some of the products generated from rice husk are: pigment, rubber filler, electronic & solar grade, an insulator, pozzolan, absorbents and in different energy application. Rice husk can also be utilized in the production of silicon-based materials. Working in the area of finding new applications for rice husk would support the following areas, directly or indirectly:

1. Affordable health
2. Energy efficiency
3. Value added agriculture
4. Sustainable energy
5. Potable water
6. Waste to wealth

Health issues would be reduced, mainly respiratory diseases caused by the smog produced when farmers burn the rice husk in the fields after reaping crops in the months of October to November. Due to wind directions in these months, in a particular direction, few areas in North India like Delhi and NCR get air-clogged due to smog. This poor air quality results in severe respiratory diseases and eye infections, especially among children and the elderly. The nutrients present within the stubble are wasted in burning, and farmers should spend on chemical fertilizers to hold soil quality. The burning of crop residue results in the ground temperature rising and the drying up of soil, necessitating extra water for irrigation. Livestock, too, is impacted through crop burning. Burning of crop residue damages other micro-organisms present within the upper layer of the soil, in addition to its organic qualities. Researchers have shown that the burning of agricultural biomass residue, or stubble burning, is a chief fitness danger. However, various researches have shown that rice stubble burning produces a large range of pollutants (RSPM, NOx and SO2) in a brief burning length, resulting in a surprising environmental impact. Rice husk burning made few geographical locations very unlivable in terms of quality of life and health safety of children and elderly, so much so that their residents have shifted to other regions. Further, people refrain from moving to such unhealthy geographic locations for jobs or study. This research is directed toward reducing the pollution by utilizing the rice husk in useful and eco-friendly products.

REFERENCES

Assureira, E (2002). Rice husk – An alternative fuel in Perú. *Carbon, 34,* 38–43.

Chungsangunsit, T, Gheewala, SH, & Patumsawad, S (2009). Emission assessment of rice husk combustion for power production. *World Academy of Science: Engineering and Technology, 53,* 1070.

Emmanuel, OO, Oliver, N, Namessan, G, & Abasiaka, A (2006). Property optimization of kaolin-rice husk insulating fire–bricks benjamin iyenagbe ugheoke. *Leonardo Electron. J. Pract. Technol, 9,* 167–178.

Giddel, MR, & Jivan, AP (2007, January). Waste to wealth, potential of rice husk in India a literature review. In *International Conference on Cleaner Technologies and Environmental Management PEC,* Pondicherry, India (Vol. 2, pp. 4–6).

Ismail, MS., Yusof, N, Yusop, MZM, Ismail, AF, Jaafar, J, Aziz, F, & Karim, ZA (2019). Synthesis and characterization of graphene derived from rice husks. *Malays J Fundam Appl Sci, 15,* 516–521.

Ismail, M. S., Yusof, N., Yusop, M. Z. M., Ismail, A. F., Jaafar, J., Aziz, F., & Karim, Z. A. (2019). Synthesis and characterization of graphene derived from rice husks. Malays., 15, 516–521.

Kumar, S, Sangwan, P, Dhankhar, RMV, & Bidra, S (2013). Utilization of rice husk and their ash: A review. *Res. J. Chem. Env. Sci, 1*(5), 126–129.

Matori, KA, Haslinawati, MM, Wahab, ZA, Sidek, HAA, Ban, TK, & Ghani, WAWAK (2009). Producing amorphous white silica from rice husk. *MASAUM Journal of Basic and Applied Sciences, 1*(3), 512–515.

Mehta, PK (1978). *U.S. Patent No. 4,105,459.* Washington, DC: U.S. Patent and Trademark Office.

Muntohar, AS (2002). Utilization of uncontrolled burnt rice husk ash in soil improvement. *Civil Engineering Dimension, 4*(2), 100–105.

Nuzzolese, AG, Gentile, AL, Presutti, V, & Gangemi, A (2016, October). Conference linked data: The scholarlydata project. In *International Semantic Web Conference* (pp. 150–158). Springer, Cham.

Ragadhita, R, Nandiyanto, ABD, Nugraha, WC, & Mudzakir, A (2019). Adsorption isotherm of mesopore-free submicron silica particles from rice husk. *Journal of Engineering Science and Technology, 14*(4), 2052–2062.

Rozainee, M, Ngo, SP, Salema, AA, Tan, KG, Ariffin, M, & Zainura, ZN (2008). Effect of fluidising velocity on the combustion of rice husk in a bench-scale fluidised bed combustor for the production of amorphous rice husk ash. *Bioresource Technology, 99*(4), 703–713.

Sarangi, M, Bhattacharyya, S, & Behera, RC (2009). Effect of temperature on morphology and phase transformations of nano-crystalline silica obtained from rice husk. *Phase Transitions, 82*(5), 377–386.

13 Clay and Ceramics as Sustainable and Green Materials to Remove Methylene Blue from Water: A Critical Analysis

Priyanka Sharma and Kushal Qanungo
Division of Chemistry, University Institute of Science,
Chandigarh University, Gharuan, Mohali, Punjab, India

CONTENTS

13.1   INTRODUCTION

Water pollution is a global phenomenon, and it adversely affects both biotic and abiotic components of the ecosystem. Furthermore, rapid industrialisation has

DOI: 10.1201/9781003127819-13

resulted in the production of different organic and inorganic pollutants that pollute the surface and groundwater, making it unsuitable for agricultural, industrial, and domestic use (Noel & Rajan, 2015; Shahadat & Isamil, 2018).

The dyeing industry is one of the heavily polluting industries. Due to the water-soluble nature of the most dyes and their high molar-extinction coefficient, they are visible, even at a very low concentration. Effluents generated from dyeing industries contain a considerable volume of dye-containing wastewater (Reisch, 1996). Due to their synthetic origin and the thermal and photo-stable nature of most dyes, they are nonbiodegradable and can exist in the environment for a long time (Ho & Chiang, 2001; Bhatia et al., 2017). These dye effluents contain many chemicals that are lethal, carcinogenic, and mutagenic to living organisms. Therefore, treatment or removal of these harmful pollutants is essential before discharging the wastewater into the environment, and thus, wastewater treatment is a great challenge for scientists and environmental engineers.

Dyes are chemicals that are used to colour different substances like paper, rubber, plastics, cosmetics, leather and textiles. They have diverse chemical composition and structures. Methylene blue (MB) is a cationic azo dye ($C_{16}H_{18}ClN_3S$). It is a solid, odourless, dark-green powder at room temperature and gives blue colour on dissolving in water. MB is used for staining cells and tissues in microbiology and pathological laboratories (Vutskits et al., 2008). Furthermore, it is also used in pharmaceutical industries, paper industries, and to dye cotton, silk, and wool. However, it can cause many health issues, like irritation and paleness of the skin (necrosis), burning sensation, shortness of breath, inflammation, nausea, vomiting, weakness, and mental confusion (Bleicher et al., 2009).

13.1.1 DYE SEPARATION METHODS

Numerous methods are employed to remove harmful pollutants/dyes from aqueous systems. These methods include physical, chemical, and biological processes like oxidation, membrane filtration, coagulation, ion exchange, aerobic degradation, and adsorption, etc. (Mondal, 2008).

Among different dye-removal methods, adsorption is one of the most preferred low-cost and flexible methods due to its easy handling and efficient dye-removal power. Adsorption involves the movement of solute (ions, atoms, molecules) from the solution phase to the adsorbent surface. The adsorbate molecules can be liquid, gas, or solid (Kandisa et al., 2016).

The process of adsorption is generally of two types: physisorption and chemisorption. In the process of physisorption or physical adsorption, the molecules of adsorbate adsorb on the surface of the adsorbent through forces like ionic interactions, hydrogen bonding, van der Waals forces, etc. Physical adsorption is generally an efficient method since it requires low activation energy; also, multilayered adsorption is possible in many cases. On the other hand, chemisorption/chemical adsorption is a process that involves the exchange of electron/ion/chemical-bond formation during adsorption (Dawood & Sen, 2014).

13.1.2 CLAY AND CERAMIC MATERIALS AS ADSORBENT

Clays are emerging as an alternative adsorbent for pollutants removal from wastewater because of their low cost, natural abundance, inorganic and ecofriendly nature (Santos et al., 2016).

Clays are hydrated aluminosilicates with a variable amount of other ions, such as Mg, Fe, alkali, and alkaline earth metals, with a particle size of less than 2 μm. Kaolinite, montmorillonite-smectite, illite, and chlorite are the major groups of clays. The kaolinite group includes kaolinite, dickite, halloysite, and nacrite minerals. The minerals pyrophyllite, talc, vermiculite, sauconite, saponite, nontronite, and montmorillonite, are present in the smectite group. Clay micas are in illite group (Adeyemo et al., 2017). Clay has several advantages, including high specific surface area, ion-exchange potential, good adsorption power, and nontoxic nature. However, very few reports show ceramics being used as an adsorbent material for MB removal (Njoya et al., 2017).

This chapter aims to give a comprehensive study of the use of clay and ceramics materials in their raw or modified forms as an adsorbent for MB removal. Different types of clays from around the world, either in natural or modified form, have been used in MB removal and have been critically analysed and discussed. These clays include: Saudi red clay (SC), natural illite clay mineral (NICM), raw bentonite (RB), plasma-modified bentonite clay (PMBC), Moroccan clay (MC), montmorillonite clay (MIC), raw ball clay beads (RBCB), modified ball clay beads (MBCB), mesoporous synthetic hectorite clay (MSHC), modified Tamazert kaolin (KT), fibrous clay mineral (FCM), calcined kamerotar clays (CKC), silonijan kaolin (SK), calcined natural clay (CNC), natural clay material (NCM), erzurum clay (EC), topkhana natural clay (TNC), modified ball clay (MBC), modified ball clay-chitosan composite (MBC-CH), and many other clay-ceramic based adsorbents. Furthermore, the adsorption behaviour of clays and ceramics has also been analysed by comparing different experimental conditions, like adsorption capacity, concentration, temperature, adsorbent dose, pH of the solution, presence of co-ions, adsorption mechanism, and kinetics involved during adsorand mental confusionption of MB molecules.

13.2 MATHEMATICAL EQUATIONS

13.2.1 CALCULATION OF ADSORPTION CAPACITY AND REMOVAL PERCENTAGE

The amount of dye adsorbed $(mg.g^{-1})$ and removal percentage (%) can be calculated using the following equation (Suteu & Malutan, 2013; Arora et al., 2019).

$$q_t = \frac{Co - Ce}{w} V$$

$$Removal \% = \frac{(C_o - C_e)}{C_o} \times 100$$

Where w is the mass of adsorbent used and V is the volume of the solution.

13.2.2 KINETICS MODELS USED FOR ADSORPTION ANALYSIS

Model	Mathematical Equation	Plot	Parameters	References
Pseudo first order (PFO)	Linear: $\ln(q_e - q_t) = \ln q_e - K_1 t$ Non-linear: $q_t = q_e(1 - e^{-K_1 t})$	$\ln(q_e - q_t)$ *vs* t	q_t = amount of dye adsorbed at time q_e = amount of dye adsorbed at equilibrium	(Moussout et al., 2018)
Pseudo second-order (PSO)	Linear: $\dfrac{t}{q_t} = \dfrac{1}{k_2 q_e^2} + \dfrac{t}{q_e}$ Non-linear: $qt = \dfrac{q_e^2 K_2 t}{q_e K_2 t + 1}$	t/q_t *vs* t	t = Contact time K_1 = Pseudo-first-order rate constant. K_2 = Pseudo-second-order rate constant	
Intraparticle diffusion	$q_t = K_{id}\, t^{0.5}$	q_t *vs* $t^{0.5}$	q_t = Amount of dye adsorbed at time t. K_{id} = Intraparticle diffusion rate constant.	(Doğan & Alkan, 2003)

13.2.3 ADSORPTION ISOTHERM MODELS USED FOR ANALYSIS

Model	Mathematical Equation	Plot	Parameters	References
Langmuir	Nonlinear: $q_e = \dfrac{q_{max}\, b C_e}{1 + b C_e}$ Linear: $\dfrac{C_e}{q_e} = \left(\dfrac{C_e}{q_m}\right) + \left(\dfrac{1}{b \cdot q_{max}}\right)$ Separation factor: $R_L = \dfrac{1}{1 + b C_0}$	q_e *vs* C_e $\dfrac{C_e}{q_e}$ *vs* C_e	q_{max} = maximum Langmuir uptake Ce = equilibrium dye concentration b = Langmuir adsorption equilibrium constant R_L = Separation factor $R_L > 1$: Unfavourable $R_L = 1$: Linear $0 < R_L < 1$: Favourable $R_L = 0$: Irreversible	(Chen, 2015)
Freundlich	Nonlinear: $\log q_e = \log K_F + \dfrac{1}{n} \log C_e$ Linear (when n = 1): $q_e = K_F C_e^{1/n}$	$\log q_e$ *vs* $\log C_e$	K_F = Freundlich constants $1/n$ = Sorption intensity	(Wang & Guo, 2020)
Henry	$q_e = K_H C_e$	q_e *vs* C_e	K_H = Henry's adsorption binding constant	(Nnenna et al., 2020)

13.3 EXPERIMENTAL FACTORS IN MB REMOVAL

13.3.1 EFFECT OF CONTACT TIME

The reaction time plays a critical role in the course of adsorption. Commonly, the quantity of dye adsorbed (mg.g^{-1}) increases with the passage of time, which indicates filling of available adsorption sites during the initial reaction period.

Clay material, like SC, NICM, PMBC, MC and cordite-based ceramics (CBC), showed maximum removal within a time range of 5 to 60 min, whereas some adsorbents, like EC, MIC, MBCB, RBCB, KT, FCM, SK, CKC, MBC, and MBC-CH, took more time (60 to 180 min) to adsorb MB from the solution phase. The use of CKC showed a time of 20 h to absorb MB and to achieve the equilibrium stage.

The adsorption of MB using EC (rich in montmorillonite and nontronite) reached equilibrium at 60 min showed a maximum adsorption capacity of 58.2 mg.g^{-1} (Gurses et al., 2006). For SC, the adsorption process was fast for the initial 5 min, and it took 40 min to reach adsorption equilibrium at a stirring rate of 150 rpm. There was a very steady rise in dye adsorption after the first 5 min, which is attributed to the negatively charged adsorbent surface, responsible for rapid electrostatic adsorption of MB at neutral pH of the solution (Khan, 2020). The CBCs with a variable composition of kaolinite, talc, and bauxite at different time intervals (0–30 min) also showed maximum adsorption for the first 5 min for different ceramic samples prepared by heating at 1300°C, 1400°C temperature (Njoya et al., 2017). MC showed promising results with a maximum adsorption rate in the first 5 min and reached equilibrium in 15 min, which indicates a fast movement of MB molecules on the surface of clay particles during initial reaction time and then the movement of dye molecules from the external surface to interlaminar regions. In contrast, kaolin took 7 h to reach equilibrium with 75% removal of MB from solution (El Mouzdahir et al., 2007). On using MIC for MB removal, equilibrium was achieved in less than 30 min, (Almeida et al., 2009).

Comparable results were also observed for MB adsorption experiments with polyamide-vermiculite nanocomposite (PVNC) with a 99% removal within a time range of 15 to 60 min with 15 mg.L^{-1} of PVNC (Basaleh et al., 2019). For Algerian palygorskite (AP), 97% of removal was observed after a contact time of 5 min with a clay mass (50 mg.L^{-1}) with a MB concentration of 10 ppm (Youcef et al., 2019).

13.3.2 EFFECT OF SOLUTION pH

Several studies have suggested that the pH of a solution is one of the major factors that control the removal capacity and the adsorption efficiency of any adsorbent during the whole adsorption process. The change in pH of the solution can change the surface characteristics of the adsorbent as well as ionisation of dye molecules. It can alter the magnitude of dissociation of different functional groups present on the available active sites of the adsorbents. Basic pH in cationic dyes commonly results in a higher removal percentage due to the presence of the negatively charged surface (Khenifi et al., 2007; Nandi et al., 2009; Karaca et al., 2013). Most of the MB adsorption studies have been carried out in a pH range of 2–12. Furthermore, the regeneration capacity of an adsorbent is influenced by the changes in pH of the regenerator solution. Commonly used reagents for altering or changing the concentration of H^+ and OH^- ions for regeneration of adsorbents are HCl and NaOH.

An increased adsorption capacity of mesoporous synthetic hectorite-alginate-beads (MSHC-AB) was found on raising the pH 3 to 10, and the highest removal (97.5%) was observed at a pH range of 12 (Pawar et al., 2018). A similar trend was also observed during MB adsorption using MIC, where both the adsorption capacity and the removal percentage were observed to be more at a basic pH range. This is ascribed to the increase in negative charge on the exterior sides of the adsorbent. MB removal using KT under basic conditions showed maximum adsorption (111 mg.g^{-1}) at a pH of 11.2, indicating a more negatively charged surface. In contrast, the lower adsorption (96 mg.g^{-1}) was observed in a pH range between 2.6 and 4.8, indicating the presence of a more positive charge on the surface of the adsorbent (BouKhemakhem & Rida, 2017). In the case of CBC, MB removal was studied within a range of 6–12, and maximum adsorption was at a pH of 6 within a contact time of 10 min (Njoya et al., 2017).

Similarly, an increased rate of adsorption was observed when the pH of the solution was changed from 4 to 12 using MBC and MBC-CH; a reverse trend was observed on decreasing the pH value (Auta & Hameed, 2014). Adsorbents like MC, SC, and EC were used to study MB adsorption, and they showed maximum adsorption at a neutral pH range. MB adsorption using CKC showed maximum adsorption at a pH of 7, and after that, remained stable, up to a pH of 12 (Duwal et al., 2016). Change in the pH (from 2–9) does not affect the adsorption capacity of TNC. The result obtained for TNC showed that the maximum percentage variation in adsorption capacity was less than 0.9% for MB solutions having different pH (Salh et al., 2020). An increase in pH of the reaction mixture from 2 to 6 for Iraqi Red Kaolin Clay (IRKC) resulted in a rise in adsorption capacity from 79.2 mg.g^{-1} to 88.3 mg.g^{-1} with pH$_{pzc}$ (pH at point of zero charge) of 7 (Jawad & Abdulhameed, 2020). A similar trend was observed for MB adsorption on PVNC when the pH was increased from 3 to 10 resulting in an increase in removal efficiency from 35% to 75%. In addition to the electrostatic forces between MB and clay surface, a strong competition between MB$^+$ and H$^+$ ions at lower pH values play an important role in MB adsorption on clay (Basaleh et al., 2019).

13.3.3 EFFECT OF MB CONCENTRATION

The initial concentration of solution provides the driving force required to move the solute particles from solution to the adsorbent surface. RB, NICM, MIC, EC, and KT showed a rise in adsorption capacity with increased MB concentration in solution. It follows that the removal mechanism, as well as withholding of the MB molecules by the adsorbent, becomes more efficient with the increase in dye concentration. This kind of adsorbent behaviour has been ascribed to the involvement of chemisorption, multilayer adsorption, or dimerization of MB molecules at a higher concentration range. Similar results were also obtained for CBC, as well as for MC.

In the case of adsorbents like MSHC and NICM, adsorption was found to be more effective with increased MB concentration, but the removal percentage was found to be decreased at higher solution concentrations. The dye uptake showed a decrease in removal capacity with the saturation of the available active sites at a higher concentration of MB (Almeida et al., 2009; Ozdes et al., 2014; Pawar et al., 2018).

For MBC-CH, a range of uptake capacity of 26.93 to 193.23 mg.g^{-1} was obtained for a concentration range of 30 to 300 mg.L^{-1}. This increase in adsorption capacity is attributed to the presence of numerous active sites on MBC-CH surface, which can easily surpass the limited number of MB molecules at low concentrations. In contrast, at a higher concentration of MB, there is competition for active sites present on the surface (Auta & Hameed, 2014). The use of bentonite alginate beads (BAB) also showed an increased rate of adsorption with the rise in MB concentration, which is attributed to the rapid diffusion of dye molecules at higher concentration (Pandey, 2019). Adsorption of MB using TNC also showed an increase in adsorption capacity from 25 mg.g^{-1} to 128 mg.g^{-1} on increasing MB concentration from 50 mg.L^{-1} to 550 mg.L^{-1} and this was accompanied by a decrease in adsorption power due to saturation of the sites of clay surface (Salh et al., 2020). Similarly, the use of IRKC showed a rise in adsorption capacity from 2.2 to 101.6 mg.g^{-1} with a change in concentration from 10 to 120 mg.L^{-1}. The increased concentration gradient provides a driving force to transfer the MB molecules to active sites.

MB removal using MSHC and MSHC-AB composite showed a high MB removal efficiency with a low initial MB concentration (5–100 mg.L^{-1}), and it showed a slight decrease with a higher initial MB concentration of 5–800 mg.L^{-1} (Pawar et al., 2018).

13.3.4 Effect of Adsorbent Dose

Usually, in the adsorption process, an increase in adsorbent dose results in an increase in removal efficiency. This is attributed to the rise in the number of existing adsorption sites at the adsorbent surface (Li et al., 2009; Saka & Sahin, 2011).

In the case of EC, an increased removal capacity (57.1 to 66.1 mg.g^{-1}) was obtained on varying the dose from 0.10 g to 0.30 g for a contact time of 120 min (Gurses et al., 2006). A variation in the dose of MC from 1.6 g.L^{-1} to 4.6 g.L^{-1} for MB concentration of 600 mg.L^{-1} resulted in an increase in the MB removal percentage at the large dosage of adsorbent (Almeida et al., 2009). An increase in MB percentage removal from 66% to 100% was observed in response to an increase in the dose of KT clay from 1 g.L^{-1} to 1.8 g.L^{-1} (El Mouzdahir et al., 2007). Similarly, for TNC, MB adsorption increased on increasing the dose from 1 g.L^{-1} to 5 g.L^{-1} (Basaleh et al., 2019), and the use of a dose of 3.4 g.L^{-1} showed a 100% removal with an adsorption capacity of 165 mg.g^{-1}. A variation in dose from 0.02 g to 0.2 g per 100 mL at a constant solution pH (5.6) and concentration (100 mg.L^{-1}) for IRKC resulted in an increased removal percentage value, from 86.2% to 99.8%, which is ascribed to availability of more surface area/active sites (Jawad & Abdulhameed, 2020).

SC as an adsorbent showed MB adsorption capacity of 50.25 mg.g^{-1} with a dose of 0.2 g in a MB solution of 50 mL. The clay showed a sharp rise in the adsorption power with a lower dose, whereas the increase in dosage showed a slight increase in the removal efficiency. It is attributed to the agglomeration of clay particles, leading to a decrease in adsorption sites and, therefore, reduction in the amount of MB adsorbed per unit mass of adsorbent and strong competition among MB molecules for a specific number of adsorption sites (Khan, 2020). An increase in the amount of NICM from 1.0 g.L^{-1} to 30.0 g.L^{-1} resulted in an increase in removal percentage from 81.4% to 99.0%. This is attributed to the increased number of clay particles, which leads to easy penetration of MB molecules onto the active adsorption sites. On the other hand, an increase in amount of NICM resulted in decreased MB uptake from 81.4 mg.g^{-1} to 3.3 mg.g^{-1}, due to the aggregation of clay particles, which results in reduction in surface area for adsorption (Ozdes et al., 2014).

Similarly, in the case of CBC, MB adsorption showed a decrease with an increase in the amount of adsorbent. It is ascribed to the agglomerations of particles at higher ceramic dose, which leads to reduced adsorption sites, leading to a decreased value of MB adsorbed per unit mass of ceramic (Njoya et al., 2017).

SK also showed a reduced value of MB adsorption with an increased adsorbent dose. This is because of a decrease in equilibrium dye concentration with an increase in the dosage of SK (Ghosh & Bhattacharyya, 2002). Similar results were also obtained for MB adsorption by FCM, which is ascribed to the micro-aggregation of fibres (Hajjaji et al., 2006).

13.3.5 Effect of Temperature

Temperature plays an essential role in the adsorption process as it provides information regarding the enthalpy and entropy changes involved during the adsorption. The rate of diffusion of solute molecules across the outer boundary, as well as in the interior pores of the adsorbent, is affected by the change in temperature. Additionally, a change in reaction temperature can alter the equilibrium adsorption capacity. The influence of temperature on the adsorption capacity of different clay and ceramic materials is discussed below.

Experimental data for adsorbents like NICM, FCM, and SK, showed that these adsorbents are excellent adsorbent for MB removal at room temperature (298K). MB removal percentage for SC was observed to increase on increasing the temperature from 25°C to 55°C (Khan, 2020).

MB removal experiment using EC under a range of reaction temperature, i.e. 20°C, 40°C and 60°C showed that the rise in temperature causes a decrease in both efficiency and effectiveness of the adsorption process.

Temperature variation (5°C–40°C) studied with NICM as an adsorbent at a constant adsorbent dose (5.0 g.L^{-1}) and concentration (100 mg.L^{-1}) resulted in a slight increase in adsorption capacity on the increase in temperature. The results have been ascribed to a decrease in viscosity of solution or increase in the mobility of dye molecules, resulting in easy diffusion of dye molecules throughout the external surface as well as in the internal pores of clay particles. The increase in adsorption capacity with increasing temperature also signifies the endothermic nature of the adsorption process (Ozdes et al., 2014).

A change in temperature from 35°C to 60°C for MIC resulted in an increase in removal percentage from 68% to 74%, as well as a rise in the amount of MB adsorbed from 245 mg.g^{-1} to 300 mg.g^{-1}, respectively. This 6% increase in removal percentage indicates the endothermic nature of the adsorption process (Almeida et al., 2009). Similarly, a variation in temperature from 30°C to 50°C for RBCB and MBCB showed that a rise in reaction temperature resulted in better dye adsorption, which has been attributed to more significant collisions between solute and solvent molecules resulting in high adsorption rate (Auta & Hameed, 2014). The MB adsorption using TK was more feasible at a higher temperature (291K, 303K, 323K), which indicates the endothermic nature of the process (BouKhemakhem and Rida, 2017).

13.3.6 EFFECT OF OTHER IONS

Co-ions have a variable effect on the adsorption behaviour of the adsorbent. The presence of co-ions may result in screening of the electrostatic interaction between the active sites and the dye molecules. In addition, the presence of ionic species could also result in an increase in the dissociation of the dye molecules, which is brought about by protonation.

In the removal of MB by MSHC-AB, the adsorption capacity has followed the order; sulphite < chloride < phosphate < bicarbonate < nitrate (Pawar et al., 2018). Results obtained for MBC and MBC-CH showed that the ionic species like SO_4^{2-} had a greater effect on MB adsorption in MBC and MBC-CH than ions like HCO_3^- and Cl^- (Auta & Hameed, 2012).

The presence of ionic species like $NaNO_3$, Na_2SO_4, $CaCl_2$, and KCl had a variable effect on MB adsorption on NICM. A higher strength of Na_2SO_4 (0 to 1.0 M) in the solution resulted in a decrease in the MB adsorption capacity (16.4 to 16.1 $mg.g^{-1}$). On the other hand, increase in strength of $NaNO_3$, KCl, and $CaCl_2$ ions from 0 to 1.0 M, resulted in an increase in the adsorption capacity; a change in q_e value from 16.4 to 16.9 $mg.g^{-1}$ for $NaNO_3$, 16.4 to 18.6 $mg.g^{-1}$ KCl and 16.4 to 18.7 $mg.g^{-1}$ in case of $CaCl_2$, respectively (Ozdes et al., 2014)

13.3.7 KINETICS AND MECHANISM

The adsorption kinetics is an essential and noteworthy factor that helps in choosing and finalising the optimal operating conditions for adsorbent-adsorbate interaction. To understand the adsorbent behaviour and to explore the reaction controlling mechanism, models like PFO, PSO, and intraparticle diffusion are suitable and have been extensively used.

For RB, the adsorption process was well explained by the PSO kinetic model as compared to PFO (Özacar & Şengil, 2006). Similarly, experiment data obtained for MSHC, NICM, RBCB or MBCB, KT, PMBC, CBC, and MC was fitted in the PSO. Correlation coefficient (R^2) for most of the adsorbents using the PSO model were higher than that of the PFO kinetic model, indicating that the MB adsorption can be described more accurately by the PSO.

In the case of FCM as an adsorbent, the MB adsorption was found to be diffusion-controlled, and the kinetics followed was PSO.

The use of BAB showed PSO kinetics and intraparticle diffusion involvement during the adsorption of MB on beads (Pandey, 2019). The use of TNC also showed

PSO kinetics and the adsorption phenomena controlled by fast film diffusion on external sites of clay and a slower diffusion on the inner sites of TNC (Salh et al., 2020). MB adsorption by IRKC showed high R^2 values followed PFO (Jawad & Abdulhameed, 2020).

13.3.8 ADSORPTION STUDIES

The adsorption isotherms are one of the most essential characteristics of an adsorbent-adsorbate interaction. The amount of dye accumulated on the adsorbent is typically analysed using adsorption isotherm models. It is a constant-temperature equilibrium correlation between the number of dye molecules per unit mass of clay (q_e) and equilibrium dye concentration (C_e). Different adsorption isotherm models are available for proper understanding of the equilibrium adsorption process and to know the mechanism of the adsorption process.

For SC, monolayer adsorption on the surface is indicated by a high correlation coefficient for the Langmuir isotherm model. Furthermore, the positive value of entropy suggests the spontaneous, endothermic, physical nature of the adsorption process, as well as randomness at the adsorbent/adsorbate interface (Khan, 2020). In the case of Pendik Bentonite (PB), the monolayer saturation capacity, (q_e), 1667 mg.g^{-1}, suggested that MB adsorption is restricted to monolayer coverage and well followed by the Langmuir isotherm (Özacar & Şengil, 2006). A comparative study between mesoporous synthetic hectorite-alginate-beads-wet (MSHC-AB-W) and powdered MSHC has shown a higher Langmuir monolayer adsorption capacity (785.45 mg.g^{-1}) for MSHC-AB-W, which signifies the differences in porosity and the number of available active sites for the adsorption of MB (Pawar et al., 2018). MB adsorption using NICM is also well demonstrated using the Langmuir isotherm study. Furthermore, the favourable nature of the adsorption process is indicated by the value R_L (0.32–0.02) and $1/n$ (0.12) for a concentration range of 50 mg.L^{-1} to 1000 mg.L^{-1}. The high value of adsorption energy (25.0 kJ mol.L^{-1}) indicates the chemical nature of adsorption process during adsorption of MB by NICM particles (Ozdes et al., 2014). Similarly, for KT, MB adsorption using Langmuir showed monolayer adsorption suggesting a homogeneous nature of the clay surface (BouKhemakhem & Rida, 2017). In case of FCM, the cation exchange process is found to be responsible for the fast diffusion of MB molecules, and it was well followed by Langmuir isotherm for a variable pH range (4–9) (Hajjaji et al., 2006). For adsorbents, MBC and MBC-CH, the adsorption process is also well followed by the Langmuir model. Comparative isotherm study for MBC and MBC-CH showed a higher value of monolayer coverage for MBC-CH than MBC (Auta &Hamced, 2014). A straight-line plot in the case of CBC and MC showed fitting of experimental data in Langmuir isotherm (Njoya et al., 2017) (El Mouzdahir et al., 2007). Langmuir and Freundlich models were well followed by MB adsorption on IRKC with a high R^2 value of 0.99 and 0.98, respectively, indicating monolayer and multilayer mode of adsorption (Jawad & Abdulhameed, 2020).

The different clay and ceramic materials used to remove MB from aqueous solutions and their associated experimental conditions are tabulated in Table 13.1.

TABLE.13.1

Comparison of Different Experimental Parameters/Result

Sl. No.	Substrate	Surface Area ($m^2.g^{-1}$)/ Particle Size	RPM	Time (min)	pH	Adsorbent Dose ($g.L^{-1}$)	MB Conc. ($mg.L^{-1}$)	Ads. Cap. (q_e) Max. ($mg.g^{-1}$)	Removal (Max.) (%)	Kinetic Model	Rate Constant ($g.mg^{-1}.min^{-1}$)	Adsorption Model	References
					Experimental Conditions				Results				
1	SC	63.15/-	150	60	7	0.1-0.7	100	50.25	100	PSO	0.552	Langmuir	(Khan, 2020)
2	PB	28/53-75 µm	500	60	7.9	1.0	100-1000	1667	978	PSO	13.09×10^{-4}	Langmuir	(Özacar & Şengil, 2006)
3	EC	30/-	90-200	15-120	5.65	0.10-0.30	10-100	58.2	–	PSO	0.00178	Langmuir	(Gurses et al., 2006)
4	MSHC	468/-	–	60-900	3-12	1-5	5-800	196	97.50	PSO	–	Langmuir	(Pawar et al., 2018)
	MSHC-AB-D	205/-						357.14		PFO		Freundlich	
	MSHC-AB-W	205/-						485.45		PFO		Freundlich	
5	NICM	3.10/-	400	60	2-10	1-30	50-1000	24.87	92.3	PSO	0.045	Langmuir	(Ozdes et al., 2014)
6	MIC	62/-	250	120	3-11	1.6-4.6	200-1500	300.30	100	PSO	456.82×10^{-2} ($g.mg^{-1}.h^{-1}$)	Langmuir	(Almeida et al., 2009)
7	MBCB	92/1-2 mm	140	1440	3-12	0.20 g/ 200 mL	30-300	100	96	PSO	2.070×10^{-4} ($g.mg^{-1}.h^{-1}$)	Langmuir	(Auta & Hameed, 2012)
8	RBCB	10/1-2 mm						34.652	53		65.90×10^{-4} ($g.mg^{-1}.h^{-1}$)		
9	KT	14/-	400	5-120	2-11.19	1.0-6.0	30-100	111	66-100	PSO	0.0013-0.004	Langmuir	(BouKhemakhem and Rida, 2017)
10	RB PMBC	64.2/- 65.3/-	100	180	2-12	0.15 g/ 200 mL	100-250	303	–	PSO	0.00247-0.0568	Langmuir	(Sahin et al., 2015)
11	FCM	115/<2 µm	10,000	100	4-9	0.3	10-25	85	–	diffusion	$(29 \pm 3) \times 10^{-3} min^{-1}$	Langmuir	(Hajjaji et al., 2006)

(Continued)

TABLE.13.1 (Continued)
Comparison of Different Experimental Parameters/Result

Sl. No.	Substrate	Surface Area (m^2/g^{-1})/ Particle Size	RPM	Time (min)	pH	Adsorbent Dose $(g.L^{-1})$	MB Conc. $(mg.L^{-1})$	Ads. Cap. (q_e) Max. $(mg.g^{-1})$	Removal (Max.) (%)	Kinetic Model	Rate Constant $(g.mg^{-1}.min^{-1})$	Adsorption Model	References
				Experimental Conditions					Results				
12	SK	–	–	180	2–10	0.8	10–25	20.49	–	Fick's law	–	Langmuir	(Ghosh & Bhattacharyya, 2002)
13	CKC	–/<63 µm	–	10–140	7–12	100 mg/ 50 mL	25–55 m mol.L^{-1}	19.47 mmol.g^{-1}	–	–	–	Langmuir	(Duwal et al., 2016)
14	MBC	–/0.5–2.0 mm	–	100	4–12	0.10 g/ 100 mL	30–300	70	–	PSO	0.629–4.870 (g.mg^{-1}.h^{-1})	Langmuir	(Auta & Hameed, 2014)
17	MBC–CH CBC	–	–	5–35	6–12	0.1–0.5 g	5–11	142 1.84	–	PSO	1.897	Langmuir	(Njoya et al., 2017)
18	MC	414/–	350	5–15	7.2	0.1 g	10–1000	135	–	PSO	0.0050	Langmuir	(El Mouzdahir et al., 2007)
19	BAB	–/2.76 mm	160	30	7	0.05% (w/v)	200–1500	2024	85	PSO	1.92×10^{-4}	Langmuir	(Pandey, 2019)
20	TNC	–/<5 µm	5000	0–360	2–9	1–5	10–250	129.8	≈100	PSO	0.00209	Langmuir	(Salh et al., 2020)
21	IRKC	–/≤250 µm	110	0–300	8	0.10 g/ 100 mL	10–120	240.4	99.8	PSO	0.18	Langmuir	(Jawad & Abdulhameed, 2020)
22	PVNC	12.307/–	4400	180	3–10	5–50 mg	15	76.42	99	PSO	0.002375	Langmuir	(Basaleh et al., 2019)
23	AP	–/5–100 µm	–	5–60	–	50g	10	57.47	97	PSO	–	Langmuir	(Youcef et al., 2019)

24	Kaolin (Halloysite) Kaolin (kaolinite)	70.90/- 6.19/-	–	5–30	3–11	25 mg/ 25 mL	5–30	29.13 26.47	–	PSO	1.05–0.01	–	(Harrou et al., 2020)
25	Ngbo Clay	-/0.75 μm	150	60	2–12	0.5–5 g/ 100 mL	100–900	–	97	PSO	377.69	Freundlich Henry	(Nnenna et al., 2020)
26	NCM	/75 μm	–	300	–	2 mg/40 mL	10–50	18.8–97.2	–	PSO	0.016	Freundlich	(Nyankson et al., 2020)
27	Pakistani Clay	–	–	10–380	2–10	–	50–300	270	–	PSO	–	Langmuir	(Rehman et al., 2021)

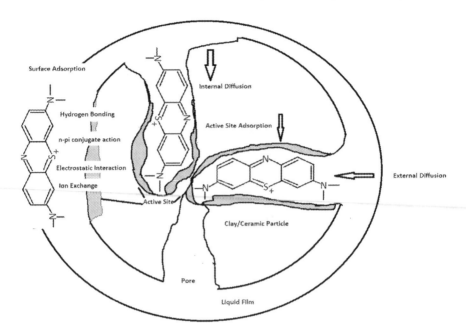

FIGURE 13.1 Graphical representation of the possible adsorption process of MB molecules on clay/ceramics.

13.3.9 A POSSIBLE MECHANISM FOR DYE ADSORPTION

The interactions involved in the removal process this text discusses can be summarized diagrammatically in Figure 13.1.

13.4 CONCLUSION AND FUTURE SCOPE

This chapter presented a summary of methylene blue removal by various clay and ceramic-based adsorbents under different experimental conditions and their adsorption mechanism. It was observed that adsorption increases during the initial hours of reaction and then remains constant on achieving equilibrium with high methylene blue removal percentage. Most of the adsorbents showed the highest adsorption at a neutral pH, whereas some of the adsorbents work well in acidic and some in basic conditions, i.e. a range from 2–12.

An increase of MB concentration has resulted in higher uptake of MB molecules, whereas changes in adsorbent dose had a variable effect on MB removal capacity. In some experimental studies, an increase in adsorbent dose resulted in increased adsorption capacity due to increased surface area. In contrast, in some studies, it resulted in decreased or constant adsorption due to agglomeration. Adsorption experiments at different temperatures showed the process is endothermic nature in general. To describe the equilibrium studies of MB adsorption, the Langmuir and Freundlich models have been adopted in most studies. The pseudo-second-order kinetics very elegantly explained the kinetics of adsorption in most of the cases.

Development and modification of clay-based adsorbents and ceramic materials with high adsorption capacities and adsorption rate, which are minimally affected by variations of pH, temperature, MB concentration, presence of other inorganic and organic impurities, are some of the focus areas for future research.

REFERENCES

Adeyemo, AA, Adeoye, IO, & Bello, OS (2017). Adsorption of dyes using different types of clay: A review. *Applied Water Science*, 7(2), 543–568.

Almeida, CAP, Debacher, NA, Downs, AJ, Cottet, L, & Mello, CAD (2009). Removal of methylene blue from colored effluents by adsorption on montmorillonite clay. *Journal of Colloid and Interface Science*, 332(1), 46–53.

Arora, C, Soni, S, Sahu, S, Mittal, J, Kumar, P, & Bajpai, PK (2019). Iron based metal organic framework for efficient removal of methylene blue dye from industrial waste. *Journal of Molecular Liquids*, 284, 343–352.

Auta, M, & Hameed, BH (2012). Modified mesoporous clay adsorbent for adsorption isotherm and kinetics of methylene blue. *Chemical Engineering Journal*, 198, 219–227.

Auta, M, & Hameed, BH (2014). Chitosan–clay composite as highly effective and low-cost adsorbent for batch and fixed-bed adsorption of methylene blue. *Chemical Engineering Journal*, 237, 352–361.

Basaleh, AA, Al-Malack, MH, & Saleh, TA (2019). Methylene Blue removal using polyamide-vermiculite nanocomposites: Kinetics, equilibrium and thermodynamic study. *Journal of Environmental Chemical Engineering*, 7(3), 103107.

Bhatia, D, Sharma, NR, Singh, J, & Kanwar, RS (2017). Biological methods for textile dye removal from wastewater: A review. *Critical Reviews in Environmental Science and Technology*, 47(19), 1836–1876.

Bleicher, RJ, Kloth, DD, Robinson, D, & Axelrod, P (2009). Inflammatory cutaneous adverse effects of methylene blue dye injection for lymphatic mapping/sentinel lymphadenectomy. *Journal of Surgical Oncology*, 99(6), 356–360.

BouKhemakhem, A, & Rida, K (2017). Improvement adsorption capacity of methylene blue onto modified Tamazert kaolin. *Adsorption Science & Technology*, 35(9–10), 753–773.

Chen, Xunjun. 2015. Modeling of Experimental Adsorption Isotherm Data. *Information*, 6(1), 14–22. 10.3390/info6010014.

Dawood, S, & Sen, T (2014). Review on dye removal from its aqueous solution into alternative cost effective and non-conventional adsorbents. *Journal of Chemical and Process Engineering*, 1(104), 1–11.

Doğan, M, & Alkan, M (2003). Adsorption kinetics of methyl violet onto perlite. *Chemosphere*, 50(4), 517–528.

Duwal, N, Joshi, S, & Bhattarai, J (2016). Study on the removal of methylene blue by calcined-Kamerotar clays as an eco friendly low cost adsorbent. *Int J Adv Res Chem Sci*, 3(11), 1–8.

El Mouzdahir, Y, Elmchaouri, A, Mahboub, R, Gil, A, & Korili, SA (2007). Adsorption of methylene blue from aqueous solutions on a Moroccan clay. *Journal of Chemical & Engineering Data*, 52(5), 1621–1625.

Ghosh, D, & Bhattacharyya, KG (2002). Absorption of methylene blue on kaolinite. *Applied Clay Science*, 20(6), 295–300.

Gurses, A, Dogar, Ç, Yalcın, M, Açıkyıldız, M, Bayrak, R, & Karaca, S (2006). The adsorption kinetics of the cationic dye, methylene blue, onto clay. *Journal of Hazardous Materials*, 131(1–3), 217–228.

Hajjaji, M, Alami, A, & El Bouadili, A (2006). Removal of methylene blue from aqueous solution by fibrous clay minerals. *Journal of Hazardous Materials*, 135(1–3), 188–192.

Harrou, A, Gharibi, E, Nasri, H, & El Ouahabi, M (2020). Thermodynamics and kinetics of the removal of methylene blue from aqueous solution by raw kaolin. *SN Applied Sciences*, 2(2), 277.

Ho, YS, & Chiang, CC (2001). Sorption studies of acid dye by mixed sorbents. *Adsorption*, 7(2), 139–147.

Jawad, AH, & Abdulhameed, AS (2020). Mesoporous Iraqi red kaolin clay as an efficient adsorbent for methylene blue dye: Adsorption kinetic, isotherm and mechanism study. *Surfaces and Interfaces*, 18, 100422.

Kandisa, RV, Saibaba, KN, Shaik, KB, & Gopinath, R (2016). Dye removal by adsorption: A review. *Journal of Bioremediation and Biodegradation*, 7(6).

Karaca, S, Gürses, A, Açışlı, Ö, Hassani, A, Kıranşan, M, & Yıkılmaz, K (2013). Modeling of adsorption isotherms and kinetics of Remazol Red RB adsorption from aqueous solution by modified clay. *Desalination and Water Treatment*, 51(13–15), 2726–2739.

Khan, M (2020). Adsorption of methylene blue onto natural Saudi Red Clay: Isotherms, kinetics and thermodynamic studies. *Materials Research Express*, 7(5), 5507.

Khenifi, A, Bouberka, Z, Sekrane, F, Kameche, M, & Derriche, Z (2007). Adsorption study of an industrial dye by an organic clay. *Adsorption*, 13(2), 149–158.

Li, K, Zheng, Z, Huang, X, Zhao, G, Feng, J, & Zhang, J (2009). Equilibrium, kinetic and thermodynamic studies on the adsorption of 2-nitroaniline onto activated carbon prepared from cotton stalk fibre. *Journal of hazardous materials*, 166(1), 213–220.

Mondal, S (2008). Methods of dye removal from dye house effluent—an overview. *Environmental Engineering Science*, 25(3), 383–396.

Moussout, H, Ahlafi, H, Aazza, M, & Maghat, H (2018). Critical of linear and nonlinear equations of pseudo-first order and pseudo-second order kinetic models. *Karbala International Journal of Modern Science*, 4(2), 244–254.

Nandi, BK, Goswami, A, & Purkait, MK (2009). Removal of cationic dyes from aqueous solutions by kaolin: Kinetic and equilibrium studies. *Applied Clay Science*, 42(3–4), 583–590.

Njoya, D, Nsami, JN, Rahman, AN, LekeneNgouateu, RB, Hajjaji, M, & Nkoumbou, C (2017). Adsorption of Methylene Blue from Aqueous Solution onto Cordierite based ceramic. *Journal of Materials and Environmental Sciences*, 8(5), 1803–1812.

Nnenna, NV, Philomena, KI, & Elijah, OC (2020). Removal of methylene blue dye from aqueous solution using modified Ngbo clay. *Journal of Materials Science Research and Reviews*, 5(2), 33–46.

Noel, SD, & Rajan, MR (2015). Impact of dyeing industry effluent on ground water quality by water quality index and correlation analysis. *Research in Biotechnology*, 6(1), 47–53.

Nyankson, E, Mensah, RQ, Kumafle, L, Gblerkpor, WN, Aboagye, SO, Asimeng, BO, & Tiburu, EK (2020). Dual application of natural clay material for decolorisation and adsorption of methylene blue dye. *Cogent Chemistry*, 6(1), 1788291.

Özacar, M, & Şengil, İA (2006). A two-stage batch adsorber design for methylene blue removal to minimise contact time. *Journal of environmental management*, 80(4), 372–379.

Ozdes, D, Duran, C, Senturk, HB, Avan, H, &Bicer, B (2014). Kinetics, thermodynamics, and equilibrium evaluation of adsorptive removal of methylene blue onto natural illitic clay mineral. *Desalination and Water Treatment*, 52(1–3), 208–218.

Pandey, LM (2019). Enhanced adsorption capacity of designed bentonite and alginate beads for the effective removal of methylene blue. *Applied Clay Science*, 169, 102–111.

Pawar, RR, Gupta, P, Sawant, SY, Shahmoradi, B, & Lee, SM (2018). Porous synthetic hectorite clay-alginate composite beads for effective adsorption of methylene blue dye from aqueous solution. *International Journal of Biological Macromolecules*, 114, 1315–1324.

Rehman, MU, Manan, A, Uzair, M, Khan, AS, Ullah, A, Ahmad, AS, & Khan, MA (2021). Physicochemical characterisation of Pakistani clay for adsorption of methylene blue: Kinetic, isotherm and thermodynamic study. *Materials Chemistry and Physics, 269*(2021), 124722.

Reisch, MS (1996). Asian textile dye makers are a growing power in a changing market. *Chemical & Engineering News, 74*(3), 10–12.

Sahin, O, Kaya, M, & Saka, C (2015). Plasma-surface modification on bentonite clay to improve the performance of adsorption of methylene blue. *Applied Clay Science, 116,* 46–53.

Saka, C, & Sahin, Ö (2011). Removal of methylene blue from aqueous solutions by using cold plasma-and formaldehyde-treated onion skins. *Coloration Technology, 127*(4), 246–255.

Salh, DM, Aziz, BK, & Kaufhold, S (2020). High adsorption efficiency of Topkhana natural clay for methylene blue from medical laboratory wastewater: A linear and nonlinear regression. *Silicon, 12*(1), 87–99.

Santos, SC, Oliveira, AF, & Boaventura, RA (2016). Bentonitic clay as adsorbent for the decolourisation of dyehouse effluents. *Journal of Cleaner Production, 126,* 667–676.

Shahadat, M, & Isamil, S (2018). Regeneration performance of clay-based adsorbents for the removal of industrial dyes: A review. *RSC Advances, 8*(43), 24571–24587.

Suteu, D, & Malutan, T (2013). Industrial cellolignin wastes as adsorbent for removal of methylene blue dye from aqueous solutions. *BioResources, 8*(1), 427–446.

Vutskits, L, Briner, A, Klauser, P, Gascon, E, Dayer, AG, Kiss, JZ, & Morel, DR (2008). Adverse effects of methylene blue on the central nervous system. *Anesthesiology: The Journal of the American Society of Anesthesiologists, 108*(4), 684–692.

Wang, J, & Guo, X (2020). Adsorption isotherm models: Classification, physical meaning, application and solving method. *Chemosphere, 127279.*

Youcef, LD, Belaroui, LS, & López-Galindo, A (2019). Adsorption of a cationic methylene blue dye on an Algerian palygorskite. *Applied Clay Science, 179,* 105145.

14 Metal Chalcogenides Based Nanocomposites for Sustainable Development with Environmental Protection Applications

Mohamed Jaffer Sadiq Mohamed
Department of Chemistry, School of Chemical Sciences and Technology, Yunnan University, Kunming, China

National Center for International Research on Photoelectric and Energy Materials, Yunnan Province Engineering Research Center of Photocatalytic Treatment of Industrial Wastewater, Yunnan Provincial Collaborative Innovation Center of Green Chemistry for Lignite Energy, Yunnan University, Kunming, China

CONTENTS

DOI: 10.1201/9781003127819-14

14.1 INTRODUCTION

Environmental emissions and energy shortages are quickly becoming significant issues in urban society. Water pollution is described as a substantial factor in the disintegration of the environment in recent years (Hao et al., 2018, and Sadiq et al., 2018a). Advanced oxidation processes are seen as high potential technologies for protecting the environment. Among the numerous advanced oxidation processes, photocatalytic technology has earned an excessive treaty of attention in water treatment due to its simple technical specifications, low-cost catalysts, and the successful elimination of environmental contaminants (Pu et al., 2017, and Sadiq et al., 2017a).

Transition metal chalcogenides are an excellent catalyst for photocatalytic degradation of organic contaminants. Among these transition metal chalcogenides, two-dimensional MoS_2 has attracted massive attention because of its superior electro-catalytic performance, chemical stability, absorption in visible frequencies, considerable carrier mobility, and direct bandgap (Chen et al., 2018, Fang et al., 2018, Yu et al., 2016, and Zhang et al., 2018). The rapid recombination of photogenerated electron-hole pairs is considered as a key issue in using MoS_2 as a catalyst in photocatalytic applications. Different techniques are used to solve this issue by enhancing the efficiency of the MoS_2 photocatalytic process, such as doping, co-deposition of noble metals, and mixing of two composite and incapacitating semiconductors with various cations and anions like nitrogen, phosphorus, sulfur, carbon, etc. (Mutyala et al., 2020, Peng et al., 2017, and Sumesh and Peter 2019). However, the above techniques may improve the photocatalytic action of MoS_2, extend the light absorption range, and overwhelm the electron-hole pair recombination process (Lin et al., 2019).

Graphene is one of the most significant exciting two-dimensional resources with excellent thermal properties and remarkable chemistry. It is also documented as an essential catalyst for photocatalytic reactions (Singh et al., 2020). Graphene has an excellent electron transport/acceptor support, which promotes the migration of photogenerated electrons, hinders the electron-hole pair recombination to a vast extent, and enhances light absorption (Sadiq et al., 2016). Graphene is currently embedded in several photocatalysts to increase its photocatalytic efficiency. In the same way, to boost its photocatalytic efficiency, graphene can also be integrated with MoS_2, taking into account the step-by-step configuration of energy levels built into the MoS_2/graphene composite (Sadiq et al., 2018b, and Yu et al., 2020).

In this study, the MoS_2/graphene nanocomposites were effectively synthesized by the hydrothermal method. The photodegradation of methylene blue dye from wastewater with the prepared MoS_2/graphene nanocomposites was examined. The as-prepared MoS_2/graphene nanocomposites exhibit improved photocatalytic efficiency relative to pure MoS_2. These results suggest that an exceptional electron acceptor improved the active transfer of photogenerated electrons of MoS_2 to the graphene, thereby inhibiting the recombination of electron-hole pairs. These effective MoS_2/graphene nanocomposites are excellent materials for catalysts in photocatalytic applications.

14.2 EXPERIMENTAL SECTION

14.2.1 MATERIALS AND METHOD

All the chemicals were procured from Sigma Aldrich, and all the experiments were used in the deionized water.

14.2.2 PREPARATION OF GRAPHENE OXIDE

Graphene oxide was prepared based on the modified Hummers process (Lavin-Lopez et al., 2016). In brief, graphite flakes (1.5 g) were mixed with concentrated sulfuric acid (45 mL) and potassium permanganate (6.0 g) and agitated for 2 hours. The blend was diluted with deionized water and hydrogen peroxide (1 mL) to complete the reaction. The obtained graphene oxide was rinsed with hydrochloric acid (10%) and deionized water until the pH was raised to 7 and dried to create graphene-oxide powder. For graphene-oxide exfoliation, it was dispersed in deionized water through sonication for 30 minutes.

14.2.3 PREPARATION OF MoS_2

MoS_2 catalytic material was prepared by the facile hydrothermal method. In brief, sodium molybdate (1.0 g), polyvinylpyrrolidone (0.03 g), and thiourea (1.2 g) were dissolved in 40 mL of deionized water. The entire blend was agitated for 2 hours at room temperature and finally treated hydrothermally at 180°C with a Teflon-lined autoclave for about 12 hours. The obtained products were washed and dried at 80°C for about 12 hours in a vacuum.

14.2.4 PREPARATION OF MoS_2/GRAPHENE NANOCOMPOSITES

MoS_2/Graphene-x (x = 0.5, 2.5, 5.0, and 10 wt.% graphene oxide) nanocomposites were prepared by the facile hydrothermal method. In brief, 1.0 g of sodium molybdate with 40 mL of polyvinylpyrrolidone (0.03 g) and thiourea (1.2 g) solution were added to the calculated amount of graphene-oxide solution. The entire blend was agitated for 2 hours to room temperature and finally treated hydrothermally at 180°C with a Teflon-lined autoclave for about 12 hours. The obtained black solid products of MoS_2/graphene were washed and dried at 80°C for about 12 hours in a vacuum. Figure 14.1 shows a schematic illustration of the synthesis of MoS_2/graphene nanocomposites.

14.2.5 CHARACTERIZATION

The synthesized nanocomposites surface morphology was studied by field-emission scanning electron microscopy (Model Carl Zeiss Ultra 55) and transmission electron microscopy (Model FEI Technai G2). The elemental composition was carried out using an X-ray photoelectron spectroscopy (Multi-lab 2000, Thermo Scientific, UK)

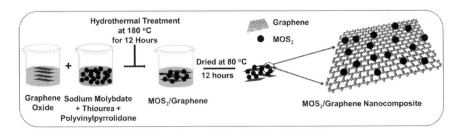

FIGURE 14.1   Schematic illustration of the synthesis of MoS₂/Graphene nanocomposites.

Mg-Kα (1253.6 eV) with a 200 W of power used as an exciting source, and 10 eV of energy used to collect data. The structural properties were characterized using an X-ray diffractometer (Rigaku, Japan) with Cu-Kα radiation ($\lambda = 0.15406$ nm) within the 2θ range of 5° to 60°. The total organic carbon was analyzed by a total organic carbon analyzer (TOC-V CSN, Shimadzu, Japan).

14.2.6   PHOTOCATALYTIC ACTIVITY MEASUREMENTS

Photocatalytic studies on the nanocomposites were conducted by determining the degradation of methylene blue solution. A pyrex glass photocatalytic reactor fitted with a water-cooled immersion tube, 100 W tungsten lamp, and the 400 nm wavelength cut-off filter was used as a visible-light treatment source. A sample suspension of 100 mL contained the methylene blue dye (10 mg/L) and the photocatalyst (10 mg) in a beaker, and the resultant sample mixture was agitated for about 30 minutes to attain an adsorption-desorption balance between the dye solution and catalysts. In photocatalytic studies, 3 mL of the reacted methylene blue solution was collected at a designated time interval. The treated methylene blue dye solution concentration was measured using UV-Vis spectroscopy (Analytik Jena) at an absorbance wavelength of about 664 nm (Sadiq et al., 2017b).

The degradation percentage of methylene blue dye was measured using the following Equation (14.1).

$$\text{Methylene blue degradation percentage} = ((C_o - C_t)/C_o) \times 100 \quad (14.1)$$

Where initial absorbance (C_o) and absorbance at a time interval (C_t) of the methylene blue dye solution, respectively.

The mineralization percentage of methylene blue was measured using the following Equation (14.2).

$$\text{Methylene blue mineralization percentage} = [(TOC_o - TOC_t)/TOC_o] \times 100$$
$$(14.2)$$

Where initial concentration (TOC_o) and concentration at a time interval (TOC_t) of the methylene blue dye solution, respectively (Sadiq et al., 2017c).

(a)   (b)

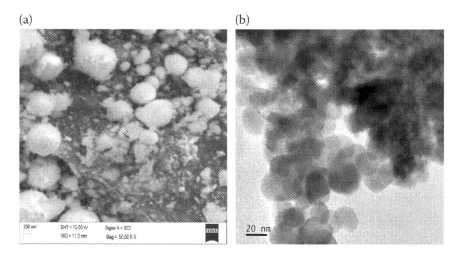

FIGURE 14.2 (a) Field-emission scanning electron microscopy image and (b) Transmission electron microscopy image of the MoS₂/Graphene nanocomposites.

14.3 RESULTS AND DISCUSSION

14.3.1 MORPHOLOGICAL STUDIES

Field-emission scanning electron microscopy and transmission electron microscopy analysis were investigated the morphology, size, and structural features of synthesized materials. The field-emission scanning electron microscopy image (Figure 14.2a) of MoS_2/graphene obtained through the hydrothermal method and MoS_2 exhibit spherical morphology with severe stacking and wrapped with fine sheet morphology graphene. To get further information on its morphology, MoS_2/graphene was thoroughly inspected using transmission electron microscopy techniques. Transmission electron microscopy image (Figure 14.2b) also indicates that the MoS_2 nanospheres are wrapped on the surface of graphene nanosheets with few layers.

14.3.2 X-RAY DIFFRACTION STUDIES

X-ray diffractometer analysis was investigated the crystalline structure of synthesized materials (Figure 14.3). The MoS_2/graphene nanocomposites (Figure 14.3b) suggest no extra impurity peaks exist. Only typical diffraction patterns of the hexagonal phase of MoS_2 (JCPDS No. 37–1492) are seen (Chen et al., 2017). For graphene oxide (Figure 14.3a), the diffraction peak appears at $2\theta = 10.5°$. However, no peak diffraction referring to graphene oxide was observed in the MoS_2/graphene nanocomposite, suggesting a reduction in graphene oxide during synthesis, removing oxygen-containing functional groups, and exfoliation (Sadiq and Bhat, 2016). It was clear that the combination of graphene and MoS_2 had a negligible impact on the crystalline structure of MoS_2.

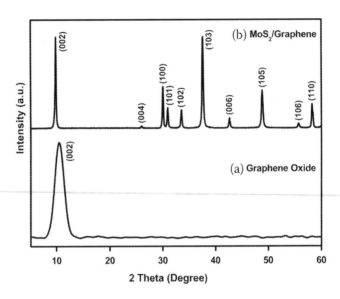

FIGURE 14.3 X-ray diffraction patterns of (a) Graphene oxide and (b) MoS2/Graphene nanocomposites.

14.3.3 X-Ray Photoelectron Spectroscopy Studies

X-ray photoelectron spectroscopy measurements were investigated the chemical status of the prepared MoS$_2$/graphene nanocomposites. X-ray photoelectron spectroscopy survey of MoS$_2$/graphene nanocomposite, as shown in Figure 14.4a. The survey spectrum shows that the elementary peaks of C, Mo, S, and no other elements are present in MoS$_2$/graphene nanocomposites. Figure 14.4b shows the high-resolution X-ray photoelectron spectroscopy spectra of the C 1s peak, which contains the binding energies at 284.8 eV, in the aromatic ring (C–C/C=C) of sp2, indicating the formation of graphene sheets (Sandhya et al., 2018). Figure 14.4c shows the high-resolution X-ray photoelectron spectroscopy spectra of Mo 3d, which corresponds to Mo 3d$_{3/2}$ at 232.8 eV and Mo 3d$_{5/2}$ at 229.7 eV of (+4) oxidation state of Mo along with this S 2s peak appeared at 226.1 eV indicates the formation of MoS$_2$ (Sabarinathan et al., 2017). Figure 14.4d shows the high-resolution X-ray photoelectron spectroscopy spectra of the S 2p peak, containing the binding energies at 163.3 eV and 161.8 eV to 2p$_{1/2}$ and 2p$_{3/2}$ (-2) oxidation state of S, respectively (Syariati et al., 2019). The above results support the presence of MoS$_2$ and graphene sheets without any other oxidation products in the MoS$_2$/graphene nanocomposites.

14.3.4 Photocatalytic Studies

The photocatalytic activity of MoS$_2$, graphene, and MoS$_2$/graphene nanocomposite was conducted in methylene blue under visible light treatment, after achieving the adsorption-desorption balance in darkness for 30 minutes. The percentage of degradation of methylene blue over time by different catalysts is shown in

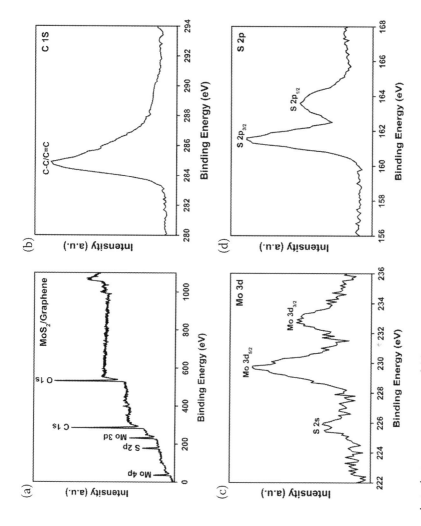

FIGURE 14.4 X-ray photoelectron spectroscopy of (a) survey spectrum, high-resolution spectra (b) C 1s, (c) Mo 3d, and (d) S 2p of MoS₂/graphene nanocomposite.

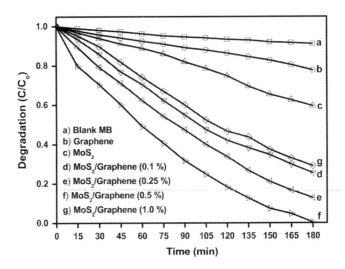

FIGURE 14.5 Degradation plots of methylene blue over various catalysts under visible light treatment.

Figure 14.5. Only degraded methylene blue without the photocatalyst are observed to 8.93% after 180 minutes. Further, it demonstrates that the self-degradation of methylene blue is insignificant and that the degradation occurs only with the aid of photocatalysts. The percentage degradation efficiency of methylene blue solutions in the existence of the graphene, MoS_2, MoS_2/graphene (0.1%), MoS_2/graphene (0.25%), MoS_2/graphene (0.5%), and MoS_2/graphene (1.0%) after 180 minutes, photocatalytic reaction was achieved to be 22.28%, 40.54%, 74.64%, 87.23%, 99.67%, and 71.06%, respectivelyg and these findings are compared with the most recent published literature, as indicated in Table 14.1.

Figure 14.6 shows that methylene blue dye photocatalytic degradation adopted pseudo-first-order kinetics on irradiation time, as indicated in Equation (14.3).

$$- \ln(C/Co) = kt \tag{14.3}$$

Where initial concentration (Co), concentration (C) at irradiation time (t), and first-order (k) rate constants (Sadiq and Bhat, 2017).

The rate constants value can be determined from the slope of the linear line. Moreover, the rate constants (Figure 14.7) was calculated from the photocatalytic reactions are 0.00122 min^{-1}, 0.00219 min^{-1}, 0.00597 min^{-1}, 0.00826 min^{-1}, 0.01423 min^{-1}, and 0.00508 min^{-1} for graphene, MoS_2, MoS_2/graphene (0.1%), MoS_2/graphene (0.25%), MoS_2/graphene (0.5%), and MoS_2/graphene (1.0%) nanocomposites, respectively. The MoS_2/graphene (0.5%) nanocomposite exhibits 11.7 times fold increments in efficiency compared to graphene and 6.5 times fold increments compared to MoS_2.

The reusability of catalytic materials provides tremendous potential for practical applications. The MoS_2/graphene nanocomposite in methylene blue degradation ratio over five consecutive cycles is shown in Figure 14.8. It can indicate that the

TABLE 14.1

Comparison of Degradation Percentage in the Photocatalytic Degradation of Methylene Blue by Various Catalysts under the Light Source

Photocatalysts	Photocatalytic Degradation of Methylene Blue Dye Under Light Source	Degradation Percentage	References
MoS_2/Graphene	Visible light	99.67%	Present work
WO_3/TiO_2	Visible light	99.0%	Abo El-Yazeed and Ahmed, 2019
Bi_2S_3/$ZnIn_2S_4$	Visible light	95.4%	Chachvalvutikul et al., 2019
CdS/TiO_2	Visible light	77.0%	Keerthana and Murugakoothan, 2019
$BaTiO_3$/GO	UV-Vis light	95.0%	Mengting et al., 2019
rGO-V_2O_5	Visible light	71.0%	Mishra et al., 2020
TiO_2/Fe_3O_4/GO	UV light	82.0%	Nadimi et al., 2019
	Visible light	76.0%	
MoS_2/TNT@CNTs	Visible light	98.7%	Qin et al., 2018
$MnTiO_3$/TiO_2	Sun light	75.0%	Suhila et al., 2020
PVDF/GO/ZnO	UV light	86.84%	Zhang, Dai, et al., 2019
Fe_3O_4@C@Ru	Sunlight irradiation	92.7%	Zhang, Yu, et al., 2019

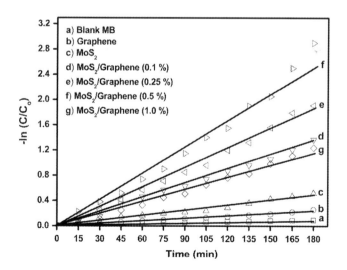

FIGURE 14.6 First-order kinetics plots of methylene blue over various catalysts under visible light treatment.

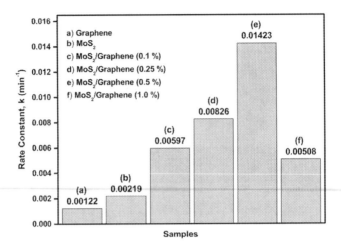

FIGURE 14.7   Rate constant of methylene blue over various catalysts under visible light treatment.

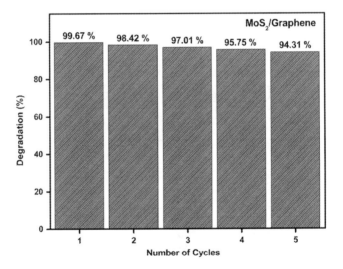

FIGURE 14.8   Stability test of MoS2/graphene nanocomposite for the photodegradation of methylene blue under visible light treatment.

most effective minimal loss of photocatalytic efficiency through the fifth cycle test. Therefore, the above findings revealed that MoS$_2$/graphene nanocomposites exhibit outstanding photocatalytic reusability and stability.

The total organic carbon content was determined to understand the products' nature during the photocatalytic degradation of methylene blue. The obtained absolute organic carbon values are used to calculate the methylene blue dye mineralization percentage using Equation (14.2) provided earlier. Figure 14.9 shows the

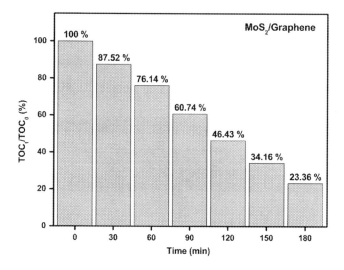

FIGURE 14.9 Mineralization curve for photodegradation of methylene blue by MoS_2/graphene nanocomposite at different intervals of time.

variation of the mineralization percentage with time by MoS_2/graphene nanocomposites. As can be observed from the plot, the total organic carbon content value reduced to 23.36%, or the mineralization level increased to 76.64% for 180 minutes of visible light treatment. The findings indicate that the entire organic carbon content is mainly transformed into carbon dioxide (CO_2) through the process (Sadiq et al., 2018a). The detected outcomes suggest that the MoS_2/graphene nanocomposite is an environmentally friendly photocatalyst.

14.3.5 MECHANISM OF THE PHOTOCATALYTIC ACTIVITY

The photocatalytic performance of MoS_2/graphene materials is primarily determined by the appropriate energy band position for light absorption, charge generation/migration, and recombination of electron-hole pairs (Zhang et al., 2016). The suggested photocatalytic mechanisms for MoS_2/graphene nanocomposites are shown in Figure 14.10. The electrons in the valence band (VB) of the MoS_2 were excited by visible light treatment at the conduction band (CB), which results in the photogeneration of holes (h^+) in the valence band and electrons (e^-) in the conduction band. These photogenerated electrons and holes are typically recombined during the photocatalytic degradation process. However, MoS_2/graphene nanocomposite photogenerated electrons on the MoS_2 conduction band tended to pass to graphene sheets. This prohibited photogenerated electrons and holes from being recombined and increased photocatalytic activity. The reduction reaction of the electrons and the oxygen molecules (O_2) took place in an aqueous solution, resulting in direct degradation of methylene blue (MB) by superoxide radical anions ($O_2 \bullet^-$). The resulting photoinduced holes were either directly oxidized with

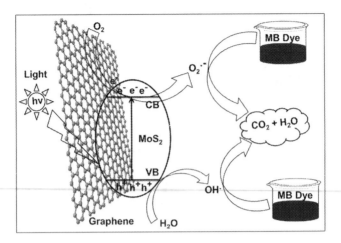

FIGURE 14.10   Proposed mechanisms for the photocatalytic degradation of methylene blue by MoS₂/Graphene nanocomposites under visible light treatment.

methylene blue or trapped in water (H_2O) to form hydroxyl radicals ($OH^•$), ensuing in the degradation of methylene blue (Yang et al., 2016).

14.4 CONCLUSIONS

To address the most critical environmental problem in the economic management of wastewater. This research was designed to use new photocatalytic materials, i.e., MoS₂/graphene nanocomposites. A series of MoS₂/graphene nanocomposite with different graphene-oxide content was effectively synthesized via a hydrothermal method. It can confirm that the MoS₂ was effectively decorated on the surface of graphene using various comprehensive characterization techniques. The photodegradation of methylene blue dye from water with the synthesized nanocomposites was also studied. We demonstrated that the MoS₂/graphene nanocomposite showed improved effective photocatalytic activity than that from pure MoS₂. Under the optimum conditions of MoS₂/graphene (0.5%), the degradation rate reaches 99.67% within 180 minutes, and the first-order rate constant is 0.01423 min^{-1}. These results suggest that an exceptional electron acceptor improved the active transfer of photogenerated electrons of MoS₂ to the graphene, thereby inhibiting the recombination of electron-hole pairs. This effective MoS₂/graphene nanocomposite is an excellent material for catalysts in various environmental applications.

ACKNOWLEDGEMENTS

The author greatly acknowledges Yunnan University, Kunming, China, for awarding Post Doctoral Research Fellowship.

REFERENCES

Abo El-Yazeed, WS & Ahmed, AI (2019). Photocatalytic activity of mesoporous WO_3/TiO_2 nanocomposites for the photodegradation of methylene blue. *Inorganic Chemistry Communications*, *105*, 102–111.

Chachvalvutikul, A, Pudkon, W, Luangwanta, T, Thongtem, T, Thongtem, S, Kittiwachana, S, & Kaowphong, S (2019). Enhanced photocatalytic degradation of methylene blue by a direct Z-scheme $Bi_2S_3/ZnIn_2S_4$ photocatalyst. *Materials Research Bulletin*, *111*, 53–60.

Chen, H, Liu, T, Su, Z, Shang, L, & Wei, G (2018). 2D transition metal dichalcogenide nanosheets for photo/thermo-based tumor imaging and therapy. *Nanoscale Horizon*, *3*, 74–89.

Chen, Z, Yan, H, Lyu, Q, Niu, S, & Tang, C (2017). Ternary hybrid nanoparticles of reduced graphene oxide/graphene-like MoS_2/zirconia as lubricant additives for bismaleimide composites with improved mechanical and tribological properties. *Composites Part A: Applied Science and Manufacturing*, *101*, 98–107.

Fang, X, Wei, P, Wang, L, Wang, X, Chen, B, He, Q, Yue, Q, Zhang, J, Zhao, W, Wang, J, Lu, G, Zhang, H, Huang, W, Huang, X, & Li, H (2018). Transforming monolayer transition-metal dichalcogenide nanosheets into one-dimensional nanoscrolls with high photosensitivity. *ACS Applied Materials Interfaces*, *10*, 13011–13018.

Hao, D, Yang, Y, Xu, B, & Cai, Z (2018). Bifunctional fabric with photothermal effect and photocatalysis for highly efficient clean water generation. *ACS Sustainable Chemistry & Engineering*, *6*, 10789–10797.

Keerthana, BGT & Murugakoothan, P (2019). Synthesis and characterization of CdS/TiO_2 nanocomposite: Methylene blue adsorption and enhanced photocatalytic activities. *Vacuum*, *159*, 476–481.

Lavin-Lopez, MP, Romero, A, Garrido, J, Sanchez-Silva, L, & Valverde, JL (2016). Influence of different improved hummers method modifications on the characteristics of graphite oxide in order to make a more easily scalable method. *Industrial & Engineering Chemistry Research*, *55*, 12836–12847.

Lin, Y, Ren, P, & Wei, C (2019). Fabrication of MoS_2/TiO_2 heterostructures with enhanced photocatalytic activity. *CrystEngComm*, *21*, 3439–3450.

Mengting, Z, Kurniawan, TA, Fei, S, Ouyang, T, Othman, MHD, Rezakazemi, M, & Shirazian, S (2019). Applicability of $BaTiO_3$/graphene oxide (GO) composite for enhanced photodegradation of methylene blue (MB) in synthetic wastewater under UV-vis irradiation. *Environmental Pollution*, *255*, 113182.

Mishra, A, Panigrahi, A, Mal, P, Penta, S, Padmaja, G, Bera, G, Das, P, Rambabu, P, & Turpu, GR (2020). Rapid photodegradation of methylene blue dye by $rGO-V_2O_5$ nanocomposite. *Journal of Alloys and Compounds*, *842*, 155746.

Mutyala, S, Sadiq, MMJ, Gurulakshmi, M, Bhat, DK, Shanthi, K, Mathiyarasu, J, & Suresh, C (2020). Disintegration of flower like MoS_2 to limply allied layer grown on spherical nanoporous TiO_2: Enhanced visible light photocatalytic degradation of methylene blue. *Journal of Nanoscience and Nanotechnology*, *20*, 1118–1129.

Nadimi, M, Saravani, AZ, Aroon, MA, & Pirbazari, AE (2019). Photodegradation of methylene blue by a ternary magnetic TiO_2/Fe_3O_4/graphene oxide nanocomposite under visible light. *Materials Chemistry and Physics*, *225*, 464–474.

Peng, W, Li, Y, Zhang, F, Zhang, G, & Fan, X (2017). Roles of two-dimensional transition metal dichalcogenides as cocatalysts in photocatalytic hydrogen evolution and environmental remediation. *Industrial & Engineering Chemistry Research*, *56*, 4611–4626.

Pu, S, Long, D, Wang, MQ, Bao, SJ, Liu, Z, Yang, F, Wang, H, & Zeng, Y (2017). Design, synthesis and photodegradation ammonia properties of $MoS_2@TiO_2$ encapsulated carbon coaxial nanobelts. *Materials Letters*, *209*, 56–59.

Qin, Q, Shi, Q, Ding, W, Wan, J, & Hu, Z (2018). Efficient hydrogen evolution and rapid degradation of organic pollutants by robust catalysts of MoS_2/TNT@CNTs. *International Journal of Hydrogen Energy, 43,* 16024–16037.

Sabarinathan, M, Harish, S, Archana, J, Navaneethan, M, Ikeda, H, & Hayakawa, Y (2017). Highly efficient visible-light photocatalytic activity of MoS_2-TiO_2 mixtures hybrid photocatalyst and functional properties. *RSC Advances, 7,* 24754–24763.

Sadiq, MMJ & Bhat, DK (2016). Novel RGO-$ZnWO_4$-Fe_3O_4 nanocomposite as an efficient catalyst for rapid reduction of 4-nitrophenol to 4-aminophenol. *Industrial & Engineering Chemistry Research, 55,* 7267–7272.

Sadiq, MMJ & Bhat, DK (2017). Novel $ZnWO_4$/RGO nanocomposite as high performance photocatalyst. *AIMS Materials Science, 4,* 158–171.

Sadiq, MMJ, Shenoy, US, & Bhat, DK (2016). Novel RGO-$ZnWO_4$-Fe_3O_4 nanocomposite as high performance visible light photocatalyst. *RSC Advances, 6,* 61821–61829.

Sadiq, MMJ, Shenoy, US, & Bhat, DK (2017a). A facile microwave approach to synthesis of RGO-$BaWO_4$ composites for high performance visible light induced photocatalytic degradation of dyes. *AIMS Materials Science, 4,* 487–502.

Sadiq, MMJ, Shenoy, US, & Bhat, DK (2017b). Enhanced photocatalytic performance of N-doped RGO-$FeWO_4$/Fe_3O_4 ternary nanocomposite in environmental applications. *Materials Today Chemistry, 4,* 133–141.

Sadiq, MMJ, Shenoy, US, & Bhat, DK (2017c). $NiWO_4$-ZnO-NRGO ternary nanocomposite as an efficient photocatalyst for degradation of methylene blue and reduction of 4-nitrophenol. *Journal of Physics and Chemistry of Solids, 109,* 124–133.

Sadiq, MMJ, Shenoy, US, & Bhat, DK (2018a). Novel $NRGO$-$CoWO_4$-Fe_2O_3 nanocomposite as an efficient catalyst for dye degradation and reduction of 4-nitrophenol. *Materials Chemistry and Physics, 208,* 112–122.

Sadiq, MMJ, Shenoy, US, & Bhat, DK (2018b). Synthesis of $NRGO$/$BaWO_4$/g-C_3N_4 nanocomposites with excellent multifunctional catalytic performance via microwave approach. *Frontiers of Materials Science, 12,* 247–263.

Sandhya, S, Sadiq, MMJ, Bhat, DK, & Hegde, AC (2018). Electrodeposition of Ni-Mo-rGO composite electrodes for efficient hydrogen production in an alkaline medium. *New Journal of Chemistry, 42,* 4661–4669.

Singh, S, Faraz, M, & Khare, N (2020). Recent advances in semiconductor–graphene and semiconductor–ferroelectric/ferromagnetic nanoheterostructures for efficient hydrogen generation and environmental remediation. *ACS Omega, 5,* 11874–11882.

Suhila, A, Aïcha, M, & Elbashir, EAS (2020). Photocatalytic degradation of methylene blue dye in aqueous solution by $MnTiO_3$ nanoparticles under sunlight irradiation. *Heliyon, 6,* e03663.

Sumesh, CK & Peter, SC (2019). Two-dimensional semiconductor transition metal based chalcogenide based heterostructures for water splitting applications. *Dalton Transactions, 48,* 12772–12802.

Syariati, A, Kumar, S, Zahid, A, Yumin, AAE, Ye, J, & Rudolf, P (2019). Photoemission spectroscopy study of structural defects in molybdenum disulfide (MoS_2) grown by chemical vapor deposition (CVD). *Chemical Communications, 55,* 10384–10387.

Yang, Y, Xu, L, Wang, H, Wang, W, & Zhang, L (2016). TiO_2/graphene porous composite and its photocatalytic degradation of methylene blue. *Materials and Design, 108,* 632–639.

Yu, X, Du, R, Li, B, Zhang, Y, Liu, H, Qu, J, & An, X (2016). Biomolecule-assisted self-assembly of CdS/MoS_2/graphene hollow spheres as high-efficiency photocatalysts for hydrogen evolution without noble metals. *Applied Catalysis B: Environmental, 182,* 504–512.

Yu, X, Zhao, G, Gong, S, Liu, C, Wu, C, Lyu, P, Maurin, G, & Zhang, N (2020). Design of MoS_2/Graphene van der waals heterostructure as highly efficient and stable electrocatalyst for hydrogen evolution in acidic and alkaline media. *ACS Applied Materials Interfaces, 12,* 24777–24785.

Zhang, D, Dai, F, Zhang, P, An, Z, Zhao, Y, & Chen, L (2019). The photodegradation of methylene blue in water with PVDF/GO/ZnO composite membrane. *Materials Science and Engineering: C*, *96*, 684–692.

Zhang, X, Lai, Z, Ma, Q, & Zhang, H (2018). Novel structured transition metal dichalcogenide nanosheets. *Chemical Society Reviews*, *47*, 3301–3338.

Zhang, L, Lan Sun, L, Liu, S, Huang, Y, Xu, K, & Ma, F (2016). Effective charge separation and enhanced photocatalytic activity by the heterointerface in MoS_2/reduced graphene oxide composites. *RSC Advances*, *6*, 60318–60326.

Zhang, Q, Yu, L, Xu, C, Zhao, J, Pan, H, Chen, M, Xu, Q, & Diao, G (2019). Preparation of highly efficient and magnetically recyclable Fe_3O_4@C@Ru nanocomposite for the photocatalytic degradation of methylene blue in visible light. *Applied Surface Science*, *483*, 241–251.

15 Renewable Energy in Smart Grid: Futuristic Power System

Hadeel Fahad Alharbi and Kusum Yadav
College of Computer Science and Engineering, University of Ha'il, Ha'il, Kingdom of Saudi Arabia

CONTENTS

15.1 INTRODUCTION

Smart grid is the future of the power sector in the 21st century. Information about consumer consumption, peak hour loads, and past failure are collected to form a robust, scalable, and self-healing power grid system. There is a full-duplex kind of transmission among the different entities involved in this system (Kabalci & Kabalci, 2017). Figure 15.1 shows the abstract representation of the smart grid system where blue color solid lines are used for transmission of power and red dotted lines are used for communication of the consumer-related information toward the power station.

As per the basic rule of science, energy can neither be created nor be destroyed. It can be converted from one form to another. Sources of energy can be categorized into two categories: renewable energy resources and nonrenewable energy resources. Different kind of renewable and nonrenewable energy resources that can be used for the generation of power are shown in Figure 15.2.

DOI: 10.1201/9781003127819-15

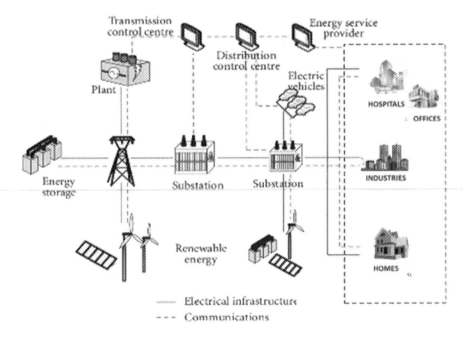

FIGURE 15.1   An abstract representation of the smart grid concept.

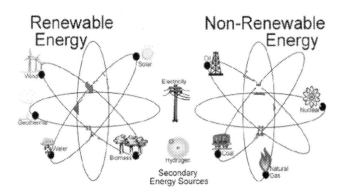

FIGURE 15.2   Renewable and nonrenewable energy resources.

Renewable energy resources are those resources that are nonexhaustive. Sun, wind, ocean waves, and flowing water are some examples of renewable resources. In contrast, nonrenewable energy resources are those whose stock is limited. Fossil fuels, oil, coal, natural gases, nuclear energy, etc., are examples of nonrenewable energy resources. Oil crisis, fluctuating oil prices, and changing political relationship have forced developing countries to find an alternate source of energy for power generation (Sinsel, Riemke, & Hoffmann, 2020).

Setup installation cost and availability are the two significant challenges associated with integrating renewable energy resources with the smart grid power system. Wind, solar energy, flowing water, and ocean waves are not available at every place and all the time.

In this chapter, the authors have discussed the details of smart grid technology and renewable energy resources in Section 15.2 and Section 15.3, respectively. Different issues and the opportunities associated with the integration of renewable energy and smart grid technology are discussed in Section 15.4. Some case studies of the successful amalgam of the smart grid system and renewable energy resources are discussed in Section 15.5. Conclusive remarks are given in Section 15.6.

15.2 SMART GRID

To ensure prosperity for the future generation, it is essential to shift from the conventional power grid to the smart power grid. This will improve energy efficiency, reduce CO_2 and other toxic gas emissions, and transition to renewable energy resources. Whether a developed or a developing country, all are investing huge money for the practical implementation of a smart power grid system. The objective of the smart grid is to maintain the balance between demand and supply while ensuring reliability, responsiveness, and optimization of resources. The significant differences between the smart grid and conventional grid are described in Table 15.1, below:

15.2.1 CHARACTERISTICS OF SMART GRID

Essential Characteristics of the Smart Grid Power System Are as Follows:

- For the enhancement of reliability, efficiency, and security of the smart grid system, digital information and control technology are used.
- Renewable energy resources are used.
- Resources are deployed in a distributed manner.
- Smart metering and communication technologies are used.
- Consumer devices are integrated with smart Internet of Things (IoT) based appliances.
- Provide the billing, consumption, and control information to the consumer timely.
- Standards for communication and IoT appliance interoperability with consumer devices should be followed (Greer et al., 2014).

TABLE 15.1

Smart vs Traditional Grid (Ourahou, Ayrir, Hassouni, & Haddi, 2020)

Conventional Grid	Smart Grid
These are analog.	It's digital.
Transmission is simplex.	Transmission is full-duplex.
Generation and distribution are centralized.	Generation and distribution are distributed at multiple locations.
IoT-based appliances are not used.	IoT-based appliances are used.
The role of the customer is passive.	The role of the customer is active.

15.2.2 COMPONENTS OF SMART GRID

A smart grid transmits power and communicates information, so multiple technologies are involved in the smart grid. Major components involved in the smart grid system are as follows:

Integrated Communication: For the transmission of information in both ways, a high-speed communication platform is required. It acts as a gel that combines other technologies with the conventional grid system.

Sensing Devices: For sensing user information for demand and peak hour estimation and generation of bills, different IoT-based sensing devices are used.

Metering Devices: To provide users timely and sufficient information about usage and charges, smart metering devices are used.

Advanced Components: These components are used in finding the electrical behavior of the grid.

Advanced Control: These components are analytical tool-enabled components that are used for the analysis of huge data. These components help in making different decisions to avoid the failure situation.

Storage: To improve the reliability of the system, power is stored in a distributed manner. For this purpose, different power storage components are required (Li, Qiao, Sun, Wan, Wang, Xia & Zhang, 2010).

15.2.3 ADVANTAGES OF SMART GRID

Following are the significant advantages of the smart grid:

Energy Efficiency: Managing the load properly and reducing line losses ensures efficient energy usage.

Green Earth: Through proper energy management, energy consumption is less relative to the traditional system, so this results in less CO_2 emission.

Customer Satisfaction: Customers are getting uninterrupted power supply at a lower cost and are aware of usage charges, resulting in greater customer satisfaction. Moreover, it allows customers to play a role in optimizing system operations.

Operational Efficiency: It uses the concept of a distributed system, resulting in better network design, and monitoring and diagnostic of the fault also becomes easy (Bayindir et al., 2016).

Besides all these advantages, some hurdles still must be addressed in the widespread adoption of smart grid concepts; these include the lack of standards, significant capital investment, unavailability of proper business model, and tje meed for a faster internet connection.

15.3 RENEWABLE ENERGY RESOURCES

There are a variety of renewable energy sources available globally. These are infinite and freely available. Details of some of the renewable energy resources are given below:

Solar Energy: Solar energy is the most widely available renewable energy resource. A photovoltaic panel is used for the conversion of solar energy into an

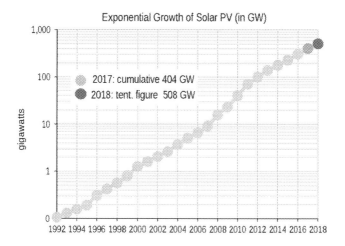

FIGURE 15.3   Solar energy generation growth.

electric current. This panel consists of a grid of silicon cells that contain electrons. On receiving light photons from sun rays, electrons inside a silicon cell get excited and start moving, generating the current. This current can be used for storage or running electric equipment (Asgher et al., 2018). Figure 15.3 shows how the generation of solar energy has exponentially increased in the last three decades.

Biomass Energy: Residue like sewerage waste, industrial waste, animal waste, plant waste, and agricultural waste are the major sources of biomass. All these contain some stored energy. When this biomass is burned, energy is released in heat, which can be converted into electricity through turbines and generators. This energy can also be used to generate biofuels (Martin, 2010).

Geothermal Energy: This is the energy generated automatically in the core of the Earth due to radioactive decay and high temperature. Water gets converted into the form of steam, and hot water and high-pressure steam come out on the Earth's surface. This steam can be used to rotate the turbine to generate mechanical energy. This mechanical energy can be converted into electrical energy (Turcotte & Schubert, 2002).

Wind Energy: The power of wind is used to rotate turbines and generate mechanical energy. Then, this mechanical energy is converted into electrical energy. For large-scale power generation, a grid of turbines is used. Wind energy generation will not harm the environment (Köktürk & Tokuç, 2017).

Figure 15.4 below shows the growth of power generation from wind energy in the last three decades.

Hydro Energy: Water is the source of energy in hydropower. Water is stored from a river through a dam. After that, it is fallen over the turbine and generates mechanical energy. Later on, this mechanical energy is converted into electrical energy. Figure 15.5 below shows the basic design of the generation of power from water.

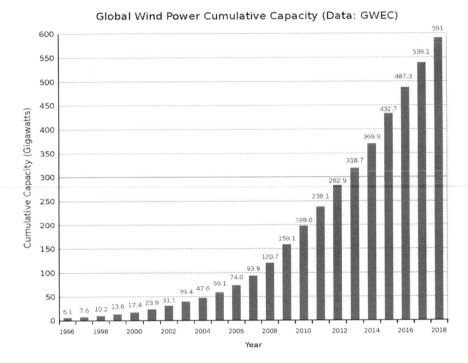

FIGURE 15.4   Wind energy generation growth.

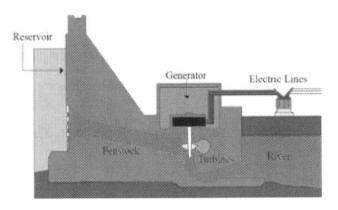

FIGURE 15.5   Process of conversion of hydro energy into power.

Figure 15.6 shows how hydro energy has contributed to the generation of power for the last 100 years. It also shows that power generation through water is still increasing.

15.4   INTEGRATION OF RENEWABLE ENERGY WITH SMART GRID

Climate change is the biggest issue in the 21st century and worldwide; all countries are working to stop this change in the climate. To handle the climate change situation,

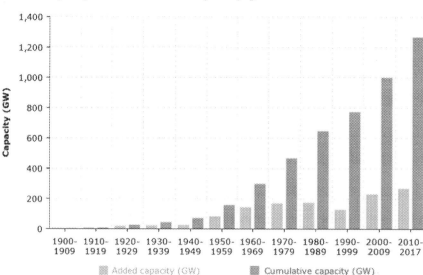

FIGURE 15.6 Hydro energy generation growth.

governments are shifting toward renewable energy resources and distributed power generation. The integration of renewable energy resources and smart grid systems will help in preventing climate change. In this section, the authors have discussed the various opportunities and challenges associated with integrating renewable energy resources and smart grids.

15.4.1 CHALLENGES

Despite all efforts to promote renewable energy as a source of energy for power generation in conventional as well as smart grid, until 2017, 73.5% of electricity production worldwide was generated through fossil fuels, with the remaining 26.5% of electricity production done through renewable energy resource (Qazi et al., 2019).

Uncertainty: Traditional power generation has handled the fluctuating customer demand by controlling the fuel used for power generation. With a renewable energy source, there is also a need to address the uncertainty about resource availability since these resources are not in control of human beings. Figure 15.7 shows the variability in power generation through the wind for 30 successive days. This shows the level of uncertainty.

Figure 15.8 also shows the uncertainty in power generation through solar energy. This is highlighting the variation in one place due to the positions of the cloud. It is also highlighting the variation in output from one geographical position to another.

Variability: System operators have to achieve a balance between demand and supply. Sometimes there is a high production of renewable energy, but demand is low, and sometimes the reverse situation is true.

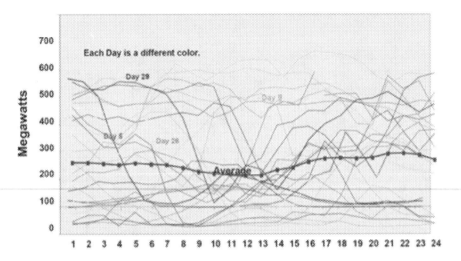

FIGURE 15.7   Power generation through the wind for 30 successive days.

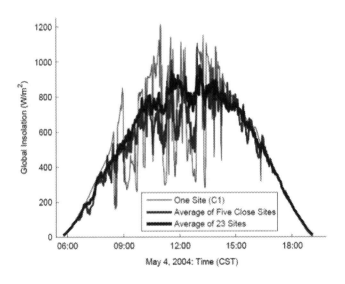

FIGURE 15.8   Power generation through solar energy on a specific day.

High Initial Cost: To accommodate renewable energy sources for power generation, there is a need for system upgrades. Moreover, renewable energy sources may be located at a distant location. This will result in an additional cost of transmission and distribution.

Distribution: Renewable energy resources are not available collectively; they are diversified and open in a distributed manner. This causes additional distribution costs.

Though a lot of research is going on in renewable energy sources, due to the lack of standardization and business model, integration is still not an easy task.

15.4.2 OPPORTUNITIES

Challenges associated with renewable energy sources can be considered as an opportunity to integrate with a smart grid system in the following ways:

Uncertainty: As a smart grid is equipped with an advanced decision support system, this will help the operator handle the uncertainty associated with renewable energy resources.

Variability: Using a smart inverter with integrated storage, excessive power generated from renewable energy sources can be stored and used in the peak demand time. This will help in handling the variability associated with renewable energy resources.

High Initial Cost: A smart grid is equipped with a smart meter, advanced energy management system, and intelligent load-calculation equipment, so additional resources will be installed only when it is required. This will defer the installation of other resources.

Distribution: A smart grid is distributed in nature, so it can efficiently utilize the distributed availability of renewable energy resources.

15.5 CASE STUDY ON SMART GRID & RENEWABLE ENERGY INTEGRATION

The following case studies of the different parts of the world have shown how a smart grid can be deployed for better technical and economic integration with uncertain and variable renewable energy sources.

Karehman wind farm located in Oland, Sweden, was designed to handle 30 MW of power. But through dynamic line rating, i.e. by collecting information about environmental air temperature, power transfer, and line temperature capacity of the line, it was increased by 60% and was able to handle 48 MW of power. This has helped them to avoid the installation of an additional transmission line. If they have installed an extra line, that would cost them around 16 million US dollars. But by just spending 7.5 lakh US dollar, they were able to deploy a dynamic rating mechanism (Babu et al., 2015).

In 2012, using a smart grid and renewable energy resource, Korea developed GAPA as an energy-independent island. The area of GAPA island was 8.5 km^2 with a population of 281 people. 450 kW diesel generator was replaced by 500 kW and 111 kW wind and solar power generation.

To manage the uncertainty and variability of renewable energy resources, they were supported by 1 MW lithium-ion batteries. This storage system helps handle the fluctuating demand and provides the initial power to start the wind turbines. An automated power-management system was used to optimize the grid operation by analyzing the electricity consumption in real time (Choi & Do, 2016). Its schematic representation is shown in Figure 15.9. This has made GAPA island carbon free.

In 2011, Ann Cavoukian of Hydro One Company in Canada had supported a case study on smart grid solutions in Ontario. The company's vision was to increase the role of renewable energy resources in the provision of electricity in Ontario. Their focus was entirely on the establishment of a smart electricity grid. The company has planned to use all renewable energy sources like solar energy, wind, hydro, and

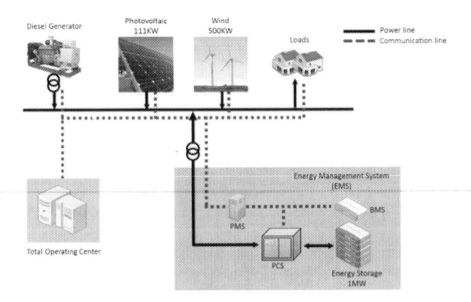

FIGURE 15.9   Schematic representation of GAPA island smart grid.

biomass in smart grid electricity distribution. The amalgam of smart technology with renewable energy resources was done to benefit customers, the company, and the environment. This plan has improved the grid distribution in terms of control, automation, and analysis. For efficient management, they have used a software called a distribution-management system at every grid distribution location. The company has provided a tool for network operations and planning to monitor the grid and information about real-time power flow (Cavoukian & Chanliau, 2013).

Austria's government has implemented multiple pilot projects of smart grid at renewable energy resource supporting locations. To test voltage control in the distribution network, they deployed a pilot project in the Salzburg region. Another pilot project was deployed in Lungau where significant wind and solar resources were available. This deployment resulted in a 20% increase in the distribution network capacity on the 30 kW line. This has deferred the need for system upgrades and provided the ability to control the voltage more accurately. It was estimated that this pilot project's deployment results in saving around 2,200,000-liter crude oil and reducing the emission of CO_2 by 5500 tons per year. At Turrach hydropower plant, they have deployed another pilot project. Before its implementation, for the grid supply to the grid, 14 km long cable was needed. But through the implementation of voltage control, this was made possible through a 50 m cable. This results in significant cost savings and improved security of transmission, increased share of renewable energy resources, and reduction in CO_2 emission (Kupzog et al., 2013).

15.6   CONCLUSION

To avoid climate change, all countries focus on those sources of energy that do not harm the environment. The usage of renewable energy in the power-generation sector is one step in this direction. Also, to utilize the available resources for power generation efficiently, it is necessary to shift from a conventional grid to a smart grid system. The situation can be more improved if the renewable energy source is integrated with a smart grid system. But there are specific challenges also. Nonavailability of renewable resources at every place and huge starting investments are the major hurdles in the smooth integration of renewable energy resources with a smart grid system. Further, it has been discussed how the challenges associated with integrating renewable energy sources and smart grid are converted into opportunities. Through case studies, it has proved that the amalgam of a smart grid system with renewable energy resources has resulted in an efficient and environment-friendly system.

As future work, the authors have planned to design a model to predict the stability of a smart grid. Also, a model of prediction of availability of renewable energy sources will act as adhesive for the integration of a smart grid with renewable energy sources.

REFERENCES

Asgher, U, Babar Rasheed, M, Al-Sumaiti, AS, Ur-Rahman, A, Ali, I, Alzaidi, A, & Alamri, A (2018). Smart energy optimization using heuristic algorithm in smart grid with integration of solar energy sources. *Energies*, 11(12), 3494.

Babu, S, Jürgensen, JH, Wallnerstrom, CJ, Hilber, P, & Tjernberg, LB (2015, November). Analyses of smart grid technologies and solutions from a system perspective. In *2015 IEEE Innovative Smart Grid Technologies-Asia (ISGT ASIA)* (pp. 1–5). IEEE.

Bayindir, R, Colak, I, Fulli, G, & Demirtas, K (2016). Smart grid technologies and applications. *Renewable and Sustainable Energy Reviews*, 66, 499–516.

Cavoukian, A, & Chanliau, M (2013). *Privacy and Security by Design: A Convergence of Paradigms*. Information and Privacy Commissioner, Ontario.

Choi, J, & Do, DPN (2016). Process and Features of Smart Grid, Micro Grid and Super Grid in South Korea. *IFAC-PapersOnLine*, 49(27), 218–223.

Greer, C., Wollman, D., Prochaska, D., Boynton, P., Mazer, J., Nguyen, C., FitzPatrick, G., Nelson, T., Koepke, G., Hefner, A., Pillitteri, V., Brewer, T., Golmie, N., Su, D., Eustis, A., Holmberg, D., & Bushby, S. (2014). NIST framework and roadmap for smart grid interoperability standards, release 3.0, Special Publication (NIST SP). Gaithersburg, MD: Gaithersburg, MD. 10.6028/NIST.SP.1108r3. https://tsapps.nist.gov/publication/get_pdf.cfm?pub_id=916755. (Accessed June 7, 2022).

Kabalci, Y, & Kabalci, E (2017). Modeling and analysis of a smart grid monitoring system for renewable energy sources. *Solar Energy*, 153, 262–275.

Köktürk, G, & Tokuç, A (2017). Vision for wind energy with a smart grid in Izmir. *Renewable and Sustainable Energy Reviews*, 73, 332–345.

Li, F, Qiao, W, Sun, H, Wan, H, Wang, J, Xia, Y & Zhang, P (2010). Smart transmission grid: Vision and framework. *IEEE Transactions on Smart Grid*, 1(2), 168–177.

Kupzog, F, Brunner, H, Schrammel, J, Döbelt, S, Einfalt, A, Lugmaier, A, & Stutz, M (2013). Results & findings from the smart grids model region Salzburg. *Salzburg, Austria: Salzburg AG*. Accessed May 4, 2015.

Martin, MA (2010). First generation biofuels compete. *New Biotechnology*, 27(5), 596–608.

Ourahou, M, Ayrir, W, Hassouni, BE, & Haddi, A (2020). Review on smart grid control and reliability in presence of renewable energies: challenges and prospects. *Mathematics and Computers in Simulation*, 167, 19–31.

Qazi, A, Hussain, F, Rahim, NA, Hardaker, G, Alghazzawi, D, Shaban, K, & Haruna, K (2019). Towards sustainable energy: a systematic review of renewable energy sources, technologies, and public opinions. *IEEE Access*, 7, 63837–63851.

Sinsel, SR, Riemke, RL, & Hoffmann, VH (2020). Challenges and solution technologies for the integration of variable renewable energy sources—a review. *Renewable Energy*, 145, 2271–2285.

Turcotte, DL, & Schubert, G (2002). *Geodynamics*. Cambridge University Press.

16 Experimental Investigation on Different Supplementary Cementitious Materials as Smart Construction Materials to Produce Concrete: Toward Sustainable Development

Bansari N. Dave
Nascent Associates, Rajkot, Gujarat, India

Damyanti G. Badagha
Civil Engineering Department, S. N. Patel Institute of
Technology & Research Centre, Umrakh, Gujarat, India

Ahmed A. Elngar
Computers & Artificial Intelligence, Beni-Suef University,
Banī Suwayf, Egypt

Pratik G. Chauhan
Civil Engineering Department, Dr. S. & S. S. Ghandhy
Government Engineering College, Surat, Gujarat, India

CONTENTS

DOI: 10.1201/9781003127819-16

16.1 INTRODUCTION

The material concrete, an artificial stone conglomerate, is made by using Portland cement, water, and aggregates. Cement has been widely used in one form or another in the construction industry for centuries. One ton of cement generation requires around 2 ton of raw materials and releases 0.95 ton ($\cong$ 1 ton) of CO_2, which spreads pollution in environment (Badagha and Modhera, 2017; Damtoft et al., 2008). Different alternatives are available and in use for cement utilization reduction nowadays (Berke, 2012).

These SCMs are basically naturally available materials or by-products from different processes. To be used in concrete, these SCMs may or may not be processed further. Few of them are pozzolans, which do not contain any cementitious characteristics on their own; instead, they react to form cementitious compounds when used with Portland cement. Alternatively, SCM like slag do exhibit cementitious properties. Supplementary cementitious materials are sometimes referred to as mineral admixtures (Badagha and Modhera, 2019).

The SCMs are often utilized in combination or individually to make concrete. As blended cement or as a separately batched ingredient, SCMs are also applied to the concrete mixture at the ready mixed concrete plant. The SCMs should essentially meet requirements of established standards, i.e. ASTM C1697-16 (2016), the quality specification for blended supplementary cementitious material. Some SCMs are listed here: fly ash, GGBS (ground-granulated-furnace-slag), metakaolin, micro silica. Unlike other mineral admixtures, micro silica and ash are additionally by-products from the industries, where their engineering values are well controlled.

Concrete that possesses high workability, high strength, and high durability is considered as high performance concrete (HPC). HPC is a concrete within which certain characteristics are produced for a particular application and setting, as defined by ACI (American Concrete Institute) (1998). ACI concept includes that durability is optional, and this has contributed to a series of HPC structures that could potentially have very long service lives, showing early in their durability-related distress.

16.2 EXPERIMENTAL PROCEDURE

16.2.1 MATERIALS

Ordinary Portland cement (OPC) of 53 grade is used to make mix design of respective grades to achieve the desired characteristics of HPC. The various properties, including physical as per IS 4031:1988, as well chemical as per IS 4032-1985, are found to satisfy the standards given in IS 12269-1987

As fine aggregates, the natural sand passing through conforming to grading Zone II of IS:383–1970 was used as an experiment after washing with pure potable water and drying in the natural sunlight. Fine aggregates and coarse aggregates used in this research work were tested, as per IS 2386-1963

Fly ash of Class F was used as per ASTM C618. Micro silica of Grade 920-D as per ASTM C1240-20, as a mineral admixture for developing HPC mixes.

16.2.2 Specimen Preparation and Test Procedure

First of all, the elements, i.e. cement FA, CA, water, and admixtures, were mixed altogether homogeneously. The materials required for making HPC cube specimens were taken in proper quantities. The quantity of material to be used for HPC was determined by weigh batching. Weigh batching was done using digital weighing machines having an accuracy of 0.005 kg.

The mix designs of higher grades of HPC M50 to M90 were done using the revised IS code method (IS: 10262-2009).

16.2.3 Tests on Hardened HPC Specimens

The concrete cube specimens after proper curing were tested for compressive strength and durability properties and are described below.

- **Compressive Strength Test:** After proper curing of 28 days of HPC cube specimens, the cubes are taken out of the tank and the surface of each cube wiped properly to let it surface dry before weighing on a digital weighing machine. Immediately after surface drying, the specimens were taken for testing in a electrically operated compression-testing machine of 3000 KN capacity. The experiment was done as per IS: 516–1959. The observations were recorded.
- **Durability Test:** High-performance concrete specimens of various grades (M50 to M90) were tested. The main tests frequently performed for durability of concrete, such as acid attack, chloride attack, sulphate attack, and permeability (RCPT), were conducted on the reference mixes of each grade of HPC in the laboratory and are discussed below.

16.3 DURABILITY TEST

Many researchers have done durability tests for different environmental effects. Some of the researchers have worked on acid attack on concrete (Reddy, Rao, and George, 2012, Swamy, Lakshmi Sudha, and Venkateswarlu, 2019, Thomas et al., 2015, Turkel, Felekoglu, and Dulluc, 2007, Zivica and Bajza, 2001). Some of the researchers studied the effect of chloride on concrete (Ardeshana and Desai, 2012, Bajad, Modhera, and Desai, 2011, Jain and Narayanan Neithalath, 2010) whereas some of the researchers studied the effect of sulphate attack on concrete (Bajad, Modhera, and Desai, 2012, Guerrero, Hernandez, and Goni, 2000, Hendi et al., 2020, Rasheeduzzafar, et al., 1994). In the experimental work, three solutions 1) acid attack (H_2SO_4) 2), chloride attack (NaCl), 3) magnesium sulphate ($MgSO_4$), are presented here, and the procedure is as followed. After preparing the HPC specimens of size 150 mm × 150 mm × 150 mm, the acid attack test was carried out. The curing of above-mentioned specimens was done for a full 28 days. After removal from the curing tank, the casted samples were stored in atmospheric temperature for two days considering the constant weight. Samples were weighed properly and engrossed in 10% solution of pH value of 2.5 for 180 days. The pH value was periodically monitored and was kept between 2–3. After 180 days of being fully merged in an acidic-environment, the samples were taken out and washed in tap water and kept in atmosphere for two days to gain constant weight. The specimens were weighed, and any loss in weight was recorded to get the percentage loss of weight.

Permeation properties are determined by the most extensively used technique in the United States, ASTM C 1202-97 (AASHTO T-277), which is based on chloride ion dispersion; it is well known as the rapid chloride-ion penetrability test (RCPT).

16.4   TEST RESULTS AND DISCUSSION

16.4.1   ACID ATTACK

Table 16.1 shows the experimental results of acid attacks on concrete specimens, which are also plotted in Figure 16.1 and Figure 16.2. Initial weight, final weight, and the loss in weight are shown in Table 16.1; the reduction in strength is also shown for the concrete before the acid attack, i.e. reference mix and after acid attack.

Figure 16.1 shows the percentage weight reduction of concrete under an acidic environment. It shows that the percentage weight loss for M80 grade concrete is 7.35, which is the highest, whereas the same is 3.24 for M90 grade concrete, which is the lowest.

Figure 16.2 shows the percentage reduction of compressive strength under an acidic environment. It indicates that the loss in compressive strength of M50 grade is highest and the same of M90 grade is least. From Table 16.1, Figure 16.1 and Figure 16.2, it is concluded that M90 shows the lowest reduction in weight after 180 days acid attack. Whereas, M80 shows highest deterioration affects under acid attack. The same table shows a vast difference in compressive strength. M50 grade shows the lowest percentage reduction in compressive strength, whereas M90 grade shows the highest percentage reduction.

16.4.2   CHLORIDE ATTACK

Table 16.2 shows the experimental results of chloride attack on concrete specimens, which are also plotted in Figure 16.3 and Figure 16.4. Initial weight, final weight,

TABLE 16.1
Experimental Result of Acid Attack

Grade	Original Weight (kg)	Final (kg)	Weight Reduction (kg)	% Weight Reduction	Average % Weight Reduction	Avg. Comp. Strength N/mm^2		
						Reference Mix	After Acidic Attack	% Reduction
M50	9.19	8.55	0.64	6.96	5.31	73.04	37.12	49.1
	9.02	8.57	0.45	4.98				
	9.02	8.665	0.36	3.99				
M60	9.08	8.54	0.54	5.95	4.9	86.01	46.92	45.4
	9.15	8.585	0.57	6.23				
	9.08	8.85	0.23	2.53				
M70	8.92	8.56	0.36	4.04	3.78	89.96	39.18	43.56
	8.87	8.535	0.334	3.76				
	8.935	8.62	0.315	3.53				
M80	8.985	8.345	0.64	7.12	7.35	92.26	54.29	41.15
	9.2	8.48	0.72	7.83				
	9	8.36	0.64	7.11				
M90	9.15	8.65	0.5	5.46	3.24	98.51	56.25	40.52
	8.99	8.965	0.025	0.28				
	9.16	8.795	0.365	3.98				

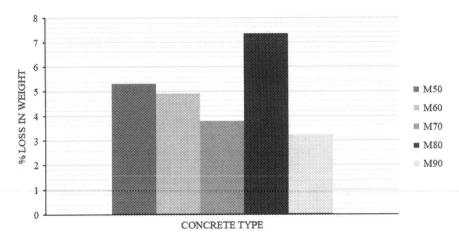

FIGURE 16.1   Percent weight reduction of concrete under acidic environment.

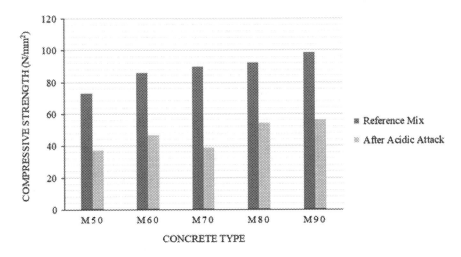

FIGURE 16.2   Percent reduction of compressive strength under acidic environment.

and the loss in weight are shown in Table 16.2. The reduction in strength is also shown for the concrete before chloride attack, i.e. reference mix and after chloride attack.

Figure 16.3 shows the percentage weight reduction of concrete due to chloride attack. It shows that the percentage weight reduction for M60 grade concrete is 1.36, which is the highest, whereas the same is 0.27 for M80 grade concrete, which is the lowest.

Figure 16.4 shows the percentage reduction in compressive strength of concrete due to chloride attack. It indicates that the loss in compressive strength of M80 grade is highest and the same of M60 grade is the lowest. From Table 16.2,

TABLE 16.2
Experimental Result of Chloride Attack

Grade	Original Weight (kg)	Final Weight (kg)	Weight Reduction (kg)	% Weight Reduction	Average % Weight Reduction	Avg. Comp. Strength N/mm²		
						Reference Mix	After Chloride Attack	% Reduction
M50	8.895	8.905	0.09	1	0.69	73.04	64.78	11.31
	8.895	8.84	0.005	0.06				
	9.025	8.935	0.09	1				
M60	8.94	8.795	0.145	1.62	1.36	86.01	84.11	2.2
	8.92	8.88	0.04	0.45				
	9.15	8.905	0.185	2.02				
M70	9.16	9.14	0.02	0.22	0.29	89.96	75.98	15.5
	9.28	9.255	0.025	0.27				
	9.23	9.195	0.035	0.38				
M80	9.08	9.05	0.03	0.33	0.27	92.26	76.18	17.43
	9.11	9.085	0.025	0.27				
	9.165	9.145	0.02	0.22				
M90	9.25	9.225	0.025	0.27	0.33	98.51	89.73	8.9
	8.475	8.465	0.01	0.12				
	9.005	8.95	0.055	0.61				

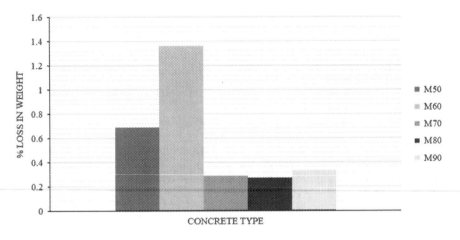

FIGURE 16.3   Percentage weight reduction under chloride attack.

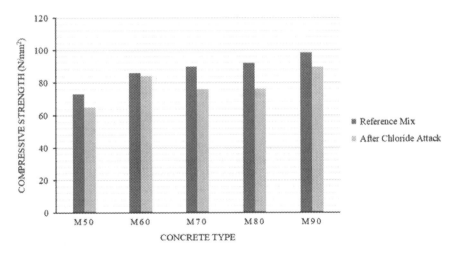

FIGURE 16.4   Percent reduction in compressive strength of concrete under chloride attack.

Figure 16.3 and Figure 16.4, it is concluded that M80 shows the lowest reduction in weight after 180 days of chloride attack, whereas M60 shows the highest deterioration effect under chloride attack. The same table shows vast differences in compressive strength. M60 grade shows the lowest percentage reduction in compressive strength, whereas M80 grade shows the highest percentage reduction.

16.4.3   SULPHATE ATTACK

Table 16.3 shows the experimental results of sulphate attack on concrete specimens, which are also plotted in Figure 16.5 and Figure 16.6. Initial weight, final weight, and the loss in weight are shown in Table 16.3. The reduction in strength is also

TABLE 16.3
Experimental Result of Sulphate Attack

Grade	Original Weight (kg)	Final Weight (kg)	Weight Reduction (kg)	% Weight Reduction (kg)	Average % Weight Reduction (kg)	Avg. Comp. Strength N/mm^2		
						Reference Mix	After Sulphate Attack	% Reduction
M50	9.245	9.05	0.195	2.11	2.23	73.04	64.78	11.31
	9	8.845	0.155	1.72				
	8.82	8.56	0.26	2.95				
M60	9.08	8.955	0.125	1.38	1.85	86.01	79.07	8.07
	8.86	8.695	0.165	1.86				
	9.05	8.84	0.21	2.32				
M70	9.19	9.16	0.03	0.33	0.31	89.96	84.87	5.65
	8.955	8.93	0.025	0.28				
	9.075	9.045	0.03	0.33				
M80	8.98	8.965	0.015	0.17	0.098	92.26	74.52	19.22
	9.11	9.1	0.01	0.11				
	9.145	9.015	0.13	0.014				
M90	9.045	9.02	0.025	0.28	0.29	98.51	77.65	21.14
	9.25	9.21	0.04	0.43				
	9.245	9.23	0.015	0.16				

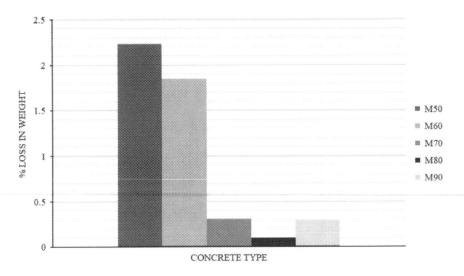

FIGURE 16.5 Percentage weight reduction under sulphate attack.

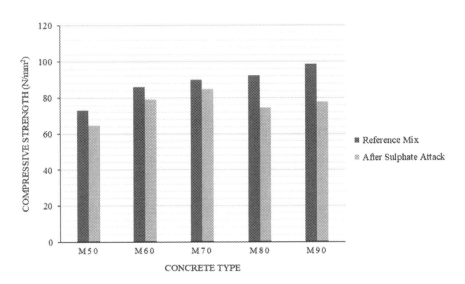

FIGURE 16.6 Percent reduction in compressive strength of concrete due to sulphate attack.

shown for the concrete before sulphate attack, i.e. reference mix and after sulphate attack.

Figure 16.5 shows the percentage weight reduction of concrete under sulphate attack. It shows that the percentage weight loss for M50 grade concrete is 2.23, which is the highest, whereas the same is 0.098 for M80 grade concrete, which is the lowest.

Figure 16.6 shows the percentage reduction in compressive strength of concrete due to sulphate attack. It indicates that the loss in compressive strength of M90 grade is the highest and the same of M70 grade is the lowest. From Table 16.2, Figure 16.3 and Figure 16.4, it is concluded that M80 shows the lowest reduction in weight after 180 days of sulphate attack, whereas M50 shows the highest deterioration effect under sulphate attack. The same table shows vast differences in compressive strength. M70 grade shows the lowest percentage reduction in compressive strength, whereas M90 grade shows the highest percentage reduction.

Figure 16.7 shows the percentage weight loss in HPC mixes for acidic, chloride, and sulphate attack on concrete specimens. It indicates that the loss in weight is higher in acid attack for all grades of concrete. Weight loss is less in chloride attack for M50, M60, and M70, whereas it is less in M80 and M90 for sulphate attack.

Figure 16.8 shows the decreased percentage in compressive strength of HPC mixes for acidic, chloride, and sulphate attack on concrete specimens. It indicates that the loss in compressive strength is less in acid attack for all grade of concrete. Compressive strength loss is higher in chloride attack for M60, M80, and M90, whereas it is higher in M70 for sulphate attack. For M50 grade concrete, compressive strength reduction is same for sulphate attack and chloride attack.

16.4.4 PERMEABILITY TEST

The test of chloride ion penetration in concrete was carried out according to ASTM C1202-97. Table 16.4 shows the results of the rapid chloride penetration test (RCPT), which indicates a very low penetrability to chloride ions, indicating that the HPC

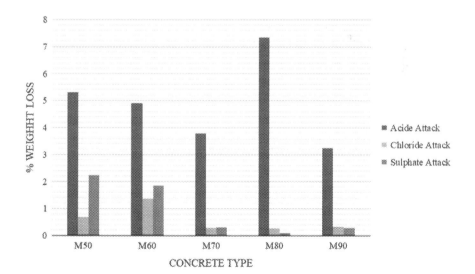

FIGURE 16.7 Percent weight loss in HPC mixes for different chemical attacks.

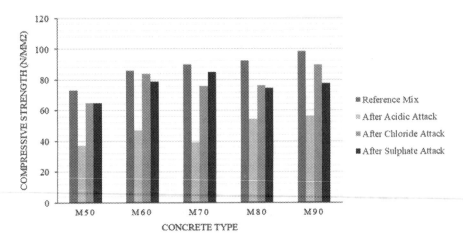

FIGURE 16.8   Percent reduction in compressive strength of HPC mixes due to different chemical attacks.

TABLE 16.4
Experimental Result of RCPT

Grade of HPC	Time of Current Passed	Curing Period	Total Charge Passed (Coulombs)	Average Value (Coulombs)
M50	360 Minutes	28 days	402	401.00
			406	
			395	
M60	360 Minutes	28 days	448	450.00
			453	
			449	
M70	360 Minutes	28 days	418	462.67
			480	
			490	
M80	360 Minutes	28 days	465	454.33
			438	
			460	
M90	360 Minutes	28 days	480	474.33
			478	
			465	

mixes possess excellent durability property. The variation in permeability values for different grades of reference HPC mixes is shown graphically in the figure. From the graph of permeability Vs grades of HPC, it is seen that for higher grades of HPC mixes, greater resistance to chloride penetration is offered by the specimens.

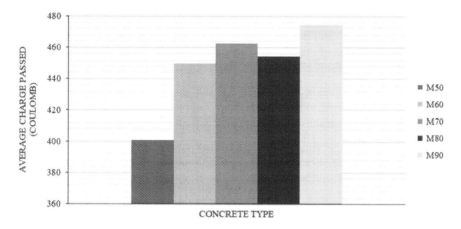

FIGURE 16.9 Variation of permeability for different grades of HPC mixes.

Figure 16.9 shows the variation of permeability for different grades of HPC mixes. It indicates the increase in charge passes with increase in grade of concrete, which means the permeability increases as the grade of concrete is higher.

16.5 CONCLUSION

In this research work, design mix is prepared using the standard method to prepare design mix. In this type of mix design, an attempt was made to prepare higher grade M50 to M90 by using SCMs like fly ash and micro silica with different proportions to replace cement and save natural resources. The properties of weight reduction, loss in compressive strength & RCPT values are investigated.

From Figure 16.7 and Figure 16.8, it is clear that different severe environmental conditions have different effects on concrete durability. The results of these tests conclude that the effect of acidic environmental is harmful for the use of SCMs because the weight loss and strength reduction are significantly high. The applications of SCMs are advisable for the concrete in aggressive chloride and sulphate atmosphere because it has less effect on concrete durability.

The present study has been done using fly ash and micro silica, so the study can be extended with GGBFS and with different dosage of hybrid SCMs. The study can also be extended using different strength tests as only compressive strength and weight loss has been considered.

REFERENCES

ACI Committee 363 (1998). Guide to Quality Control and Testing of High-Strength Concrete, 363.2R-98, American Concrete Institute, Farmington Hills, Michigan, 18.

Ardeshana A, and Desai A (2012). Durability of fiber reinforced concrete of marine structures. International Journal of Engineering Research and Application (IJERA), 2 (4), 215–219. http://www.ijera.com/

ASTM C1697 – 16 (2016). Standard Specification for Blended Supplementary Cementitious Materials, ASTM International, West Conshohocken, PA, www.astm.org

ASTM C1240 – 20 (2020). Standard Specification for Silica Fume Used in Cementitious Mixtures, ASTM International, West Conshohocken, PA, www.astm.org

Badagha D, and Modhera C (2017). M55 grade concrete using industrial waste to minimize cement content incorporating CO2 emission concept: An experimental investigation. Materials Today: Proceedings, 4(9), 9768–9772. 10.1016/j.matpr.2017.06.264

Badagha D, and Modhera C (2019). Parametric Experimental Studies on Sustainable Concrete Containing Waste under Different Curing Conditions, International Journal of Recent Technology and Engineering (IJRTE), 7(6S2), 1–5. https://www.ijrte.org/wp-content/uploads/papers/v7i6s2/F10000476S219.pdf

Bajad M, Modhera C and Desai A (2011). Influence of a fine glass-powder on strength of concrete subjected to chloride attack. International Journal of Civil Engineering and Technology (IJCIET), 2(2), 01–12.

Bajad M, Modhera C, and Desai A (2012). Resistance of Concrete Containing Waste Glass Powder Against MgSO4 Attack. IUP Journal of Structural Engineering; Hyderabad, 5(3), 28–47.

Berke N (2012). Handbook of Environmental Degradation of Materials, Elsevier Inc. 10.101 6/B978-1-4377-3455-3.00011-0

Damtoft JS, Lukasik J, Herfort D, Sorrentino D, and Gartner EM (2008). Sustainable development and climate change initiatives, Cement and Concrete Research, 38, 115–127. 10.1016/j.cemconres.2007.09.008

Guerrero A,Hernandez MS and Goni.S (2000). The role of the fly ash pozzolanic activity in Simulated Sulphate radioactive liquid waste. Waste Management, 20, 51–58.

Hendi A, Behravan A, Mostofinejad D, Kharazian H, and Sedaghatdoost A (2020). Performance of two types of concrete containing waste silica sources under MgSO4 attack evaluated by durability index. Construction and Building Materials, 241, 118–140. 10.1016/j.conbuildmat.2020.118140.

IS:516-1959 (2004). Methods of Tests for Strength of Concrete, Bureau of Indian Standards, Manak Bhavan, Bahadur Shah Zafar Marg, New Delhi, https://bis.gov.in/

IS:4031-1998 (2005). Methods of Physical Tests for Hydraulic Cement, Bureau of Indian Standards, Manak Bhavan, Bahadur Shah Zafar Marg, New Delhi, https://bis.gov.in/

IS:4032-1985 (2005). Methods of Chemical Analysis of Hydraulic Cement, Bureau of Indian Standards, Manak Bhavan, Bahadur Shah Zafar Marg, New Delhi, https://bis.gov.in/

IS:12269-1987 (2004). Specification For 53 Grade Ordinary Portland Cement, Bureau of Indian Standards, Manak Bhavan, Bahadur Shah Zafar Marg, New Delhi, https://bis.gov.in/

IS:10262-2009. Concrete Mix Proportioning – Guidelines, Bureau of Indian Standards, Manak Bhavan, Bahadur Shah Zafar Marg, New Delhi, https://bis.gov.in/

IS:383-1970 (2002). Specification for Coarse and Fine Aggregates from Natural Sources for Concrete, Bureau of Indian Standards, Manak Bhavan, Bahadur Shah Zafar Marg, New Delhi, https://bis.gov.in/

IS:2386-1963 (2002). Methods of Test for Aggregates for Concrete, Bureau of Indian Standards, Manak Bhavan, Bahadur Shah Zafar Marg, New Delhi, https://bis.gov.in/

Jain J, and Narayanan Neithalath N (2010). Chloride transport in fly ash and glass powder modified concretes – Influence of test methods on microstructure. Cement and Concrete Composites, 32, 148–156.

Rasheeduzzafar OSB, Amoudi S, Abduljauwad M, and Maslehuddin (1994). Magnesium-sodium sulfate attack in plain and blended cements. Journal of Materials in Civil Engineering, 6(2), 201–222. 10.1061/(ASCE)0899-1561(1994)6:2(201)

Reddy, BM, Rao, HS, and George, MP (2012). Effect of hydrochloric acid on blended cement (fly ash based) and silia fume blended cements and their concretes. International Journal of Science and Technology, 1(9), 467–480.

Swamy N, Lakshmi Sudha J, and Venkateswarlu D (2019). Consequence of concrete due to acids. International Journal of Recent Technology and Engineering (IJRTE), 8(2), 5896–5899. 10.35940/ijrte.B1093.078219

Thomas B, Gupta R, Mehra P, and Kumar S (2015). Resistance to acid attack of cement concrete containing discarded tire rubber. Ukieri Concrete Congress – Concrete Research Driving Profit and Sustainability, 1235–1244.

Turkel S, Felekoglu B and Dulluc S (2007). Influence of various acids on the physico–mechanical properties of pozzolanic cement mortars. Sadhana, 32(6), 683–691.

Zivica, V and Bajza, A (2001). Acid attack on cement based materials – A review, part 1, Construction and Building Materials, 15, 331–340.

17 Solar Thermal Power Generation: Application of Internet of Things for Effective Control and Management

Subhra Das

Solar Engineering Department, Amity University Haryana, Gurugram, Haryana, India

CONTENTS

17.1 INTRODUCTION

Solar thermal power generation uses concentrating solar thermal collectors to concentrate beam radiation onto a receiver to heat working fluid at a temperature suitable to convert thermal energy effectively to electricity. Solar thermal power plants, when operated in combination with fossil fuel-based power units or with adequate storage, can provide firm and dispatchable power. Though the principle of producing high temperature by concentrating solar radiation has been known for centuries, this technology has been used for commercial power generation only for the last half century. This technology has true potential to replace fossil fuels for power generation and mitigate climate change by reducing greenhouse gas emissions, but various issues, mainly the cost economics, have constrained its growth over the last ten years. Governments have taken initiatives to increase renewable energy proportion in the energy mix. On 11 January 2010, the government of India had launched National Solar Mission (NSM), setting a target of 20 GW solar power generation by the year 2022. In 2015, the target for solar power was increased to 100 GW. As per Annual Report 2018–2019 of the Ministry of New & Renewable Energy, solar power projects under NSM phase I were allotted in two batches. In August 2010 under NSM Batch I, 150 MW solar photovoltaic and 470 MW solar thermal projects were allotted, and under NSM Batch II 350 MW solar photovoltaic projects were allotted through a reverse-bidding process. Under the bundling scheme in phase I of NSM, 533 MW solar PV and 200 MW solar thermal projects have been commissioned. This shows that of all the solar power projects installed in India, only 27% are ST projects. This shows that solar thermal power projects could not attract solar developers as compared to solar PV projects.

Concentrated solar power (CSP) plants are developed in an arid and semi-arid climate where solar radiation is plentiful. These places are characterized by a high level of direct normal irradiance (DNI). In India, a total of 84,229 km^2 of wastelands with wind speed greater than 4 m/s spread across 14 states with expected DNI value greater than 1800–2000 kWh/m^2/year has been identified (Sharma et al., 2015). The extreme climate of these regions poses significant challenges for the installation of solar thermal power plants in terms of water scarcity, dust accumulation, strong wind, high temperature, electricity transmission, and energy supply security (Xu et al., 2016).

Dust accumulation on concentrators can cause significant degradation of the performance of CSP (Boddupalli et al., 2017). Water is essential: 20–40 gallons/MWh of water is required to clean mirrors, 30–60 gallons/MWh of water is required for steam-cycle processes, 0–900 gallons/MWh of water is required for cooling systems for steam-cycle processes, and 750–900 gallons/MWh of water is required for wet cooling systems (Bracken et al., 2015). Surface water sources are far away from CSP plants, which are mostly located in desert areas; hence, these plants use groundwater to fulfil their water requirement. The huge requirement of water for cleaning or power generation in solar thermal power plants, and its dependency on groundwater, poses a question on its sustainability in the absence of long-term studies on natural recharge of groundwater and factors affecting natural recharge. As a result, poor planning of groundwater usage may lead to overdraft, water-quality degradation, and impact on other unseen environmental factors.

Hundreds of concentrators are connected in series and parallel connections in a typical STE power plant. The life of the STE plant is at least 30 years, with minimum degradation in performance, which is a major advantage of this plant over the solar PV power plant. In addition, most of the components of the STE solar field can be recycled and reused.

An STE power plant consists of solar field, thermal storage, and power-generation blocks. The major components of solar field are solar thermal collectors, receivers, inlet, and outlet piping assembly. Energy flows in with solar radiation in the solar field, which is concentrated onto the receiver by concentrators. Heat-transfer fluid flowing through the receivers removes some of the heat absorbed by the receiver, and the rest is lost to the ambient from the surface of the receiver. Unlike solar photo-voltaic panels, CSPs work only with direct normal irradiance (Sharma et al., 2015).

Various factors affect the objectives of the solar thermal power project by increasing the project budget or the time of delivery, or by affecting the system performance. These factors have an associated uncertainty related to their occurrence and are described as risks associated with the project. In this chapter, we first present an overview of political, environmental, social, and technological risks associated with solar thermal power generation projects. The *political risks* are related to energy policies or treaties made governments adopt during a period and are country specific. The energy policy gives a legal framework for recourse or remedy; fiscal policy related to taxation, allowances, subsidies, or grants; regulations pertaining to grid connection, health & safety, environmental factors (Turney and Fthenakis, 2011), land-use planning, and access (Ong et al., 2013). Increasing government subsidies in renewable energy (RE) has a positive impact on RE projects since it encourages people to invest in RE projects, thereby having a positive impact on business. But reducing or withdrawing subsidies may lead to a decline in the number of customers, and hence, losses in business.

Environmental risks address the impact of the project on natural resources (Bracken et al., 2015; Stults, 2015), damages to flora and fauna due to cleaning up of project site (Das, 2018; Lieberman et al., 2014), pollution caused during construction & operation of the project, and the impact of wastes from the project during its life cycle. Most CSP plants are in desert areas having a scarcity of surface water. Plants depend upon groundwater to meet water requirements, which leads to the question of sustainability of such plants (Bracken et al., 2015). Huge land-area requirements for the CSP projects also have an impact on wildlife and vegetation. The total land-area requirement for a CSP project may vary from 3.5–4.0 acres/GWh/yr based on the CSP technology used. Since a large land area is needed for installing a solar thermal power plant, wastelands are best suited for this purpose (Sharma et al., 2015). Utilizing wastelands for power generation solves the problem of land crisis for humans, but it becomes a threat for wildlife. Often, 1,000 acres of land are cleared of all vegetation to construct the facility (Bannur, 2018). This causes the loss of habitats of wildlife in that area, and there is a disturbance caused to their natural living conditions, such as food availability, loss of hiding spots, preying strategy, etc., due to human intervention (Kumarankandath and Goswami, 2015). The entire land area is cleaned and labelled to install the collectors and other components. The area is devoid of vegetation, and, in some cases, vegetation can be grown only up to a specific height so that it does not cast a shadow. The impact of alteration in vegetation over the large land area due to the installation of STP plant is not studied thoroughly. To make renewable energy environmentally friendly, renewable energy-generation projects should be carefully planned by studying their impacts on wildlife, natural resources, and the environment. Alternate arrangements should be proposed analysing the loss of the natural habitat and its impact on natural resources. Certain other factors are unique to solar thermal power plants, viz. the reflecting surfaces of mirrors appear as a water body to flying birds. Deaths and injuries to birds colliding with mirrors are potential problems. These hazards are common for central tower-type power plants; birds get burnt by the intense sun rays while they come down to catch their prey. These hazards can be a potential threat to the solar power plant, as witnessed by the 392 MW Ivanpah solar thermal power plant costing $2.2 million, which faced the threat of closure because of huge protests by environmentalists (Das, 2018).

Social risks are associated with labours and the behaviour of the public toward the project. The analysis done by Skill Council for Green Jobs in the year 2016 shows that in countries like India, where there are very few solar thermal power plants installed and operational, there is a limited scope of finding people having expertise in CSP technology. Risks associated with the availability of reliable skilled manpower affect the commissioning of these projects significantly. Also, acceptance of projects by the local community reduces the risks of damage due to theft or vandalism (Ikejemba and Schuur, 2018).

Technological challenges can be broadly associated with three areas: performance, services, and project designing & construction. Some of the key factors of technical risks are unavailability of reliable DNI data for the site, incorrect design calculation (Bannur, 2018; Kumarankandath and Goswami, 2015), incorrect specifications of components, unavailability of raw material suppliers/ manufacturers, cost and time associated with project commissioning, decommissioning, abandonment and cleaning, unavailability of finance for the project, reliability of operations & maintenance facility, reliability of components, and the underperformance of the plant. One of the major challenges of CSP technology is the access to reliable DNI data for the location; this data is needed to estimate land and mirror area requirements. An unreliable DNI data may lead to faulty design considerations and underperformance of the plant.

Parabolic trough technology dominates entire CSP technology based on installed capacity. It uses mineral oil as a heat-transfer fluid and uses Rankine cycle with low temperature (<400°C) steam turbines to generate power and have relatively low efficiency (Kearney et al., 2003). Direct steam-generating towers and the usage of molten salt as a heat-transfer medium are two technological advancements made in CSP technology to achieve high temperatures and better efficiencies at low cost (Spelling et al., 2015). Both technologies have associated drawbacks based on the physical and chemical characteristics of the heat-transfer medium. Molten salt or mineral oil systems suffer from freezing problems if the temperature of the heat-transfer medium drops too low below crystallization temperature. This results in parasitic power consumption for heat tracing.

Unskilled manpower possesses a risk of system failure as O&M staff may not be able to identify/detect the actual reason for faults occurring in the system. As reported in Helioscsp solar thermal energy news, globally, the total installed capacity of concentrated solar power technology was 6451 MW by the end of 2019. Consequently, there is a huge requirement of trained manpower for the operation & maintenance of the CSP plant.

Most solar thermal power projects are installed in an area away from local townships, and hence, solar developers face a major challenge to get reliable energy sources to power the UPS or parasitic loads. This challenge may lead to freezing of HTF in the entire pipeline in the solar field during winter, which may result in

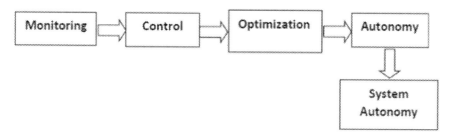

FIGURE 17.1   Maturity model for IoT.

shutting down the plant, as reported by IITB (2009) in its report on the national solar thermal power plant.

Internet of Things (IoT) can play a significant role in reducing uncertainties associated with solar thermal power generation, especially in dealing with technological risks. IoT was first introduced by the Massachusetts Institute of Technology in 1999, and since then, IoT has been used in various processes for real-time monitoring of data and for establishing efficient communication & feed-backing processes to increase overall system performance.

The maturity model for IoT is shown in Figure 17.1, which shows the first two steps as monitoring & control of equipment & processes (Ramamurthy and Jain, 2017). The next steps are optimizing system parameters based on environmental conditions, followed by equipment running autonomously and adjusting to input parameters. The final step is communicating with other system components to optimize performance and finally establishing system autonomy.

In the 1990s, the power sector introduced supervisory control and data acquisition to automate processes. This is a simple version of IoT. The automation includes collecting data for underlying processes from logic controllers, analysing & interpreting data, and communicating instructions for controlling the processes. IoT can play a vital role in the power sector by making it more efficient or smart. IoT is a network that integrates the actions of all users involved in the process and efficiently delivers the desired output. It has a rapid controllable two-way flow of information and automated system control & delivery (IRENA, 2019). Wireless communications system plays an important role in IoT framework. It connects sensor devices to communication gateways to perform two-way communication of data between each block of IoT (Motlagh et al., 2020). The requirement of huge financial investment has been identified as one of the challenges in the adoption of IoT in the Asia and Pacific energy sector. Energy policy in different countries and the availability of skilled manpower to handle IoT systems in the power sector are some other challenges that need to be addressed (Ramamurthy and Jain, 2017). Financial instruments like results-based lending, sovereign lending, and other lending provisions from banking sectors can help in the modernisation of the power sector by adopting IoT. India has taken a major step in standing up to face the challenges of IoT in the power sector by setting up stations for monitoring renewable generations, issuing regulations for the use of smart meters, and formulating standards for IoT and standards for smart grids (Kumar, 2017).

Smart meters used in renewable energy projects are an example of IoT, which delivers near real-time consumption and generation data. A high level of control & monitoring of the system can be achieved by complementing operational technology (like SCADA and smart meters) by channels to share information & communicate with other blocks in the CSP processes. IoT in the power sector can improve utilization efficiency by integrating infrastructural resources with electrical power systems by a reliable communication channel. Its introduction in smart grids will efficiently provide real-time monitoring, fault detection, assistance for maintenance & control for generation, transmission & distribution of electricity, and other aspects of the power grid. A piece of evidence for IoT adoption leading to improvement in system efficiency was reported by GE. It reported an increase in average conversion efficiency from 33% to 49% by introducing digital power plant systems for gas and coal plants in 1999.

17.2 SOLAR THERMAL POWER GENERATION: A BRIEF DESCRIPTION

A solar thermal power plant can be divided into three main blocks, namely, solar field, thermal energy storage, and power block. Energy from the sun is trapped in the solar field using concentrating collectors. The concentrators track the sun and reflect beam radiation incident on its surface to the receiver, which is placed at focus in case of point focus collector or along the focal line in case of line focus collector. The receiver absorbs incident solar radiation and transfers a part of it to heat-transfer fluid, and the rest is lost to the ambient. The temperature of heat-transfer fluid can be raised to 400°C and 2000°C, respectively, for line and point focus systems. Thermal oil, molten salts, liquid metals like sodium, etc., are used as working fluids in the absorber to collect heat from the incident/reflected solar radiation. The hot fluid is stored in the thermal storage tank and is utilized to generate superheated steam at high pressure to turn a steam turbine, coupled with a generator, to generate electricity in the power block. The majority of uncertainty associated with solar thermal power plants concerns Block I due to uncertainty in the availability of solar radiation, which varies over the year due to the revolution of Earth and over the day due to rotation of the Earth.

17.3 MODEL DESCRIPTION FOR CSP PLANT

In the present discussion, we shall consider a concentrating parabolic trough power plant to design the IoT model.

Inputs to the model include:

- Specifications of each component of the three blocks, including sensors & controllers.
- Date & time of monitoring.
- Description of the plant location, which includes latitude, longitude, azimuth angle of the site, its climate, its annual average temperature, and solar radiation availability.

Real-time monitoring

The following data is collected at 15-minute intervals:

- Global solar radiation, direction normal radiation, wind speed, ambient temperature, humidity.
- Temperature of heat transfer fluid (HTF) and mass flow rates in the three blocks. The temperature of the glass covers of the receiver.
- Tracking errors.

17.3.1 Modelling Solar Field

The solar field comprises of arrays of parabolic trough collectors, header pipes for hot and cold fluids, heat-transfer fluid (HTF), main HTF pump, auxiliary heating system, main controller, sensors, and solar field controller. The main controller is used to determine operation mode by taking feedback from various sensors, including the incident angle of solar radiation, zenith angle, beam irradiance, input and output temperature of HTF from solar field, the temperature of the storage tank, and turbine status.

17.3.1.1 Angle of Incidence of Solar Radiation

The angle of incidence θ for beam radiation can be computed using the following relations for five different tracking modes, as tabulated in Table 17.1.

The tracking requirement is decided based on the angle of incidence by the controller. The frequency of adjustment of the collector will depend upon its concentration ratio.

17.3.1.2 Model to Predict Direct Normal Irradiance

Direct normal irradiance (DNI) on titled surface at a future time step t_p is estimated using the model proposed by Das (2020):

$$DNI_T(t_p) = DNI_T(t_n) + p\Delta I(t_n) \quad (17.6)$$

$$\text{where } t_p = t_n + ph \text{ and } h = t_n - t_{n-1} \quad (17.7)$$

$$\Delta I(t_n) = \frac{1}{2}[(DNI_T(t_n) - DNI_T(t_{n-1})) + (DNI'_T(t_{n+1}) - DNI'_T(t_n))] \quad (17.8)$$

where $DNI'_T(t_{n+1})$ and $DNI'_T(t_n)$ are estimated DNI on a tilted surface in future & current time step, respectively. Here, t_n, t_{n-1} and t_{n+1} represent current, previous, and future time steps, respectively.

There are many models to estimate DNI on the tilted surfaces, including the Meinel model, Ineichen's Model, etc. In the present discussion, the Ineichen model is used to compute direct normal irradiance on the tilted surface at time step t_n using the following relation (Shen et al., 2018):

TABLE 17.1
Different Tacking Modes and Corresponding Angle of Incidence of DNI (Sukhatme and Nayak, 2012)

Tracking Mode	Description	Angle of Incidence, cos θ	
Mode I	The focal axis is placed along E-W direction and is horizontal. Concentrator is rotated about the horizontal E-W axis and is adjusted once every day. In this case, at solar noon the solar beam is normal to the collector aperture plane.	$sin^2\delta + cos^2\delta \cos\omega$	(17.1)
Mode II	The focal axis is placed along E-W direction and is horizontal. Concentrator is rotated about horizontal E-W axis and with continuous adjustment. In this case, angle of incidence of solar beam with aperture plane is minimum.	$\sqrt{1 - cos^2\delta \sin^2\omega}$	(17.2)
Mode III	The focal axis is placed along N-S direction and is horizontal. Concentrator is rotated about horizontal N-S axis and is continuously adjusted. In this case, angle of incidence of solar beam with aperture plane is minimum.	$\sqrt{\cos^2\delta \, sin^2\omega + (\sin\varphi \sin\delta + \cos\varphi \cos\delta \cos\omega)^2}$	(17.3)
Mode IV	The focal axis is placed along N-S direction and inclined at an angle equal to the latitude. Concentrator is rotated about an axis parallel to the earth's axis. In this case, at solar noon the aperture plane is due south.	$\theta = \delta$	(17.4)
Mode V	The focal axis is placed along N-S direction and is inclined. Concentrator is continuously rotated about an axis parallel to the focal axis and about a horizontal axis perpendicular to this axis. In this case, solar radiation is always incident normally on the aperture plane.	$\theta = 1$	(17.5)

$$DNI\,(t_n) = b\,I_0 \frac{\exp[-0.09.\ AM.\ (TL-1)]\cos\theta}{\cos\theta_z} \qquad (17.9)$$

$$\text{where } b = 0.664 + 0.163/\,f_h \qquad (17.10)$$

$$f_h = \exp[-altitude/8000]\text{ and} \qquad (17.11)$$

TL is Linke turbidity, which can be computed using the following relation (Linke, 1922):

$$I = I_0 \exp(-\delta_{cda} \times TL \times AM) \qquad (17.12)$$

where I represent global solar radiation at a given time step, I_o is extraterrestrial radiation and δ_{cda} is the optical thickness of a water and aerosol-free atmosphere, which is expressed as (Linke, 1922):

$$\delta_{cda} = 0.128 - 0.054\,\log(AM) \qquad (17.13)$$

AM is air mass, which takes a value equal to 1 when the sun is at zenith and equal to 2 for zenith angle equal to 60°. Air mass is computed using the following relation for zenith angle from 0° to 70° at sea level (Duffie and Beckman, 1980):

$$AM = 1/\cos\theta_z \qquad (17.14)$$

17.3.1.3   Useful Energy Gain in Collector

The useful energy gain in the collector at the current time step t_p is obtained from the energy balance on the absorber surface as:

$$Q_u(t_p) = S(t_p) - Q_L(t_p) \qquad (17.15)$$

where S(tp) and Q_L(tp) are absorbed solar radiation and energy losses from the collector at time tp. The absorbed solar radiation S at time step (t_p) is given by (Lüpfert et al., 2004):

$$S(t_p) = DNI_T(t_p).\ A_{net}.\ \eta_o \qquad (17.16)$$

A_{net} represents the net solar field collector area and η_o represents the optical efficiency of the collector. The optical efficiency of the collector can be obtained using the relation (Rohani et al., 2017):

$$\eta_o(\theta) = K(\theta).\ \cos(\theta).\ \eta_{opt,o}.\ \eta_{end}(\theta).\ \eta_{shad}(\theta).\ \eta_{clean} \qquad (17.17)$$

$\eta_{opt,o}$ is the peak optical efficiency of the collector. The angle of incidence of solar radiation θ can be obtained using Equations (17.1)–(17.5) based on the mode of tracking mechanism. A correction factor, K(θ) for incidence angle-dependent optical properties of a collector called incidence angle modifier is computed using the following relation (Rohani et al., 2017):

$$K(\theta) = 1 - 5.25097 \times 10^{-4} \times \left(\frac{\theta}{\cos\theta}\right) - 2.859621 \times 10^{-5} \times \left(\frac{\theta^2}{\cos\theta}\right) \quad (17.18)$$

Row shading factor $\eta_{shad}(\theta)$ can be computed using the relation (Wang et al., 2016):

$$\eta_{shad} = |\cos R| \frac{L_{space}}{w} \quad (17.19)$$

where L_{space} is the central distance between two parabolic trough concentrators, w is the width of aperture and R is the sun-tracking angle. Minimum and maximum values of η_{shad} are 0 and 1 respectively; corresponding to full shade and no shade rows. For single-axis tracking with horizontal axis, the sun tracking angle R can be computed using the relation (Marion and Dobos, 2013):

$$R = tan^{-1}\theta_z \sin(\gamma_s - \gamma_a) \quad (17.20)$$

where γ_s and γ_a are solar azimuth angle and axis azimuth angle, respectively. The end loss factor $\eta_{end}(\theta)$ is expressed as (Wang et al., 2016):

$$\eta_{end}(\theta) = 1 - \frac{f \tan\theta}{L} \quad (17.21)$$

where f and L are the focal length and length of the parabolic trough collector, respectively. The cleanliness factor η_{clean} is expressed as the ratio of power absorbed from solar radiation by the absorber in certain operating conditions to that with perfectly clean mirror and absorber under same operating conditions (Rohani et al., 2017). η_{clean} can be computed using the measured data.

The energy loss from the receiver at any time t_p can be obtained using the relation (Yaghoubi et al., 2013):

$$Q_L(t_p) = \pi D_{co} L h_c\left(T_{co}(t_p) - T_a(t_p)\right) + \varepsilon_c \pi D_{co} L\sigma\left(T_{co}^4(t_p) - T_{sky}^4(t_p)\right) \quad (17.22)$$

where D_{co} represents the outer diameter of cover; L is the length of the receiver; T_{co} is the temperature of the outer surface of cover; T_a & T_{sky} is ambient & sky temperature, respectively; ε_c is the emissivity of cover and σ is Stefan Boltzman constant. The convective heat-transfer coefficient h_c can be calculated using the relation (Yaghoubi et al., 2013):

$$h_c = 4D_{co}^{-0.42}V^{0.5} \quad (17.23)$$

The simulation model for the solar field is described using a block diagram shown in Figure 17.2. It consists of two blocks viz control & optimization block and monitoring block. The two blocks work in sync by interacting with each other. The monitoring block monitors the daily weather conditions, like DNI, ambient temperature, wind

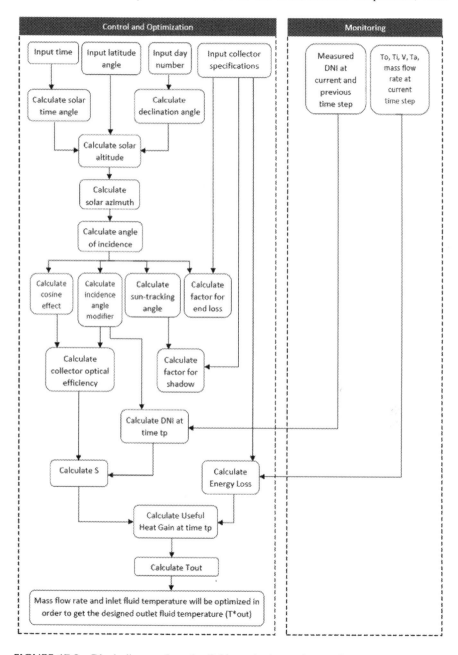

FIGURE 17.2 Block diagram for solar field monitoring and control.

speed, and the controllable parameters of the solar field, including inlet and outlet fluid temperature, mass flow rate, the temperature of the collector; it then transfers this information to the control unit. The control & optimization unit estimates the DNI and outlet fluid temperature at future time steps based on collector specifications, site details, and the data collected from the monitoring unit. The control unit then suggests a possible change in mass flow rate or inlet fluid temperature to attain the designed outlet temperature from the solar field.

17.3.1.4   Outlet Fluid Temperature

The outlet fluid temperature at time t_p can be estimated as a function of HTF temperature in storage tank T_1, useful heat gain, mass flow rate, and specific heat of HTF using the relation:

$$T_{out}(t_p) = T_l(t_n) + \frac{Q_u(t_p)}{\dot{m}c_p}$$ (17.24)

17.3.2   Modelling Thermal Energy Storage

In the present case, liquid thermal energy storage has been considered, which receives thermal energy from solar parabolic trough collectors from Block I and discharges energy to the boiler for generating steam in the power block, as shown in Figure 17.3.

The hot HTF after transferring heat to the water in the boiler is circulated back to the storage tank, which mixes with the fluid inside the tank, thereby lowering its

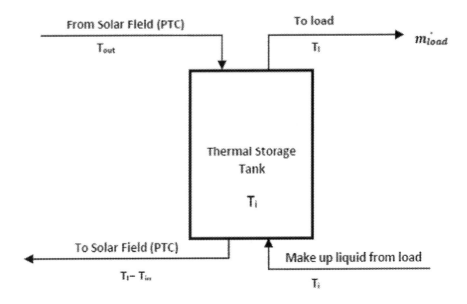

FIGURE 17.3   A well-mixed sensible heat liquid storage tank.

temperature. Thus, the fluid is continuously heated by circulating it through the parabolic trough collector array during operation and works in a closed loop.

Consider a well-mixed liquid at uniform temperature T_1 in the tank whose temperature varies with respect to time. Energy balance on the tank gives the following equation (Sukhatme and Nayak, 2012):

$$\left[\left(\rho V c_p\right)_l + \left(\rho V c_p\right)_t\right]\frac{dT_l}{dt} = \dot{m}c_p(T_{out} - T_l) - \dot{m}_{load}c_p(T_l - T_i) - (UA)_t(T_l - T_a) \quad (17.25)$$

where $\left(\rho V c_p\right)_l$ and $\left(\rho V c_p\right)_t$ represent heat capacity respectively of liquid in the tank and tank material, $\dot{m}$ is the mass flow rate of HTF in the solar field, T_l is HTF temperature in the storage tank which is same as inlet fluid temperature of the collector; T_{out} is outlet fluid temperature from the solar field; T_a is ambient temperature; T_i is the temperature of HTF that enters storage tank from the load or makeup tank $(T_i \ll T_l)$ and $(UA)_t$ is the product of overall heat transfer coefficient and surface area of the tank. In Equation (17.25), the first term on the left represents useful heat gain from the solar field, the second term represents energy transferred to load, and the third term represents energy losses from the storage tank.

17.3.3 Power Block

The power block of a utility-scale CSP system consists of equipment that converts thermal energy collected from the solar field into useful mechanical energy and electrical energy using a conventional steam Rankine cycle. The power output and heat inputs are functions of HTF temperature, condenser pressure, and flow rates. The power cycle efficiency is defined as:

$$\eta_{cycle} = \frac{Work\ output}{Heat\ input} \quad (17.26)$$

17.4 APPLICATIONS OF IOT

A block diagram for IoT enabled solar thermal power generation system is shown in Figure 17.4.

The IoT framework consists of the main controller, which communicates with each of the three blocks through two subcontrollers viz solar field (SF) controller and power block (PB) controller. The model consists of sensors and controllers, which are connected to each of the three blocks. The sensors provide real-time monitoring of meteorological data and time-varying system parameters to the main controller. The main controller communicates this information to SF and PB controller, which are used to optimize processes associated with the solar field, thermal storage, and power block.

The IoT model enables the system to predict DNI in a future time step based on real-time meteorological data, as explained in Section 17.3. The estimated value for DNI can be used to predict useful heat gain in the solar field at a future time step.

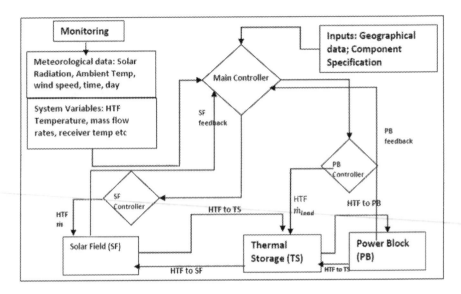

FIGURE 17.4   IoT in solar thermal power generation.

SF controller computes the optimal values of control parameters like mass flow rate and inlet fluid temperature of HTF required to obtain desired output from the solar field and communicates the same to the main controller for further action. The controllers in the power block control the flow of HTF and steam in the power block through thermal storage, boiler, turbine, and condenser to ensure optimal cycle efficiency.

The main controller, subcontrollers & sensors fitted in each block control the system variables and monitor the system performance. Constant feedback from each block is sent to the main controller; this helps in detecting any kind of fault in the system components or operational errors, thereby improving the reliability of the system. IoT helps in reducing man-made errors to a great extent and accessing data from anywhere through the internet, thus providing improved flexibility of tracking and feedback process.

17.5   CONCLUSION

The application of the Internet of Things (IoT) in solar thermal power generation will increase the reliability of the system. It will enable real-time monitoring of system variables, thereby estimating output from each of the three blocks. Based on estimated values, it would be able to optimize controllable variables like mass flow rate, inlet fluid temperature, sun-angle tracking, and many more. Based on the feedback from the real-time monitoring block, the system can generate an alert for the requirement of cleaning of collectors or the requirement of operations & maintenance of components in these blocks. This will save time and cost involved in the O&M of these systems and reduce wastage of water. IoT would help the project team to get reliable real-time feedback on the system; as a result, actions can be planned and executed without any delay. This will help in enhancing the performance and reliability of solar thermal power plants.

Thus, it can be concluded that the application of the Internet of Things in solar thermal power generation will revolutionize power plants.

REFERENCES

A Report on National Solar Thermal Power Plant. IIT Bombay. http://www.ese.iitb.ac.in/ ~NSTPP/sites/default/files/National%20Solar%20Thermal%20Power%20Plant_Report. pdf. Retrieved on 12 August 2018.

Annual Report MNRE 2018–2019. https://mnre.gov.in/

Bannur S, 2018. Concentrated solar power in India: Current status, challenges and future outlook. *Current Science*, 115(2).

Boddupalli N, Singh G, Chandra L, & Bandopadhyay B, 2017. Dealing with dust -Some challenges and solutions for enabling solar energy in desert regions. *Solar Energy*, 150, 166–176.

Bracken N, Macknick J, Tovar-Hastings A, Komor P, Gerritsen M, & Mehta S, 2015. *Concentrating Solar Power and Water Issues in the U.S. Southwest*. Technical Report of Joint Institute of Strategic Energy Analysis.

Das S, 2018. An Overview of Risk Management in Solar PV Projects. in Gaji, M.M. Eng. & Verma, A. Dr (eds.), *Development of Solar Power Generation and Energy Harvesting*, Chapter 5, Daya Publishing House, 47–56.

Das S, 2020. Short term forecasting of solar radiation and power output of 89.6 kWp solar PV power plant, *Materials Today: Proceedings*, 39, 1959–1969. doi: 10.1016/j.matpr.2 020.08.449

Duffie JA, & Beckman WA, 1980. *Solar Engineering of Thermal Processes Solar Engineering*. 1st edition John Wiley and Sons, Inc., Hoboken, New Jersey.

Helioscsp Solar Thermal Energy News. http://helioscsp.com/concentrated-solar-power-had-a-global-total-installed-capacity-of-6451-mw-in-2019/#:~:text=a%20thermochemical%20reaction.-, Concentrated%20solar%20power%20had%20a%20global%20total%20installed%20capacity %20of,from%20382%20MW%20in%202018.&text=Among%20the%20larger%20CSP %20projects,technology%20without%20thermal%20energy%20storage

Ikejemba ECX, & Schuur PC, 2018. Analyzing the impact of theft and vandalism in relation to the sustainability of renewable energy development projects in sub-Saharan Africa. *Sustainability*, 10, 814. doi:10.3390/su10030814

International Renewable Energy Agency (IRENA), 2019. *Internet of Things: Innovation Landscape Brief*. https://www.irena.org/media/Files/IRENA/Agency/Publication/2019/ Sep/IRENA_Internet_of_Things_2019.pdf?la=en&hash=0576FFAC16E131D9AB017B 118655933FD892A6C6

Kearney D, Herrmann U, Nava P, Kelly B, Mahoney R, Pacheco J, Cable R, Potrovitza N, Blake D, & Price H, 2003. Assessment of molten salt heat transfer fluid in a parabolic trough solar field. *Journal of Solar Energy Engineering*, 125(2), 170–176. doi: 10.1115/1.1565087

Kumar S, 2017. *IoT in Indian Electricity Transmission & Distribution Sectors*. https:// pronto-core-cdn.prontomarketing.com/581/wp-content/uploads/sites/2/2017/06/6_IoT-in-Indian-Electricity-Transmission-_-Distribution-Sectors.pdf

Kumarankandath A, & Goswami N, 2015. *The State of Concentrated Solar Power in India. A Roadmap to Developing Solar Thermal Technologies in India*. Centre for Science and Environment. https://shaktifoundation.in/wp-content/uploads/2014/02/State-of-CSP-in-India_low-res.pdf. Retrieved on 12 August 2018.

Lieberman E, Lyons J, & Tucker D, 2014. *Making Renewable Energy Wildlife Friendly. Defenders of Wildlife*. https://defenders.org/sites/default/files/publications/making_renewable_energy_wildlife_friendly.pdf Retrieved on 9 March 2019.

Linke F, 1922. Transmissions-Koeffizient und Trubungsfaktor. Beitr. *Phys. fr. Atmos*, 10, 91–103.

Lüpfert E, Herrmann U, Price H, Zarza E, & Kistner R, 2004. Towards standard performance analysis for parabolic trough collector fields. *SolarPaces Conference*, Mexico, Oaxaca.

Marion WF, & Dobos AP, 2013. *Rotation Angle for the Optimum Tracking of One-Axis Trackers*. Technical Report, NREL/TP-6A20-58891, July.

Motlagh NM, Mohammadrezaei M, Hunt J, & Zakeri B, 2020. Internet of things (IoT) and the energy sector. *Energies*. file:///C:/Users/Admin/Downloads/energies-13–00494.pdf

Ong S, Campbell C, Denholm P, Margolis R, & Heath G, 2013. *Land-Use Requirements for Solar Power Plants in the United States Technical Report*, National Renewable Energy Laboratory (NREL), June. https://www.nrel.gov/docs/fy13osti/56290.pdf

Ramamurthy A, & Jain P, 2017. *The Internet of Things in the Power Sector Opportunities in Asia and the Pacific*. Sustainable Development Working Paper Series. ADB Publishers.

Rohani S, Fluri TP, Dinter F, & Nitz P, 2017. Modelling and simulation of parabolic trough plants based on real operating Data. *Solar Energy*, 158, 845–860.

Sharma C, Sharma AK, Mullick SC, & Kandpal TC, 2015. Assessment of solar thermal power generation potential in India. *Renewable and Sustainable Energy Reviews*, 42, 902–912.

Shen Y, Wei H, Zhu T, Zhao X, & Zhang K, 2018. A data-driven clear sky model for direct normal irradiance. *IOP Conference Series, Journal of Physics*, 1072.

Skill Council for Green Jobs. September 2016. *Skill Gap Report for Solar, Wind and Small Hydro Sector*. http://sscgj.in/wp-content/uploads/2016/06/SCGJ-skill-gap-report.pdf

Spelling J, Gallo A, Romero M, & González-Aguilar J, 2015. A high-efficiency solar thermal power plant using a dense particle suspension as the heat transfer fluid. *Energy Procedia*, 69, 1160–1170.

Stults ES, 2015. *Minimizing Water Requirements for Electricity Generation in Water Scare Areas*. Thesis submitted to Faculty of the Worcester Polytechnic Institute. https://web.wpi.edu/Pubs/ETD/Available/etd-050415–092758/unrestricted/estults.pdf

Sukhatme SP, & Nayak JK, 2012. *Solar Energy, Principles of Collection and Storage*. 3rd edition. Tata McGraw Hill Education Pvt Ltd.

Turney D, & Fthenakis V, 2011. Environmental impacts from the installation and operation of large-scale solar power plants. *Renewable and Sustainable Energy Reviews*, 15, 3261–3270.

Wang J, Wang J, Bi X, & Wang X, 2016. Performance simulation comparison for parabolic trough solar collectors in China. *International Journal of Photoenergy*, 2016, 16 pages, Article ID 9260943, Hindawi Publishing Corporation.

Xu X Vignarooban K, Xu B, Hsu K, & Khannan AM, 2016. Prospects and problems of concentrating solar power technologies for power generation in the desert regions. *Renewable and Sustainable Energy Reviews*, 53, 1106–1131.

Yaghoubi M, Ahmadi F, & Bandehee M, 2013. Analysis of heat losses of absorber tubes of parabolic through collector of Shiraz (Iran) solar power plant. *Journal of Clean Energy Technologies*, 1(1), January, 33–37.

Index